# Governing Biobanks

Biobanks are proliferating rapidly worldwide because they are powerful tools and organisational structures for undertaking medical research. By linking samples to data on the health of individuals, it is anticipated that biobanks will be used to explore the relationship between genes, environment and lifestyle for many diseases, as well as the potential of individually tailored drug treatments based on genetic predisposition. However, they also raise considerable challenges for existing legal frameworks and research governance structures. This book critically examines the current governance structures in place for biobanks in England and Wales. It shows that the technologies, techniques and practices involved in biobanking do not always conform neatly to existing legal principles and frameworks that apply to other areas of medical research. Using a socio-legal approach, including interview data gathered from the scientific community, this book provides unique insights and makes recommendations about appropriate governance mechanisms for biobanking in the future. It also explores the issues around the secondary use of information, such as consent and how to protect privacy, when biobanks are accessed by a number of different third parties. These issues have relevance both within England and Wales and to a wide international audience, as well as for other areas where large datasets are used.

# Governing Biobanks

## Understanding the Interplay between Law and Practice

Jane Kaye
Susan M C Gibbons
Catherine Heeney
Michael Parker
and
Andrew Smart

·HART·
PUBLISHING
OXFORD AND PORTLAND, OREGON
2012

Published in the United Kingdom by Hart Publishing Ltd
16C Worcester Place, Oxford, OX1 2JW
Telephone: +44 (0)1865 517530
Fax: +44 (0)1865 510710
E-mail: mail@hartpub.co.uk
Website: http://www.hartpub.co.uk

Published in North America (US and Canada) by
Hart Publishing
c/o International Specialized Book Services
920 NE 58th Avenue, Suite 300
Portland, OR 97213-3786
USA
Tel: +1 503 287 3093 or toll-free: (1) 800 944 6190
Fax: +1 503 280 8832
E-mail: orders@isbs.com
Website: http://www.isbs.com



British Library Cataloguing in Publication Data
Data Available

ISBN: 978-1-84113-905-0

Typeset by Compuscript Ltd, Shannon
Printed and bound in Great Britain by
Lightning Source UK Ltd

In memory of my mother, Margaret Kaye,
6 March 1931–22 February 2011
*Jane Kaye*

For Jeremy and Mummie
*Susan MC Gibbons*

# *Preface*

The idea for the Governing Genetic Databases project originated because of a discussion with a scientist talking about his research. In early 2004, Jane Kaye and Michael Parker went to see Professor Kwiatkowski from the Wellcome Trust Centre for Human Genetics in Oxford (WTCHG) about a funding application for a project to be called the Malaria Genomic Epidemiology Network (MalariaGEN) that he was proposing to submit to the Bill and Melinda Gates Foundation and the Wellcome Trust. MalariaGEN was going to bring together more than 30 partner institutions in 21 countries (in both developing and developed countries) and to involve the collection of very large numbers of DNA samples combined with clinical data from malaria-endemic countries in Africa and South East Asia. At around the same time, Mike Parker, Andrew Smart and Jane Kaye were also involved in the Oxford Genetics Knowledge Park where the ethical, legal and social implications of the collection of DNA samples, with the possibility of the information being used for a number of secondary research projects was also being identified as an important area requiring further research. The legal research that Sue Gibbons had carried out as part of the ELSAGEN project had highlighted the complexity of the law that applied to biobanks in the UK. Together, these projects reflected developments that were happening not just in the UK, but across the world, and the important and difficult questions they raised provided the background motivation for the Governing Genetic Databases Project which has resulted in this book.

The work carried out for the Governing Genetic Database project was funded by Wellcome Trust grant number WT 076070/Z/04/Z. In carrying out this research Michael Parker, Jane Kaye and Andrew Smart were also funded through the Oxford Genetics Knowledge Park by the UK Departments of Health and of Trade and Industry. Jane Kaye was funded by a Wellcome Trust Fellowship (grant number WT 081407/Z/06/Z). During the period of the project, Michael Parker's contribution was also supported by grants from the Wellcome Trust (number 087285/Z/08/Z) and the European Commission (LSHM-CT-2007-037273). During the writing of the book Catherine Heeney has been supported by Governing Genetic Databases Project Grant Wellcome Trust Code WT 076070/Z/04/Z, European Commission LSHB-CT-2006-037319 and by a CSIC JAE. doc Fellowship. Jane Kaye would like to thank Patrick Ky and Patrick Woolley for their assistance with Chapters One and Two. We would also like to thank the team at Hart Publishing and Sarah Newton for her copy editing.

# Contents

*Preface* .......... vii

**Part I Setting the Scene**

1. From an Idea to a Project .......... 3
*Jane Kaye, Susan MC Gibbons, Catherine Heeney, Michael Parker and Andrew Smart*

2. Embedding Biobanks in a Changing Context .......... 30
*Jane Kaye*

**Part II Key Legal and Sociological Findings**

3. Mapping the Regulatory Space .......... 51
*Susan MC Gibbons*

4. Dynamic Networks of Practice .......... 93
*Catherine Heeney*

5. Governance in Practice: Patterns of Awareness and Engagement .......... 115
*Susan MC Gibbons and Andrew Smart*

6. General Attitudes to Governance .......... 140
*Susan MC Gibbons and Andrew Smart*

7. Attitudes to Particular Laws and Governing Bodies .......... 163
*Susan MC Gibbons and Andrew Smart*

8. Preferences for Governance .......... 202
*Susan MC Gibbons and Andrew Smart*

9. Enacting Governance—The Case of Access .......... 232
*Catherine Heeney and Andrew Smart*

## Part III Reflections, Ramifications and Recommendations

10. Reflections on Practice and Governance........................................261
*Jane Kaye*

11. Ethics and the Governance of Biobanks........................................282
*Catherine Heeney and Michael Parker*

12. Conclusions........................................302
*Jane Kaye, Susan MC Gibbons, Catherine Heeney, Michael Parker and Andrew Smart*

*Appendices*........................................327

*Index*........................................353

# Part I

# Setting the Scene

# 1

# From an Idea to a Project

*Jane Kaye, Susan MC Gibbons, Catherine Heeney, Michael Parker and Andrew Smart*

## HISTORY OF PROJECT

THIS BOOK PROVIDES a snapshot of a period of time from 2005–2009 when a large number of biobanks were being established in the UK and elsewhere. When this project commenced, advances in computer technology and high-throughput DNA sequencing had led to an increase in the amount of genetic research being conducted in the UK and the possibility of being able to establish resource collections.[1] Such advances had also led to a rapid increase in the number of collections of DNA samples that were being amassed, either for individual projects or specifically as resources for multiple other researchers to use.[2] These advances signalled the change in research focus from genetics to the study of the complexity of the whole genome, and also a change in practice to the investment in infrastructure projects or platforms. The results of the Human Genome Project, which involved the sequencing of the whole genome, had recently been completed in 2001. Large population biobanks—such as the Icelandic Health Sector Database (1999), the Estonian Genome Project (2000), and UK Biobank (2001)—had gained considerable attention because of their ambitious plans that tested many of the existing principles of medical research practice and data protection law. In addition, funders were supportive of proposals to establish disease-specific biobanks, as well

[1] FP Perera and IB Weinstein, 'Molecular Epidemiology: Recent Advances and Future Directions' (2000) 21(3) *Carcinogenesis* 517 (and references cited therein); R Tutton and O Corrigan, 'Introduction: Public Participation in Genetic Databases' in R Tutton and O Corrigan (eds), *Genetic Databases: Socio-ethical Issues in the Collection and Use of DNA* (London, Routledge, 2004).

[2] JV McHale, 'Regulating Genetic Databases: Some Legal and Ethical Issues' (2004) 12 *Medical Law Review* 70; I Hirtzlin, C Dubreuil, N Préaubert, J Duchier, B Jansen, J Simon, PL de Faria, A Perez-Lezaun, B Visser, GD Williams and A Cambon-Thomsen, 'An Empirical Survey on Biobanking of Human Genetic Material and Data in Six EU Countries' (2003) 11 *European Journal of Human Genetics* 475.

as encouraging the collection of DNA samples as 'add-ons' to existing longitudinal studies.

Despite this growth in the number of collections being established, there were no clear or accepted models for the organisational, procedural or security mechanisms of biobanks within the UK. Also, there was a lack of clarity in terminology with a number of different terms being applied to these kinds of research repositories, both within the UK and elsewhere. In the UK, 'genetic database' was the most commonly used term. This was a situation in which practice, terminology and standards were in a state of development and still emerging, and there was considerable uncertainty within the research community. Researchers and clinicians had to 're-invent the wheel' in many cases when creating biobanks. They drew on the shared cultural practices of their professions and past experience in areas such as pathology, clinical trials and epidemiology, as a basis to develop structures and procedures for managing these collections. However, none of those areas had dealt with some of the specific issues raised by genetics. This lack of precedent created uncertainty and led to different procedures being established in different institutions or in different disease areas of medicine. This also led to regional and national differences which are now having an effect on the ability of researchers to share samples and data, particularly within global collaborations.

The legal framework in the UK also provided no answers to the questions and difficulties that researchers were experiencing and with which they were having to struggle. We had concerns that researchers could be at risk of operating unlawfully, as practices appeared to be emerging in a different direction to the law, but also without any close or well-informed consideration of the legal requirements. The law that applied to medical research and genetics in the UK was, and still is, a complex, inconsistent, fragmented, confusing and incomplete regulatory framework.[3] The legal research that Sue Gibbons had done as part of the ELSAGEN project had demonstrated this.[4] Unlike many other jurisdictions in Europe, in the UK

[3] D Price, 'From Cosmos and Damien to Van Veltzen: the Human Tissue Saga Continues' (2003) 11 *Medical Law Review* 1; C Mannhalter, 'Collection, Storage and Use of Genetic Data: Issues of Gene Banks', paper presented at conference on UNESCO Universal Declaration on the Human Genome and Human Rights: present status—future perspectives (June 2003); J Bohannon, 'UK Researchers Hope for Clarity in Tissue Use' (2002) 298 *Science* 1867; J Black, 'Regulation as Facilitation: Negotiating the Genetic Revolution' in R Brownsword, W Cornish and M Llewellyn (eds), *Law and Human Genetics: Regulating a Revolution* (Oxford, Hart Publishing, 1998); R Brownsword, 'Regulating Human Genetics: New Dilemmas for a New Millennium' (2004) 12 *Medical Law Review* 14.

[4] European Commission Framework 5, Quality of Life grant, 'Ethical, Legal and Social Aspects of Genetic Databases: A European Comparison (ELSAGEN)' with partners in Iceland, Sweden, Estonia and the UK, Contract number QLG6-CT-2001-00062, 2002–2005.

there was no specific legal instrument dedicated to biobanks.[5] Thus, it was very difficult to ascertain precisely what laws did govern biobanks within the UK at that time,[6] and also what was required of researchers and biobank operators, as practice itself was still evolving. At the same time, there were—and still remain—no universally accepted definitions of 'genetic databases' or 'biobanks' to be found in the law, guidelines or codes of practice within the UK.[7] There was also uncertainty whether these new emerging practices, procedures and standards were being developed in compliance with legal requirements.[8] This was largely to do with the fact that the legal requirements were insufficiently clear, too diffusely spread, excessive, and did not apply specifically to this new area of DNA collections.

Compounding the confusing state of existing UK law, and uncertainty over how it should be applied to biobanks, genetic information also has characteristics[9] that test traditional legal principles. For example, the law made, and still does, a distinction between 'information' and 'biological samples'. Different rules and governance regimes apply to each. Often, these regimes impose differing standards—for example, in relation to participant consent. Yet, DNA/genetic material is *both* information *and* bodily sample. It can be difficult to determine, for legal purposes, when it should be treated as one or the other. The familial nature of genetic information also has implications for determining privacy interests and who should have a right to access the results of genetic tests or research.[10] There is also a need to understand how the complex notion of 'public interest' should be

[5] J Kaye and P Martin, 'Safeguards for Research Using Large Scale Dna Collections' (2000) 321 *British Medical Journal* 1146; P Martin and J Kaye, 'The Use of Large Biological Sample Collections in Genetics Research: Issues for Public Policy' in P Glasner (ed), *Reconfiguring Nature: Issues and Debates in the New Genetics* (Aldershot, Ashgate, 2004).

[6] H Newiss, 'Genetic Databanks: How Secure Is Your Information?' (2001) 1(6) *Genetics Law Monitor* 1; J Strobl, E Cave and T Walley, 'Data Protection Legislation: Interpretation and Barriers to Research' (2000) 321 *British Medical Journal* 890.

[7] McHale, 'Regulating Genetic Databases', above n 2; Black, 'Regulation as Facilitation', above n 3; Brownsword, 'Regulating Human Genetics', above n 3; P Martin, 'Genetic Governance: The Risks, Oversight and Regulation of Genetic Databases in the UK' (2001) 20 *New Genetics and Society* 157.

[8] PR Ferguson, 'Legal and Ethical Aspects of Clinical Trials: the Views of Researchers' (2003) 11 *Medical Law Review* 48.

[9] R Mackenzie, 'Paradigms of Author/Creator Property Rights in Intellectual Property Law: Ethical Implications for the Acquisition, Access, and Control of Genetic Information' in AK Thompson and RF Chadwick (eds), *Genetic Information: Acquisition, Access, and Control* (New York, Kluwer Academic/Plenum Publishers, 1999); M Richards, 'How Distinctive is Genetic Information?' (2001) 32 *Studies in History and Philosophy of Biological and Biomedical Science* 663.

[10] BM Knoppers, 'Genetic Information and the Family: Are We Our Brother's Keeper?' (2002) 20(2) *Trends in Biotechnology* 85; D Bell and B Bennett, 'Genetic Secrets and the Family' (2001) 9 *Medical Law Review* 130; GT Laurie, 'Obligations Arising from Genetic Information: Negligence and the Protection of Familial Interests' (1999) 11(2) *Child and Family Law Quarterly* 109.

construed and protected[11] if the human genome is to be regarded as the common heritage of humankind. By the early 2000s, these characteristics had triggered a heated debate as to whether genetic information should be treated as 'special' or 'exceptional' and whether its use should be the subject of separate regulation and governance structures; a debate which remained hotly contested and largely unresolved in 2004.[12]

In addition, population biobank proposals, such as the Icelandic Health Sector Database, had led to an extensive international debate over the principles that should be applied to biobanks.[13] This debate had highlighted in particular the need to consider the familial nature of genetic information and the perceived risk to privacy that biobanks may present.[14] In short, the principal issues raised by the debate were: consent, especially for secondary research purposes;[15] feedback to participants;[16] benefit-sharing;[17] participation

[11] L Beecham, 'BMA Annual Representative Meeting: Debate Needed on Balance Between Patient Confidentiality and Needs of Research' (2004) 329 *British Medical Journal* 7457; K Korts, S Weldon and ML Guðmundsdóttir, 'Genetic Databases and Public Attitudes: A Comparison of Iceland, Estonia and the UK' (2004) 8(1/2) *TRAMES* 131; V Árnason and G Árnason, 'Informed Democratic Consent? The Case of the Icelandic Database' (2004) 8(1/2) *TRAMES* 164; Kaye and Martin, 'Safeguards for Research', above n 5; G Laurie, *Genetic Privacy: A Challenge to Medico-legal Norms* (Cambridge, Cambridge University Press, 2002).

[12] S Holm, 'There is Nothing Special About Genetic Information' in AK Thompson and RF Chadwick (eds), *Genetic Information: Acquisition, Access, and Control* (London, Kluwer Academic/Plenum Publishers, 1999); LO Gostin and JG Hodge, 'Genetic Privacy and the Law: An End to Genetics Exceptionalism' (1999) 40 *Jurimetrics* 21; TH Murray, 'Genetic Exceptionalism and "Future Diaries": Is Genetic Information Different from Other Medical Information?' in MA Rothstein (ed), *Genetic Secrets: Protecting Privacy and Confidentiality in the Genetic Era* (New Haven, CT, Yale University Press, 1997).

[13] BM Knoppers, 'Of Populations, Genetics and Banks' (2001) 1 *Genetics Law Monitor*; Árnason and Árnason, 'Informed Democratic Consent' above n 11.

[14] HDC Roscam Abbing, 'Central health Database in Iceland and Patient's Rights' (1999) 6 *European Journal of Health Law* 363; JR Gulcher, K Kristjánsson, H Gudbjartsson and K Stefánsson, 'Protection of Privacy by Third-Party Encryption in Genetic Research in Iceland' (2000) 8 *European Journal of Human Genetics* 739.

[15] J Kaye, 'Abandoning Informed Consent: The Case of Genetic Research in Population Collections' in R Tutton and O Corrigan (eds), *Genetic Databases: Socio-ethical Issues in the Collection and Use of DNA* (London, Routledge, 2004); T Caulfield, REG Upshur and A Daar, 'DNA Databanks and Consent: A Suggested Policy Option Involving an Authorization Model' (2003) 4(1) *BMC Medical Ethics* 1; E Wright Clayton, KK Steinberg, MJ Khoury, E Thomson, L Andrews, MJ Ellis Kahn, LM Kopelman and JO Weiss, 'Informed Consent for Genetic Research on Stored Tissue Samples' (1995) 274(22) *Journal of the American Medical Association* 1786; G Williams and D Schroeder, 'Human Genetic Banking and the Limits of Informed Consent' in P Glasner (ed), *Reconfiguring Nature: Issues and Debates in the New Genetics* (Aldershot, Ashgate, 2004).

[16] S Eriksson, 'Should Results from Genetic Research be Returned to Research Subjects and Their Biological Relatives? (2004) 8(1/2) *TRAMES* 46.

[17] Korts, Weldon and Guðmundsdóttir, 'Genetic Databases and Public Attitudes', above n 11; AC da Rocha, 'Ethical Aspects of Human Genetic Databases: Distinctions on the Nature, Provision, and Ownership of Genetic Information' (2004) 8(1/2) *TRAMES* 34.

in decision-making;[18] protecting privacy;[19] access;[20] ownership;[21] and intellectual property[22] (especially patents and copyright).[23]

However, it was unclear whether the principles that were being proposed for the population biobanks that were being developed as research resources would or should also be applicable to smaller collections established for specific research projects focusing on particular diseases. In addition, it was not appropriate simply to transplant such principles into the UK context without considering carefully their potential implications and fit.[24] For instance, a blanket requirement that all research collections should have an independent oversight body, such as that proposed at the time for UK Biobank, clearly would have been unduly burdensome if applied to a collection of several hundred samples, maintained by one research group, for a single research project. Accordingly, the principles being developed within the context of the population biobanks debate needed to be assessed carefully before being applied to other types of collections, and sound reasons needed to be identified and articulated before such principles were accepted, modified or rejected.

In 2005, the Human Tissue Act 2004 was still to be fully implemented. The Human Tissue Authority, which had the authority to issue codes of practice and licences to regulate tissue collections, had just been established. At that time, it remained uncertain whether (or how far) the scope of its authority would extend to genetics and biobanks. Now that the Act is fully operational, it is evident that extracted DNA falls largely outside its scope. While the regulation of biobanks by such a body as the Human Tissue Authority was seen by many as a means to bring some much needed

[18] RR Sharp and MW Foster, 'Involving Study Populations in the Review of Genetic Research' (2000) 28(1) *Journal of Law, Medicine and Ethics* 41.

[19] S Alpert, 'Privacy and the Analysis of Stored Tissues', commissioned paper, National Bioethics Advisory Commission *Research Involving Human Biological Materials: Ethical Issues and Policy Guidance* (Rockville MD, The Commission, January 2000) vol II, A-1; Laurie, *Genetic Privacy*, above n 11; MA Rothstein, 'Genetic Secrets: A Policy Framework' in MA Rothstein (ed), *Genetic Secrets: Protecting Privacy and Confidentiality in the Genetic Era* (New Haven, CT, Yale University Press, 1997).

[20] BM Knoppers, M Hirtle, S Lormeau, CM Laberge and M Laflamme, 'Control of DNA Samples and Information' (1998) 50 *Genomics* 385.

[21] Rocha, 'Ethical Aspects of Human Genetic Databases', above n 17.

[22] T Caulfield, 'Regulating the Commercialization of Human Genetics: Can We Address the Big Concerns?' in AK Thompson and RF Chadwick (eds), *Genetic Information: Acquisition, Access, and Control* (New York, Kluwer Academic/Plenum Publishers, 1999).

[23] HT Greely, 'Informed Consent and Other Ethical Issues in Human Population Genetics' (2001) 35 *Annual Review of Genetics* 785; HT Greely, 'Human Genomics Research: New Challenges for Research Ethics' (2001) 44(2) *Perspectives in Biology and Medicine* 221; S Wilson, 'Population Biobanks and Social Justice: Commercial Or Communitarian Models? a Comparative Analysis of Benefit Sharing, Ownership and Access Arrangements' (2004) 8(1/2) *TRAMES* 80.

[24] JE Wylie and G Mineau, 'Biomedical Databases: Protecting Privacy and Promoting Research' (2003) 21 (3) *TRENDS in Biotechnology* 113.

clarity and certainty to the situation, this would not be an easy task—not least, as no typology had ever been developed of the various kinds of human biobanks used for research purposes in this jurisdiction. (The EUROGENEBANK Project did carry out a survey of 'biobanks', but this was based only on information from 12 respondents from across the UK.[25]) Also, no research had been conducted either into the question of what a biobank would look like if it were to be constructed and run in conformity with the legal requirements, or, beyond the appropriate limits of law, the ethical principles that should apply to different types of biobanks. Without such research, we felt that the regulatory bodies would be left in a quandary, having to attempt to regulate the unknown. An anticipated output of our project, then, was to provide an evidence base for such authorities to develop guidelines for governing biobanks within England and Wales, which would be based on a clear understanding of both the theoretical and practical issues that practitioners were facing.

## OTHER PROJECTS

Prior to the research reported in this book, there had been no research projects of this size focused specifically on the governance of biobanks within the UK. Some studies had been conducted on biobanks within Europe. The ELSAGEN[26] project (2002–2005), for example, compared four population biobanks that had been or were in the process of being established: namely, the Icelandic Health Sector Database (1999), the Estonian Genome Project (2000), the Swedish UmanGENE and UK Biobank (2001). The approach of the ELSAGEN project was to carry out empirical research into public attitudes to these biobanks and to compare the legal frameworks and regulatory bodies in each jurisdiction, whilst also focusing on the ethical concerns around population biobanks in these countries.[27] As a member of this consortium, we could draw upon the legal research. The European Commission also had funded a number of projects that focused on the ethical, legal and social issues of biobanking. The EUROGENEBANK Project[28] involved a quantitative survey to determine the numbers and types of biobanks that were in existence and what they were being used for in six European countries, as a first step towards a comprehensive audit of biobanks. More recent projects on biobanking concerns are GeneBanC

[25] Hirtzlin et al, 'An Empirical Survey on Biobanking', above n 2.

[26] European Commission Framework 5, Quality of Life grant, 'Ethical, Legal and Social Aspects of Genetic Databases', above n 4.

[27] V Árnason et al (eds), *Your Genes in a National Bank? Ethical, Legal and Social Concerns* (Cambridge, Cambridge University Press, 2008).

[28] Hirtzlin et al, 'An Empirical Survey on Biobanking', above n 2.

(2006–2009)[29] and PRIVILEGED (2007–2009)[30] which focus on privacy and data protection issues, and Tiss.EU (2008–2011) that is concerned with the legal frameworks for samples and tissue across the European Union.[31]

Governance in genomics was a little explored area in the literature at this time with only a Wellcome Trust report[32] and a paper[33] by Paul Martin on the specific topic of governance and biobanks in the UK. Since then, governance of biobanks has become a topic of greater interest and concern.[34] There have been a number of papers using legal analysis that have concentrated on how current legal requirements might apply to biobanks within the UK and Europe.[35] There have also been a number of empirical studies that have sought the opinions of research participants and the general

[29] http://www.genebanc.eu/ (accessed 30 January 2011).

[30] http://www.privileged.group.shef.ac.uk/ (accessed 30 January 2011).

[31] http://www.tisseu.uni-hannover.de/index.php (accessed 30 January 2011).

[32] P Martin and J Kaye, *The Use of Biological Sample Collections and Personal Medical Information in Human Genetics Research* (London, Wellcome Trust, 1999).

[33] Martin, 'Genetic Governance', above n 7.

[34] See, eg, J Gerards and H Janssen, 'Regulation of Genetic and Other Health Information in a Comparative Perspective' (2006) 13 *European Journal of Health Law* 339 [esp s 4]; M Majumder, 'Cyberbanks and Other Virtual Research Repositories' (2005) 33 *Journal of Law Medicine and Ethics* 31; M Deschenes and C Salle, 'Accountability in Population Biobanking: Comparative Approaches' (2005) 33 *Journal of Law Medicine and Ethics* 40 [esp s 3]; M Rothstein, 'Expanding Ethical Analysis of Biobanks' (2005) 33 *Journal of Law Medicine and Ethics* 89 [esp p 97]; D Winickoff, 'Partnership in UK Biobank: A Third Way for Genomic Property' (2005) 33 *Journal of Law Medicine and Ethics* 440 [esp 445–50]; H Gottweis and K Zatloukal, 'Biobank Governance: Trends and Perspectives' (2007) 74 *Pathobiology* 206; S Wallace et al, 'Governance Mechanisms and Population Biobanks: Building a Framework For Trust' (2008) 6(2) *GenEdit* 11; B Salter and M Jones, 'Biobanks and Bioethics: The Politics of Legitimation' (2005) 12 *European Public Policy* 710; SMC Gibbons, 'Regulating Biobanks: A Twelve-Point Typological Tool' (2009) 17 *Medical Law Review* 313; J Kaye and M Stranger (ed), *Principles and Practice in Biobank Governance* (Aldershot, Ashgate, 2009).

[35] BM Knoppers, 'Biobanking: International Norms' (2005) 33 *Journal of Law Medicine and Ethics* 7; A Cambon-Thomsen, E Rial-Sebbag and BM Knoppers, 'Trends in Ethical and Legal Frameworks for the Use of Human Biobanks' (2007) 30 *European Respiratory Journal* 373; LB Andrews, 'Assessing Values to Set Policies for Consent, Storage, and Use of Tissue and Information in Biobanks' in K Dierickx and P Borry (eds), *New Challenges for Biobanks: Ethics, Law and Governance* (Antwerp and Oxford, Intersentia, 2009); DE Winickoff, 'Biosamples, Genomics, and Human Rights: Context and Content of Iceland's Biobanks Act' (2001) 4 *Journal of Biolaw and Business* 11–17; M Anderlik, 'Commercial Biobanks and Genetic Research: Ethical and Legal Issues' (2003) 3 *American Journal of Pharmacogenomics* 203–15; T Caulfield, 'Tissue Banking, Patient Rights, and Confidentiality: Tensions in Law and Policy' (2004) 23 *Medical Law International* 39–49; G Richardson, 'The Banking of Embryonic Stem Cells: The Legal and Ethical Framework in the UK' (2004) 20 *Law Human Genome Review* 147–60; J Kaye, 'Do we Need a Uniform Regulatory System for Biobanks Across Europe?' (2006) 14 *European Journal of Human Genetics* 245–46; SMC Gibbons et al, 'Lessons from European Population Genetic Databases: Comparing the Law in Estonia, Iceland, Sweden and the United Kingdom' (2005) 12(2) *European Journal of Health Law* 103; J Kaye, HH Helgason, A Nõmper et al, 'Population Genetic Databases: A Comparative Analysis of the Law in Iceland, Sweden, Estonia and the UK' (2004) 8 *TRAMES* 15–34.

public about biobanking in general.[36] In addition, there have been studies that (like our study) have focused on biobanks in particular and especially the views of researchers.[37] However, there have been no studies that have drawn from the theoretical literature on regulation and combined this with empirical analysis to focus on researcher perspectives in the field of biobanking. Essential background materials for this research that we have conducted, and on which we report in this book, have been the studies that looked at scientists and how they share data,[38] as well as the theoretical research done on boundary objects.[39] Such studies have been instrumental in informing the analysis in Chapter Nine and our concluding chapter. We have also drawn upon other studies that have looked at stem cell banks and regulation from the perspectives of researchers.[40] These studies have provided a basis from which we have been able to develop and compare the findings for our own study—particularly, for thinking about the role of

[36] A Kettis-Lindbald et al, 'Genetic Research and Donation of Tissue Samples to Biobanks. What Do Potential Sample Donors in the Swedish General Public Think?' (2005) 16 *European Journal of Public Health* 433; M Levitt and S Weldon, 'A Well Placed Trust?: Public Perceptions of the Governance of DNA Databases' (2005) 15 *Critical Public Health* 311; G Haddow, S Cunningham-Burley, A Bruce and S Parry, 'Generation Scotland: Consulting Publics and Specialists at an Early Stage in a Genetic Database's Development' (2008) 18(2) *Critical Public Health* 139–49; H Busby, 'Biobanks, bioethics and concepts of donated blood in the UK', (2006) 28(6) *Sociology of Health and Illness* 850–65.

[37] A Smart et al, 'Social Inclusivity vs Analytical Acuity? a Qualitative Study of UK Researchers Regarding the Inclusion of Minority Ethnic Groups in Biobanks' (2008) 9(2) *Medical Law International* 169–90; S Hjörleifsson, R Strand and E Schei, 'Health as A Genetic Planning Project: Enthusiasm and Second Thoughts Among Biomedical Researchers and Their Research Subjects' (2005) 3(1) *Genomics, Policy and Society* 52–65; F Milanovic, D Pontille and A Cambon-Thomsen, 'Biobanking and Data Sharing: A Plurality of Exchange Regimes' (2007) 23(1) *Genomics, Society and Policy* 17–30; I Hirtzlin et al, 'An Empirical Survey on Biobanking', above n 2; A Capron, A Mauron and B Elger, 'Ethical Norms and the International Governance of Genetic Databases and Biobanks: Finding from an International Study' (2009) 19 *Kennedy Institute of Ethics Journal* 101.

[38] SJ Ceci, 'Scientists' Attitudes Toward Data Sharing, (1998) 13 *Science Technology and Human Values* 45–52; S Hilgartner and SI Brandt-Rauf, 'Data Access, Ownership and Control: Toward Empirical Studies of Access Practices' (1994) 15 *Knowledge* 355–72; C Hine, 'Databases as Scientific Instruments and their Role in the Ordering of Scientific Work' (2006) 36 *Social Studies of Science* 269–98.

[39] SL Star and J Griesemer, 'Institutional Ecology, Translations' and Boundary Objects: Amateurs and Professionals in Berkeley's Museum of Vertebrate Zoology, 1907–39' (1989) 19(3) *Social Studies of Science* 387–420; SL Star, 'This is Not a Boundary Object: Reflections on the Origins of a Concept' (2010) 35(5) *Science, Technology and Human Values* 602–17; M Callon, 'Some Elements of a Sociology of Translation: Domestication of the Scallops and the Fishermen of St Brieuc Bay' in J Law (ed), *Power, Action and Belief;* Sociological Review Monograph 32 (London, Routledge and Kegan Paul, 1985); T Gieryn, 'Policing STS: A Boundary-Work Souvenir from the Smithsonian Exhibition on "Science in American Life" (1996) 21(1) *Science, Technology, and Human Values* 100–15.

[40] N Stephens, P Atkinson and P Glasner, 'The UK Stem Cell Bank: Securing the Past, Validating the Present, Protecting the Future' (2008) 17(1) *Science as Culture* 43–56; SP Wainwright, C Williams, M Michael, B Farsides and A Cribb, 'Ethical Boundary-Work in the Embryonic Stem Cell Laboratory' (2006) 28(6) *Sociology of Health and Illness* 732–48.

regulation in fostering legitimacy and how scientific practice develops and is conceptualised by researchers.

## CONDUCTING THE RESEARCH

Against this background, our aim was to carry out an interdisciplinary analysis that integrated our separate areas of expertise in law, sociology and ethics to provide a view of the relationship between working practices and systems of oversight in this complex and fast-moving area of biomedicine. The legal research would map and analyse the UK 'regulatory space', in order to help us to understand the content, mechanisms and strengths/deficiencies of relevant legal, regulatory and guidance frameworks. The sociological research would collect and analyse interview data collected from professionals in the field, in order to provide a basis for understanding their working practices, and the concerns that they face in relation to various forms oversight. The ethical analysis would then provide a basis for thinking about the normative principles that might be applied to our findings. The aim was to bring together each of these elements in order to reflect upon the intersection between the formal oversight mechanisms that are 'on the books' and what was happening 'in practice' at a particular time. This information, we anticipated, would be useful for policymakers and regulatory bodies that were, and still are, trying to establish appropriate systems of oversight, and also for biomedical scientists in the field and scholars in law, sociology and ethics.

### Clarifying Key Parameters and Definitions

To undertake this work it was necessary to address a number of preliminary (and ongoing) issues relating to the parameters of the project. The jurisdictional focus on 'England and Wales' (rather than the UK) was a relatively straightforward reflection of differences in the laws that apply in Scotland and Northern Ireland, particularly following devolution. It was more difficult, however, to pin down definitions relating to other key concepts, in particular governance and regulation, and 'biobanks' and 'genetic databases'. Below we discuss the stance that our project took toward these ideas.

#### *Governance and Regulation*

Defining the kinds of oversight that might be included in our sociological and legal analyses required us to enter a contested terrain about the nature of 'governance': how to define it and how to distinguish it from 'regulation'. There are no universal definitions for these terms; how they are defined is largely dependent upon the differing theoretical perspectives, research methodologies and assumptions of those who may be using them.[41] The way that they are conceptualised and described also varies according to the context being studied (for example, utility companies, and the public or financial sector). Alongside this diversity, there has also been a move away from traditional state-centred definitions of governance and regulation, towards broader understandings of the involvement of private and non-state actors. This has resulted in both broad and narrow definitions of 'governance' and 'regulation' that encompass the law-making powers of parliaments through to the more general understanding of 'the intentional activity of attempting to control, order or influence the behaviour of others'.[42]

Traditionally, the term 'governance' was associated with government and the exercise of power held by political leaders.[43] For example, some commentators have understood governance as 'the capacity of government to make and implement policy, in other words, to steer society'.[44] These understandings—referred to as 'old governance' approaches—see national governments steering from the top down, and concentrate on the degree of control that the government is able to exert over social and economic activities.[45] However, the term has now developed a broader meaning which extends to processes and actors outside the narrow realm of government.[46] An example of this is from a United Nations Development Program report in which governance is defined as:

> The exercise of political, economic and administrative authority to manage a nation's affairs. It is the complex mechanisms, processes, relationships and institutions through which citizens and groups articulate their interests, exercise their rights and obligations and mediate their differences.[47]

Thus, no longer is the focus solely on governmental organs as being the only relevant actors and institutions in the authoritative articulation of the rules of the game.[48] Instead, the concern of the various versions of this newer approach is on the role of networks—the structures, processes and

[41] See G Stoker, 'Governance as Theory: Five Propositions' (1998) 50 *International Social Science Journal* 17, 17 who makes this point about governance; and Jordan and Levi-Faur, who make the same point about regulation in 'The Politics of Regulation in the Age of Governance' in J Jordan and D Levi-Faur (eds), *The Politics of Regulation: Institutions and Regulatory Reforms for the Age of Governance* (Cheltenham, Edward Elgar, 2004) 1, 1.

[42] J Black, 'Critical Reflections on Regulation' (2000) 27 *Australian Journal of Legal Philosophy* 1, 19.

[43] A Mette Kjær, *Governance* (Oxford, Polity, 2004) 1.

[44] *Ibid* 10.

[45] Kjær, *Governance*, n 43 above, 11.

[46] Kjær, *Governance*, n 43 above, 111.

[47] United Nations Development Program, *Reconceptualising Governance* (Discussion Paper 2, 1997) 9.

[48] Kjær, *Governance*, above n 43, 3.

relationships involved in exercising authority in the pursuit of common goals.[49]

'Regulation' is often referred to as an identifiable and discrete mode of governmental activity. However, like 'governance', the term has no single agreed meaning. Baldwin and Cave have suggested that definitions of regulation may be broadly placed into three categories: targeted rules; deliberate state intervention; and all forms of societal control or influence.[50] In its narrowest sense of regulation as targeted rules, the focus is on 'the promulgation of rules [by the state], accompanied by some mechanism, typically a public agency, for monitoring and promoting compliance with these rules'.[51] Lawyers (and, to a lesser extent, economists) tend to favour narrow definitions of 'regulation' owing to a concern with the state's monopoly over the coercive power of the law.[52] A broader understanding of regulation is deliberative state intervention, often found in the political economy literature, which captures 'all efforts of state agencies to steer the economy'[53] or 'all state actions designed to influence industrial or social behaviour'.[54] This approach is not just concerned with traditional rule-making but also encompasses other modes of influence such as the use of economic incentives (eg, taxation and subsidies), contractual powers, deployment of resources, franchises and the supply of information (eg, disclosure requirements).[55] The third category identified by Baldwin and Cave (all forms of societal control or influence) takes us closer to the newer definitions governance, as regulation is seen as more than simply 'law', and law is only one of a number of relevant regulatory mechanisms. Understandings of regulation in this category regard all such mechanisms (whether intentional or incidental)[56] as affecting behaviour and, therefore, as having regulatory status, whether these are actions of the state or derive from other sources, such as market forces.[57]

In this study we use both of these terms but in different ways. *Governance* will be used in its broadest form (covering multiple actors, activities and mechanisms) while *regulation* will be used in a more narrow sense. Our approach reflects a definition of governance as being

[49] Kjær, *Governance*, above n 43, 3–4.

[50] R Baldwin and M Cave, *Understanding Regulation: Theory, Strategy, and Practice* (Oxford, Oxford University Press, 1999) 1–2.

[51] R Baldwin, C Scott and C Hood, 'Introduction' in R Baldwin, C Scott and C Hood (eds), *A Reader on Regulation* (Oxford, Oxford University Press, 1998) 1, 3.

[52] B Morgan and K Yeung, *An Introduction to Law and Regulation* (Cambridge, Cambridge University Press, 2007) 4.

[53] C Hood, H Rothstein and R Baldwin, *The Government of Risk* (Oxford, Oxford University Press, 2001) 23.

[54] Baldwin and Cave, *Understanding Regulation*, above n 50, 2.

[55] Baldwin and Cave, *Understanding Regulation*, above n 50, 2.

[56] Baldwin and Cave, *Understanding Regulation*, above n 50.

[57] Kjær, *Governance*, above n 43, 2.

> [a] multifaceted compound situation of institutions, systems, structures, processes, procedures, practices, relationships and leadership behaviour in the exercise of social, political, economic, and managerial/administrative authority in the running of public or private affairs.[58]

In contrast, we treat regulation as being narrower in scope, applying just to the formal structures of law and legally constituted regulatory bodies that play a role in regulating biomedical research. The relative breadth of these definitions is particularly important to our project, as we aim to reflect upon the relationship between narrowly-defined regulation (eg, laws and guidance 'on the books') and broadly defined governance (eg, the processes and practices reported by biomedical science practitioners).

### *Biobanks and Genetic Databases*

A second complex problem for setting the scope of our research was to try to define what constitutes a 'genetic database' and what such an entity should be called. Multiple terms have appeared in the literature describing collections of genetic (or other biological) materials and/or accompanying health-related information. In a desire to understand the diversity of practices that might fall under the aegis of our study, we did not impose a strict definition at the outset of our work, lest it restrict the scope of our legal research or supplant the understandings of those whom we intended to interview in the sociological research. Nevertheless, at least a working definition was necessary to start the sociological research process.[59] For the purposes of recruiting potential interviewees, we therefore adopted the initial position that a 'genetic database' was something that satisfied at least two of the following criteria: it supported the production of DNA data (due to the availability of associated biological data); it contained information on DNA (eg, results from tests); it contained genomic sequence data; it contained genotype and phenotypic data; it had an associated collection of samples from which DNA could be extracted; it involved some IT system for managing and storing the data. In contrast, the legal research initially focused on how the law and regulatory bodies applied to broad areas—such as 'research' and 'genetics'—to provide a basis for how this might apply to genetic databases.

In 2005, 'genetic database' was the most commonly used term in the UK and so it was the one that we adopted at the outset of our sociological

[58] John-Mary Kauzya, *Local Governance Capacity Building for Full Range Participation: Concepts, Frameworks, and Experiences in African Countries' Experiences in African Countries*, Background Paper, 4th Global Forum on Reinventing Governance 2002, 2 available at www.unpan.org/DPADM/GlobalForum/4thGlobalForum/tabid/604/Default.aspx#Link_Press (accessed 2 June 2010).

[59] Hirtzlin et al, 'An Empirical Survey on Biobanking', above n 2.

research. When Catherine Heeney approached potential participants to be involved in the sociological arm of the project, they were invited to discuss their involvement with 'genetic databases' in an interview, and (as is shown in Chapter Four) they expressed different ideas about what such an entity might be. In the rest of Europe at this time, however, the term 'biobank' was more common. In the intervening years, 'biobank' has become the most common term used internationally, including in the UK, and particularly in relation to 'population-based' collections of biological (including genetic) samples. Given this shift in usage, in this book we use the term 'biobank' as our default, even though, at times, we use 'genetic database' where it is more accurate for describing our research processes and findings.

## Sociological Research Methodology

The sociological research aimed to chart the diversity of 'genetic databases' in England and Wales, and to explore with the biomedical professionals who created, used or managed them their awareness, attitudes, and experiences of governance. The objectives were to

a) devise a typological account of biological sample/information collections that could be defined as 'genetic databases' based on the characteristics that our respondents used to describe them;
b) describe the practices and procedures used to manage and govern these collections; and
c) deport the concerns and needs in respect to governance voiced by those who created, used or managed the collections.

Andrew Smart designed the research using an exploratory qualitative methodology.[60] At the time when our study was proposed and designed, the existing research on governance had adopted a survey-style approach which had provided a broad picture of key issues and practices.[61] There was, however, a paucity of in-depth accounts that considered the complexities and nuances of what professionals did in practice (and why they did it). An exploratory qualitative approach was therefore adopted as it would enable us to chart this ground by allowing the study's respondents to define and explain pertinent issues from their own points of view.[62]

[60] EG Guba and YS Lincoln, 'Competing Paradigms in Qualitative Research' in NK Denzin and YS Lincoln (eds), *Handbook of Qualitative Research* (Thousand Oaks, Calif, Sage, 1994) 105–17.
[61] Hirtzlin et al, 'An Empirical Survey on Biobanking', above n 2.
[62] S Arthur and J Nazroo, 'Designing Fieldwork: Strategies and Materials' in J Ritchie and J Lewis (eds), *Qualitative Research Practice. A Guide for Social Science Students and Researchers* (London, Sage, 2003) 109–37.

A purposive (non-random) sampling strategy, which sought to reflect heterogeneity within the field, was used to ensure that a range of participants with relevant views and experiences would be included in our study.[63] Our recruitment strategy was largely dictated by a paucity of information on the existence and location of 'genetic databases' and the people who ran them in England and Wales. As such, a snow-balling strategy was adopted in order to identify and recruit potential participants, whereby each interviewee was asked to provide the names of other people or studies that might be relevant to our project.[64] At the start of the research process, Catherine Heeney conducted four 'scoping' interviews[65] with individuals who were known to be involved with a variety of genetic studies and collections, from large multi-site international collaborations, sizable epidemiological projects and large disease-based studies, through to small Mendelian studies. These interviewees also represented a range of disciplines and specialisations.[66] The sample of respondents in the study started from the recommendations of these interviewees. To ensure that our sample included a wide range of the manifold types and sizes of database, Catherine Heeney used *Mindmanager* software to create a visual map of types of organisations that were known or were likely to have genetic databases in a range of sectoral and organisational settings.[67] She continued to approach potential respondents until she ceased to identify novel variations of database.

Semi-structured interviews were chosen as the research method that best suited the study aims, as such interviews enable both focus and flexibility in data collection.[68] The team collaborated in the design of an interview schedule. This was informed by the data gathered in the 'scoping interviews' which, for example, provided insights into aspects of practice and current terminology. Given our research aims, interviews which included open-ended questions offered a greater chance of gathering in-depth responses and reflections from practitioners, allowing interviewees to discuss their attitudes and experiences of governance in this area and drawing out what was particularly important to them.[69] The interview schedule also

[63] J Ritchie, J Lewis and G Elam, 'Designing and Selecting Samples' in J Ritchie and J Lewis (eds), *Qualitative Research Practice: A Guide for Social Science Students and Researchers.* (London, Sage, 2003) 77–108.

[64] *Ibid.*

[65] This data did not form part of the dataset for the study.

[66] See appendix 5 below.

[67] Knowledge of the existence of these entities was arrived at in a similar way as suggestions about who should be interviewed—via knowledge of the field coming from funders, the research team (who sought to keep up with current developments in the field) and the scoping interviews. Knowledge of a project could be the,'way in' to request an interview with particular individuals and conversely awareness of the work of particular researchers could lead to a focus on the particular research projects that they were involved in. Unsurprisingly the larger and more visible projects were most likely to be included in the former way.

[68] S Kvale, *InterViews* (London, Sage 1996).

[69] *Ibid.*

included a number of more 'closed' questions designed to collect descriptive information about the 'genetic database' itself, as well as direct questions about the laws, regulation and guidance and about interactions with 'actors' involved in governance.

Catherine Heeney conducted 49 interviews with a broad cross-section of practitioners, drawn from a range of professional backgrounds, areas of expertise, experience, positions, responsibilities, and organisational or institutional affiliations. The interviews took place between May of 2006 and November 2007 and were carried out face-to-face at the respondents' workplaces (except for one which was conducted by telephone). For the purposes of identifying respondents and communicating in our finding, we have labelled each respondent according to two primary characteristics—their 'professional background' and the type of organisation at which they were working at the time of interview.[70] We fully recognise that even these apparently straightforward matters are not always clear. As will be shown (in Chapter Four), many people have complex professional histories or overlapping areas of expertise, and organisational affiliations can be multiple. For the sake of clarity, however, we have allocated respondents into a small number of categories that represent their professions and types of organisation using the following labels:

> ***Profession:*** Bioinformatician; Clinical Scientist; Research Scientist; Epidemiologist; IT Specialist; Other health professional (each one is specified)
> ***Organisation:*** The respondents worked primarily at: University (Uni); University linked to NHS Hospital (Uni/Hos); Repository (Repository); NHS Trust or Hospital (NHS); Pharmaceutical Company (Pharma).

A few of these labels may benefit from further explanation. The essence of the division between 'Clinical Scientist' and 'Research Scientist' is that we use the former to denote those respondents who had (or had previously had) routine contact with patients (usually in NHS settings). We have used 'IT Specialist' to label those respondents who primarily created and maintained information technology infrastructures; while 'Bioinformaticians' (who often had substantial expertise in ICT) were biomedical scientists or researchers exploiting the power of ICTs. We have used the term 'Repository' to refer collectively to organisations that included specialist biobanks (created to collect new research samples and data), and networks (of people and technologies) created to facilitate or promote sharing of existing research samples and/or data.

[70] See appendix 5.

The interviews were audio-recorded and transcribed verbatim. A thematic coding scheme was developed iteratively.[71] Catherine Heeney read the first five interviews and drew out a list of common and interesting themes relating to our research aims, which were refined and organised in discussions with Andrew Smart. This process was repeated twice more with two further sets of five transcripts. By the end of this process, four broad codes, each containing numerous narrower sub-codes, were agreed: 'Context', 'Guidance,' 'Ethics in Practice' and 'Typology'. Catherine Heeney applied this coding scheme to the interview transcripts using the *N6* qualitative data analysis software package and made the results available to the team for further analysis. The 'Typology' code aimed to capture descriptive data about the practices associated with 'genetic databases'. The various sub-codes formed the 'structural dimensions'[72] under which the observed characteristics of 'genetic databases' are described and discussed by Catherine Heeney in Chapter Four. The 'Guidance' code contained the following sub-codes: 'Advisory Bodies'; 'Guidance' (including written guidance from organisations which were not, strictly speaking, advisory bodies); 'Implicit Guidance' (including standards of behaviour absorbed during medical training, as well as informal chats with colleagues); and 'Laws'. The data captured under these codes were collectively analysed by the whole team, and Sue Gibbons and Andrew Smart developed this work to form the basis of Chapter Five. The 'Context' code included four sub-codes that captured data on the broader political, social and scientific environment, including one entitled 'Attitudes to Governance' (a large sub-code comprised of 494 data items). Data coded here were further analysed by Sue Gibbons and Andrew Smart and form the basis for Chapters Six to Eight. The 'Ethics in Practice' code aimed to capture data relating to the practices that interviewees described as having an ethical significance. There were four sub-codes: 'Confidentiality Arrangements'; 'Consent'; 'Record Management'; and 'Access Arrangements'. This latter sub-code was also large, comprising 336 data items which were further analysed by Catherine Heeney and Andrew Smart to form the basis for Chapter Nine.

Our decision to use exploratory qualitative research, based on a purposive heterogeneous sample and a snow-balling recruitment strategy, introduces an associated range of methodological limitations. A common concern is the necessity for interpretative judgments to be made about interview data. Such judgments are unavoidable. However, as described above, the development of the coding framework and subsequent analyses were formed through discussions between team members, in which matters of interpretation

[71] R Gomm, *Social Research Methodology. A Critical Introduction* (Basingstoke, Palgrave MacMillan, 2004) 189.

[72] I Chompalov and W Shrum, 'Institutional Collaboration in Science: A Typology of Technological Practice' (1999) 24 *Science, Technology, and Human Values* 338–72.

were held open to debate and approval. A second concern relates to generalisation from samples of respondent selected non-randomly, and that may be relatively small in size and heterogeneous. We made attempts to ensure diversity in our sample, not least because charting such diversity was one of our core aims. Nevertheless, we make no claims that our findings about the actions and opinions of our respondents are statistically representative of others involved in 'genetic databases'. A different form of generalisability—theoretical generalisability—can be invoked to underpin the relevance of our research.[73] Our evidence documents the existence of particular actions and opinions, and creates a basis for theorising that similar patterns may exist elsewhere (thus providing the grounds for further research to discover the extent of this). Our research thus charts some relatively unknown waters and lays a foundation for further exploration.

The size and constitution of the sample also restricts our ability *systematically* to disaggregate the respondents' views and practices by factors such as the type of research that their 'genetic database' was used for, the number of samples, how long it had existed and so forth. Where such intersections are notable, attention will of course be drawn to them, but as warranting future investigation rather than as conclusive evidence of differences. We have, as noted above, labelled each respondent according to two primary characteristic that represent their professions and types of organisation. We recognise that labeling respondents in this way risks imposing these characteristics as analytical devices, and we guard against jumping readily to conclusions based on the divisions that the labels impose between, say, 'Bioinformaticians' and 'Clinical Scientists', or between people who worked primarily in Universities compared to those working in Repositories. As will be shown in Chapter Four, there is enormous diversity and complexity in this field of activity, which serves as a warning against drawing firm differentiations between different respondents when considering their views and experiences.

Ethical approval for the study was obtained via the COREC process prior to the commencement of fieldwork. During this process a number of documents were consulted, including the Guidelines from the British Sociological Association.[74] Interviewees upon first contact, usually by email, were given a description of the project and its aims, followed by an information sheet and a letter inviting their participation. The information sheet contained an explanation relating to the fact that data would be used for research purposes only and that any further dissemination of data would be in non-identifiable form. This information was also given verbally at the start of

[73] J Mason, *Qualitative Researching* (London, Sage, 1996).

[74] British Sociological Association, *Statement of Ethical Practice* (2002), available at <http:// www.britsoc.ac.uk>.

interviews, and interviewees were given the opportunity to ask for further information before being asked to sign a consent form.

### *The Legal Research*

The purposes of the legal research were

a) to examine the existing framework of laws and regulations applicable to biobanks in England and Wales;
b) to identify the bodies that were responsible for the regulation and oversight of biobank activities within England and Wales; and
c) to explore the nature and possible shortcomings within the existing legal governance mechanisms, rules and frameworks.

The first task conducted by Sue Gibbons was to identify, collate and critically analyse the many laws, guidelines and governance structures that applied to biobanks in England and Wales at the time of our study. At that time, no comprehensive accounts existed either of the laws that applied to biobanks in the UK or of potential regulators. The legal and quasi-legal materials covered included: legislation; regulations; common law and equitable doctrines; guidelines, circulars and codes of practice; and international instruments (including EC legislation). From this compilation it was possible to get a sense of the various bodies that inhabited the regulatory space for medical research. This, in turn, generated an interest in focusing on the powers, remits and responsibilities of the various bodies (eg, the Human Tissue Authority, the Information Commissioner and research ethics committees) that presently regulated research and the use of human tissue and information.

One of the aims of the legal research was to explore theoretical approaches to governance and the key substantive issues that surround biobanking. We reviewed the general literature on theories of governance and regulation and then focused on the literature on regulating biobanks. Intrinsic to this research was ensuring an understanding of the approaches and experiences of governing biobanks in other countries, and the international debate over the issues and principles that arise in respect of population biobanks. From this investigation, we also identified the concept of the 'regulatory space'[75] as being a useful conceptual tool and framing device to start to disentangle the complexity of the regulatory framework that applied (and still applies) to medical research (see Chapter Three). This would provide a basis for further analysis of the way in which the relevant bodies interacted, as well

[75] See, eg, L Hancher and M Moran, 'Organizing Regulatory Space' in R Baldwin, C Scott and C Hood, *A Reader on Regulation* (Oxford, Oxford University Press, 1998) esp 148–72.

as providing a firm conceptual basis for comparison with the empirical research findings.

Like the empirical research, in the legal research we faced similar issues of how to define and narrow the scope of our enquiry. We had hoped to focus on the law that applied specifically to biobanks in England and Wales. However, that inevitably required a broader understanding of the law and the other ways in which medical research was regulated in the UK. Keeping up to date with all of the changes in this field from 2005 until 2009 was a huge task as the law in this field was in a constant state of change. In addition, regulatory bodies were continually being created, disbanded or merged, or their powers were being expanded, reduced or transferred. It is worth noting that, with the election of the coalition government in 2010, many of the advisory and other relevant bodies that were part of the regulatory framework during this time have been or are scheduled to be disbanded, and more changes to the legislative framework are anticipated. It was also evident when we started the project that it was not possible to capture all of the bodies having influence in regulating biobanks purely by focusing on those bodies with statutory powers. It was only by mapping all of the bodies that were operating in the regulatory space, both nationally and at a European level, that we were able to get a sense of the most influential bodies, and how medical research was being governed in England and Wales.

### *The Ethical Analysis*

The ethical analysis sought to engage with both the empirical examples and conceptual framework provided in Chapter Four, as well as drawing on the practical experience of Michael Parker providing ethical support and analysis for a wide variety of projects involving genetics research. The ethical issues drawn out in Chapter Eleven resulted, therefore, from discussions on the interconnected and open nature of genetic databases, and the resources and personnel associated with these entities. The aim of combining bioethics with social science methods and theory in this way was to ensure that the ethical issues presented took into account practices encountered empirically and the conceptual framework developed in Chapter Four to capture these developments. An important characteristic of practice, as captured by the empirical work for the GGD project, is that the boundaries of the entities in question, biobanks or 'genetic databases, are open and subject to change due to numerous influences and interactions, which means that the features of genetic databases are not always predictable. Given this context the task was to consider what were the key ethical areas and how best to engage with the dynamic nature of practice. The ethical concerns or issues which were raised by reflection upon practice were contextualised within the wider context of ongoing developments affecting the biomedical research community. The description of types of collaborations and technical structures described were built up from the observations made by Michael Parker and drew on knowledge of a variety of research projects. This supported a broad consideration of developments relevant to the potential and current ethical issues associated with genetic databases. The chapter, therefore, drew upon the results from the empirical work carried for the GGD in conjunction with an overview of relevant developments in order to define and develop the ethically interesting aspects of practice in this field and reflect upon possible ways of dealing with these.

*Bringing Different Disciplines Together*

As this research drew together different disciplinary approaches, one of the challenges we faced, both individually and as a team, was to become familiar with the starting points of each academic discipline, but, at the same time, also to allow the concerns of each discipline to be explored through the research. We had to ensure that the research could have credibility when judged by its disciplinary peers but also that it added to the aims of the project as a whole and was interdisciplinary. This was a challenge, but also very stimulating, as the lawyers and ethicist learnt about 'boundary objects',[76] while the sociologists and ethicist learnt about the distinctions between national and European law. To enable this conceptual and informational sharing and enrichment to happen, we met as a team on a regular basis and had a series of research days where we presented our respective findings and compared different interdisciplinary perspectives. By bringing together and assessing the findings from the two major prongs of our research (legal and empirical), our aim was to be in a position to provide a snapshot of how a new area and approach for medical research was being governed in England and Wales.

To ensure that we were in touch with important issues, we established an expert advisory panel for this project. Overseeing the project were the following experts: Dr Elizabeth Fisher (regulation and law), Prof David Vaver (intellectual property law), Prof Dominic Kwiatkowski (population genetics), Dr Sue Dopson (organisational sociology), Prof Rory Collins (epidemiologist, Co-Director of the Clinical Trials Service Unit, University of Oxford), Dr Jane Green (epidemiologist) and Dr Fiona Douglas (clinical geneticist). Meetings of this panel at the start and towards the close of the project were held to assist with the conceptualisation and presentation of the research.

[76] Callon, 'Some Elements of a Sociology of Translation', above n 39.

## ASSESSING A GOVERNANCE STRUCTURE

One of our aims was to be in a position to make an assessment of the current governance structure for biobanks in England and Wales. To be able to evaluate whether a system of oversight is good, acceptable or in need of reform, as we are trying to do in this book, clear criteria or benchmarks are required. There are many different ways of thinking about 'good regulation'. This section will set out the criteria that can be found in regulatory guidelines, such as those articulated by the Better Regulation Task Force (BRTF), which are used in the analysis in various parts of the book and in the final chapter. The approach of academics, such as Baldwin and Cave, is also a useful starting point in understanding the elements of legitimacy that are required for 'good regulation'. Attempting to define legitimacy can be a thankless task. As Stone says, 'We know it exists as a force that holds societies together, but we cannot give very satisfactory explanations of how to create it or why it is sometimes very strong and sometimes seems to disappear.'[77] Even so, it is generally agreed that '[a] legitimate authority is one that is recognized as valid or justified by those to who it applies'.[78] The approach of Baldwin and Cave was designed for the evaluation of governmental processes, and identifies the evaluation as being one of *legitimation*. It attempts to map the 'sorts of reasons' that persuade people to accept regulatory decisions, rather than conducting an exercise in moral reasoning in order to evaluate whether a decision is morally correct.[79] Further, the criteria suggested by Baldwin and Cave summarise in more general terms many of those criteria expressed by the BRTF. While these principles are theoretical in nature, they have provided a useful basis for thinking about the practical nature of governance and regulation in this field as we have progressed.

### Elements Required for Legitimacy

To evaluate a system of oversight, Baldwin and Cave suggest that the evaluation can be performed by looking at five dimensions of legitimation: legislative mandate; accountability; due process; expertise; and efficiency.[80] We will look at these in turn.

[77] D Stone, *Policy Paradox: The Art of Political Decision-Making* (New York and London, Norton, 2002) 285.

[78] V Bekker, G Dijkstra, A Edwards and M Fenger, 'Governance and the Democratic Deficit: Introduction' in V Bekker, G Dijkstra, A Edwards and M Fenger (eds), *Governance and the Democratic Deficit; Assessing the Democratic Legitimacy of Governance Practice* (London and New York, Ashgate, 2007) 7.

[79] Morgan and Yeung, *An Introduction to Law and Regulation*, above n 52, 222.

[80] Baldwin and Cave, *Understanding Regulation*, above n 50, ch 6.

### *Legislative Mandate*

The legislative mandate criterion states that regulatory action deserves support when it is authorised by Parliament.[81] Regulators have legitimacy if they have fulfilled their mandate which they have been given by Parliament, which is elected by the people. The difficulty is that statutes often give regulators broad discretion and powers of discretion in implementing their mandates.[82] Furthermore, statutes may set out objectives that exist in tension with each other, or which may even be in conflict.[83] However, there is an increasing recognition that the nature of the state is undergoing considerable change, particularly with globalisation pressures, and that the role of the state has moved on from being the instigator and actor of regulation to having more of a steering role.[84] Therefore, in this role, Parliament may delegate decision-making to other bodies, or take a 'hands-off' approach—thereby allowing areas of activity to be subject to levels of interference or scrutiny by Parliament.

### *Accountability*

This basic principle requires that a regulatory agency be accountable for its interpretation of its mandate to a representative body.[85] As long as such oversight is in place, it means that delegation by Parliament is acceptable. However, if the regulator is not accountable to Parliament or another elected institution, this can be criticised on the basis that is it unrepresentative; and, if accountability is to a court, then questions of judicial competence arise.[86] Scott has argued that accountability can also be carried out by bottom-up mechanisms such as markets, grievance mechanisms or consultations with users, as well as horizontal checks and balances that are not parliamentary-based. Auditors, third-party accreditation standards, and supervision by public interest groups all are relevant.[87] Such an extension acknowledges the increasingly decentred nature of modern regulation.[88]

### *Due Process*

The third dimension outlined by Baldwin and Cave is due process, which is premised on the assertion that public support is cultivated if regulatory

[81] Baldwin and Cave, *Understanding Regulation*, above n 50, 78.

[82] Baldwin and Cave, *Understanding* Regulation, above n 50, 78.

[83] Baldwin and Cave, *Understanding Regulation*, above n 50, 78.

[84] Bekker et al, 'Governance and the Democratic Deficit', above n 78; A-M Slaughter, 'Global Government Networks, Global Information Agencies and Disaggregated Democracies' (2003) 24 *Michigan Journal of Internationl Law* 1041–74.

[85] Baldwin and Cave, *Understanding Regulation*, above n 50, 78–79.

[86] Baldwin and Cave, *Understanding Regulation*, above n 50, 79.

[87] C Scott, 'Accountability in the Regulatory State' (2000) 27 *Journal of Law and Society* 38, 41–42.

[88] Morgan and Yeung, *An Introduction to Law and Regulation*, above n 52, 229.

agencies adopt procedures that are fair, accessible and open.[89] This involves improving the clarity of regulatory objectives and setting out guidelines detailing the regulatory visions and compliance guidelines.[90] The aims are to reduce uncertainty and to improve the consistency and predictability of regulation. Attention is therefore focused on equality, fairness and consistency of treatment, as well as on the levels of participation that regulatory decisions and policy processes allow to the public, consumers and other affected parties.[91] The limitations of this criterion are that further guiding principles are necessary in order to explain when procedures are fair, accessible and open. For example, who should be able to participate? What are the procedures for participation? In what manner, and to what extent, can interested parties participate?[92] Additionally, trade-offs may also have to be made against the effective implementation of the mandate; for example, greater public participation may lead to less effective decision-making and delay implementation of the regulatory regime.[93]

*Expertise*

Many regulatory functions require the exercise of expert judgment. This is particularly the case where the decision-maker must consider and balance competing opinions, options and/or values.[94] Where this need arises, the regulator may claim support on the basis of his or her expertise and the nature of the task at hand, rather than providing reasons or justification.[95] There are a number of problems with this criterion. These include the fact that it makes it difficult to judge whether the decisions are appropriate or effective; its failure to explain why *this* particular issue requires expert judgment; a natural distrust by lay persons to experts; and the potential for experts to be biased.[96]

Efficiency

Finally, according to Baldwin and Cave's rubric, a regulator may claim support on the basis of acting efficiently. This argument may refer either to productive efficiency, in the sense that the legislative mandate is being implemented at the least possible cost, or that the regulation leads to results

[89] Baldwin and Cave, *Understanding Regulation*, above n 50, 79.
[90] Baldwin and Cave, *Understanding Regulation*, above n 50, 322.
[91] Baldwin and Cave, *Understanding Regulation*, above n 50, 79.
[92] Scott, 'Accountability', above n 87.
[93] Baldwin and Cave, *Understanding Regulation*, above n 50, 79.
[94] Baldwin and Cave, *Understanding Regulation*, above n 50, 80.
[95] Baldwin and Cave, *Understanding Regulation*, above n 50, 80.
[96] Baldwin and Cave, *Understanding Regulation*, above n 50, 80.

that are efficient.[97] The general problem with this criterion is that wealth maximisation alone does not provide a full account of regulatory justification and evaluation, because such arguments do not justify any particular distribution of rights within society.[98] Debate still exists concerning the conflict between efficiency and social or distributional objectives.[99]

### Better Regulation Principles

The five dimensions described by Baldwin and Cave could be used to assess the legitimacy of a state or regime, as well as an area of regulation. In contrast, the better regulation principles as outlined by the UK Better Regulation Commission identify five principles of good regulation which, it recommends, regulators should bear in mind when devising, implementing, enforcing and reviewing regulations. The BRTF is used in assessments made by the Cabinet Office and largely applies to government bodies, but does not have the same scope as the dimensions formulated by Baldwin and Cave.

By way of background, from the mid-1990s, the UK government actively pursued a 'better regulation' agenda. Its implementation now affects public authorities across the board. This includes prominent actors within the biobanking regulatory space.[100] During March 2005, the Hampton Review made a range of recommendations for reducing regulatory burdens without compromising regulatory outcomes, essentially by promoting more efficient approaches to regulatory inspection and enforcement.[101] Among other things, it identified various 'principles' that should apply throughout the regulatory system. Paraphrased, the main Hampton Principles require regulators to: allow economic progress; concentrate resources based on comprehensive risk assessments; give 'authoritative, accessible advice easily and cheaply'; limit inspections and visits; streamline the information required from regulatees; impose 'proportionate and meaningful sanctions' on persistent violators; and be both independent in their decision-making and 'accountable for the efficiency and effectiveness of their activities'.[102]

Also in March 2005, the Better Regulation Commission reported to the UK Government. It made eight recommendations for cutting the administrative

[97] Baldwin and Cave, *Understanding Regulation*, above n 50, 81.

[98] S Kelman, 'Cost-Benefit Analysis: An Ethical Critique' (1981) 5(1) *Regulation* 33; A Kronman, 'Wealth Maximisation as a Normative Principle' (1980) 9 *Journal of Legal Studies* 227; Baldwin and Cave, above n 50, 76–77.

[99] Baldwin and Cave, *Understanding Regulation*, above n 50, 81.

[100] Notably, the Human Tissue Authority, Information Commissioner, and Human Fertilisation and Embryology Authority. See ch 3 below.

[101] P Hampton, 'Reducing Administrative Burdens: Effective Inspection and Enforcement' Final Report, (London, HM Treasury, March 2005).

[102] *Ibid* 7, Box E2.

costs of regulation to businesses.[103] Among other things, the Commission concluded that any 'appropriate framework of regulation' must pass five 'tests', as embodied in its so-called 'five principles of good regulation'.[104] Those principles stipulate that regulation should be: transparent; proportionate; accountable; consistent; and targeted only at cases in which action is needed. This fivefold set of 'good governance' principles is now enshrined in the Legislative and Regulatory Reform Act 2006 (LRRA).[105] The accompanying statutory Regulators' Compliance Code[106] stipulates how any organisations exercising regulatory functions under the LRRA must go about developing their general policies and principles for exercising those functions, setting standards, and giving general guidance. The Regulators' Compliance Code is both based on the Hampton Principles and upholds the LRRA's five good governance principles.

The 'better regulation' principles are: proportionality; accountability; consistency; transparency and targeting.[107] They will be used as the basis for assessing the governance structure for biobanking in our final chapter.

1. **Proportionality** requires that: 'Regulators should only intervene when necessary. Remedies should be appropriate to the risk posed and costs identified and minimised'.[108] For example, this involves: considering all alternative options for achieving policy objectives; no 'gold plating' of EC Directives; and consideration of educational rather than punitive approaches to enforcement.
2. **Accountability** is explained in the following terms: '*Regulators must be able to justify decisions and be subject to public scrutiny*'.[109] This means that: proposals should be published and those affected consulted before decisions are made; reasons should be given for final decisions; the establishment of standards against which regulators and enforcers can be judged; well-publicised accessible, fair and effective complaints and appeals procedures; and lines of accountability between regulators and enforcers to Ministers, Parliaments and assemblies, and the public.
3. **Consistency** means that: '*Government rules and standards must be joined up and implemented.*'[110] Therefore: regulators should be consistent with each other and work with each other; regulation should

103 Better Regulation Task Force, 'Regulation—Less is More: Reducing Burdens, Improving Outcomes: A BRTF Report to the Prime Minister' (London, March 2005).

104 *Ibid* Annex B.

105 Legislative and Regulatory Reform Act 2006, esp ss 2 and 21.

106 Department for Business, Enterprise and Regulatory Reform, *Regulators' Compliance Code: Statutory Code of Practice for Regulators* (17 December 2007).

107 Better Regulation Task Force, *Principles of Good Regulation* (revised 2007) available at <http://archive.cabinetoffice.gov.uk/brc/publications/principlesentry.html>.

108 *Ibid.*

109 Better Regulation Task Force, *Principles of Good Regulation*, above n 107.

110 Better Regulation Task Force, *Principles of Good Regulation*, above n 107.

be predictable in order to give stability and certainty to those being regulated; enforcement agencies should apply regulations consistently across the country; and new regulations should take account of other existing or proposed regulations (domestic, EU or international).

4. **Transparency** is described as follows: '*Regulators should be open and keep regulations simple and user-friendly*.'[111] This means that: policy objectives should be clearly defined and communicated to interested parties; consultation must take place before proposals are developed to ensure that stakeholder views and expertise are taken into account; those being regulated should be made aware of their obligations, with law and best practice distinguished; those being regulated should be given time and support to comply; and the consequences of non-compliance should be clear.
5. **Targeting** means that: '*Regulation should be focused on the problem and minimize side effects*.'[112] The key considerations in this regard are: regulation should avoid a scattergun approach; regulators should adopt a 'goals-based' approach, with enforcers and those being regulated given flexibility in deciding how to meet targets; enforcers should predominately focus on those whose activities give rise to the most serious risks; and regulations should be reviewed to determine whether they are still necessary and effective.

## THE STRUCTURE OF THIS BOOK

The findings from this study must be seen as a snapshot of peoples' experiences of the governance system for biobanks over a specific period of time (2005–2008) and in one jurisdiction, England and Wales. This was a time when biobanks were starting to proliferate but there was no clear or dedicated framework as to how they should be governed. The picture that we have gained is one of complexity and diversity, with biobanks being established and utilised in very nuanced and specific contexts. To add to this complexity, each of these contexts—whether they be clinical or research orientated, disease-specific or more generic—has a set of values and practices that affect the choices made about the adoption and enactment of governance, even though there are some identical governance structures—such as national legislation or the Information Commissioner's Office—that should have influence over the field generally. Despite this diversity and complexity, and the fact that qualitative interviews findings cannot reliably be used as a definitive statement about situations, we can report on the concerns

[111] Better Regulation Task Force, *Principles of Good Regulation*, above n 107.
[112] Better Regulation Task Force, *Principles of Good Regulation*, above n 107.

that our respondents had about the governance mechanisms for biobanks. This study can provide us with broader insights into the ways in which new technologies and methodologies within health interact or are absorbed into existing governance structures. It also provides an understanding of the role of law in this context and how the regulatory framework applies to medical research.

This book is structured in three parts, with the first setting the scene and outlining the background to biobanking and genetics. It outlines the rationale for the establishment of biobanks and documents the influences on scientific practice. The second part contains the bulk of the analysis from the mapping of the regulatory space, and the legal and empirical research, and is the largest section of the book. The third part pulls together and synthesises these findings to provide a basis for the final concluding chapter, wherein we reflect on our findings about the governance of biobanks in England and Wales. In the final chapter we provide an analysis that is based on the better regulation principles as a foundation for thinking about some of the advantages and disadvantages of the way in which biobanks have been, and are, governed and regulated in England and Wales. This analysis also has the potential to provide insights into the ways in which new areas of research or approaches are accommodated within existing governance frameworks, and the ways in which they might better be supported.

# 2

# Embedding Biobanks in a Changing Context

*Jane Kaye*

THE LAST FIVE years have seen considerable growth in the biobanking field, with the establishment of new biobanks, such as population biobanks, but also with the development of plans to link existing clinical or research collections of samples on a national and international level.[1] It is very difficult to disentangle the influences that are responsible for these developments as they are all interrelated and influence each other. However, one project that has had a significant effect on the development of sequencing technology, computer science and the way of 'doing science' in genomics is the Human Genome Project (HGP). The advances in sequencing technologies and in bioinformatics spurred by this project have increased the ability to integrate, compile and compare large datasets. This technology would not have advanced without having specific scientific problems to address, that required specialist solutions. But in turn, these technological solutions have enabled different kinds of scientific questions to be asked, and in doing so have had an influence on scientific practice. The open-access policies formulated after the HGP, by funding bodies in collaboration with the scientific community, have also encouraged and supported the development of infrastructure to support genomics. Coupled with this, are some fundamental changes that have occurred in the regulatory framework within the UK. The purpose of this chapter is to provide an understanding of the changing context in which biobanks are embedded, focusing in particular on the period of the project—2005–2009—and to show how events within and outside of the scientific context have had an effect on this field.

[1] Eg, the BBMRI: www.bbmri.eu.

## THE HUMAN GENOME PROJECT

The antecedents of the ways that biomedical research and biobanks are currently organised and executed can be found in the approach and principles that were developed in the HGP. The HGP commenced in 1990, and was finally completed in 2003, with the aim of mapping the whole human genome. The purpose of the Human Genome Project was to identify all the approximately 20,000–25,000 genes in human DNA and determine the sequences of the 3 billion base pairs that make up human DNA. This ambitious project was first discussed in 1988 in the USA and commenced in 1990 under the leadership of James Watson. The draft sequence of the human genome was announced on 26 June 2000.[2] It was finally completed in 2001 and the full sequence was published in *Nature* on 15 February 2001.[3] It involved many funding bodies and the sequencing of the genome was carried out by a number of key research institutions across the globe. The majority of the sequencing effort was carried out by the National Institutes of Health (NIH) and the Department of Energy (DOE) in the USA, with a contribution of 60 to 70 per cent of this sequence, and the remainder coming from the effort of other key international partners, such as the Sanger Centre, funded by the Wellcome Trust.[4] The two key principles of the HGP were inclusion—that the collaboration would be open to centres from any nation—and the second principle was rapid and unrestricted data release.[5] It was this big vision approach that was novel in the biomedical sciences—both in the task that it was trying to achieve and the manner in which it was done. Without understanding the HGP and how it operated it is not possible to understand the rationale that lies behind biobanks, as the HGP was the forerunner to many of the organisational practices that now exist in genomics. The project achieved three significant things—an approach based on common endeavour and sharing, the development of new sequencing technologies and computational science and a knowledge platform to launch new scientific projects and resources.

[2] C Macilwain, 'World Leaders Heap Praise on Human Genome Landmark' (2000) 405 *Nature* 983–84.

[3] (2001) 409 *Nature* 745–964.

[4] FS Collins, A Patrinos, E Jordan, A Chakravarti, R Gesteland, L Walters and the members of the DOE and NIH planning groups, 'New Goals for the U.S. Human Genome Project: 1998–2003' (1998) 282 *Science* 682–89.

[5] International Human Genome Consortium, 'Initial Sequencing and Analysis of the Human Genome' (2001) 409 *Nature* 860–921; Human Genome Project: www.ornl.gov/sci/techresources/Human_Genome/home.shtml.

## A NEW APPROACH

The HGP marked a change from hypothesis-led projects to one large, international, collaborative project focused on a common purpose involving many teams of researchers who had previously been competitors for funding. The vision of the HGP attracted 'a remarkable collection of scientists, representing many countries, many different disciplines, and many levels of seniority, coalesced around these shared goals'.[6] The initial success of the project was due to the fact that it was organised through international interactions on a scientist-to-scientist level. 'Had we at first tried to organize this project through heads of state and ministers of health, the HGP would not have worked so easily and efficiently'.[7] This was the first time that a project in this field had been organised on a global level involving so many different groups. The HGP provided a precedent that is now an increasingly common characteristic of genomic science, where large interdisciplinary consortia led by scientists are continuingly being united under a shared common goal. Projects that have followed the HGP are the Encyclopedia of DNA Elements (ENCODE), the Human Epigenome Project, the HAPMAP Project and the 1000 Genome Project.

The HGP built upon the expertise and knowledge that had been developed by smaller projects in sequencing smaller genomes.[8] However, sequencing something as big as the human genome was a different order of magnitude and required a different kind of organisation. As James Watson wrote in 1990, the 'cottage industry' approach of sequencing in small bespoke labs with 'small groups of individuals, each working at a different site seems unlikely to succeed'.[9] The ambitious vision of the HGP and the scale of the enterprise required a new approach to carrying out biomedical science that had not been tried before. Those leading the endeavour argued that the only way to complete it within the anticipated 15 years time scale was to bring together resources and people from around the globe. In the end over 2,000 scientists in 20 centres in six countries—China, France, Germany, Great Britain, Japan and the United States—were involved in the HGP. However, embarking on a project of this kind also creates a number of tensions—a number of which are still evident today in this field (see Chapter Nine).

One of the early concerns expressed by Watson was whether self interest would trump this wider endeavour. Writing in 1990, Watson expressed as concern as to whether scientists would join in this collaborative effort and

[6] FS Colins, M Morgan and A Patrinos, 'The Human Genome Project: Lessons from Large-Scale Biology' (2003) 300 *Science* 286–90, 286.

[7] *Ibid*, 288.

[8] International Human Genome Consortium, 'Initial Sequencing', above n 5.

[9] JD Watson, 'The Human Genome Project Past, Present, and Future' (1990) 248 *Science* 44–49, 45.

be prepared to sequence a whole chromosome and not just locate the genes within them.

> It would be naive to expect that any extensive human sequence data will be released by a sequencing group until it has reasonable time to explore its implications. However, making the sequences widely available as rapidly as practical is the only way to ensure their full value will be realised and is the only acceptable way to handle information produced at public expense.[10]

The second major tension was whether funding bodies and the general public in the US would require that this should be a national project. One of the key ideas underlying the project was that of moving beyond a solely national perspective. 'The nations of the world must see that our human genome belongs to the world's people as opposed to its nation.'[11] This ethos of sharing promulgated through the HGP is the key principle behind biobanking, as a biobank is a resource established specifically to be made available to other researchers no matter where they are located.

## Open-Access Policies

The twin principles of collaboration and open access to the data generated in the HGP were supported by policies developed by the scientific community. The Bermuda Agreement (1996) was the first document to lay out the principles for open access in the field of genomics and applied specifically to the sequence data that was being produced in the HGP. The key idea being promoted in the Bermuda Agreement was that the pre-publication genome sequence 'should be freely available and in the public domain in order to encourage research and development and to maximise its benefit to society'.[12] The other benefit of this policy is that it stopped individual scientists hoarding data but also prohibited companies such as Celera patenting the genes that had been identified through the HGP.[13] The Bermuda Agreement was followed by the Fort Lauderdale Agreement in 2003,[14] which also applied to sequence data but went further by laying out a plan of 'tripartite responsibility' for sequence producers, users and funders. It called for the establishment of 'community resources' to achieve rapid and open data release. It was the first statement that articulated the benefits of the

[10] *Ibid*, 46.

[11] Watson, 'The Human Genome Project Past', above n 9, 49.

[12] The Bermuda Principles, First International Strategy Meeting on Human Genome Sequencing (Bermuda, 25–28 February 1996), www.ornl.gov/sci/techresources/Human_Genome/research/bermuda.shtml#1.

[13] E Marshall, 'Bermuda Rules: Community Spirit, With Teeth' (2001) 291 Science 1192.

[14] *Sharing Data from Large-scale Biological Research Projects: A System of Tripartite Responsibility* (Fort Lauderdale, Wellcome Trust) www.genome.gov/Pages/Research/WellcomeReport0303.pdf.

approach that had been developed through the HGP. The Fort Lauderdale agreement stated that

> community resource data sets benefit the users enormously, giving them the opportunity to analyse the data without the need to generate it first. The data sets are, in general, much larger, *richer* and of higher quality than individual laboratories could normally generate.[15]

Such datasets have been presented as the 'drivers of progress in biomedical research' and therefore they should be 'made immediately available for free and unrestricted use by the scientific community to engage in the full range of opportunities for creative science'.[16]

These documents together set out the key principles which now dominate thinking and practice regarding open access to genome sequence data in North America and the UK. The open-access principles underlying these developments have since been applied by national funding bodies beyond projects which generate sequence data to other areas of biomedical research. The Toronto Agreement which was the sequel to the Fort Lauderdale and Bermuda, also endorsed the principle of open access to sequence data but suggested that this should be extended further to include clinical data. Examples of such policies are those of the National Institutes of Health (NIH 2003),[17] Genome Canada (2005),[18] and the UK Medical Research Council (MRC 2006).[19] All of these organisations now make data sharing a requirement of some of the funding they provide in genomics. These policies have been developed with scientists and have created a climate in which data sharing is becoming more the norm, rather than just being applied to the large sequencing projects.

## ADVANCES IN SEQUENCING TECHNOLOGY

The HGP also required a change in the way of executing science as 'at first everyone did everything', but then there was a move to specialists who focused on one part of the sequencing process.[20] As Watson had previously predicted 'sequencing facilities would have to be created that are far greater

[15] *Sharing Data from Large-scale Biological Research Projects*, above n 14, 3.

[16] *Sharing Data from Large-scale Biological Research Projects*, above n 14, 3.

[17] NIH Data Sharing Policy and Implementation Guidance (updated: 5 March 2003) <http://grants.nih.gov/grants/policy/data_sharing/data_sharing_guidance.htm>.

[18] Genome Canada Data Release and Resource Sharing Policy www.genomecanada.ca/medias/PDF/EN/DataReleaseandResourceSharingPolicy.pdf.

[19] Medical Research Council Policy on Data Sharing and Preservation www.mrc.ac.uk/PolicyGuidance/EthicsAndGovernance/DataSharing/PolicyonDataSharingandPreservation/index.htm#P16_1333.

[20] J Sulston and G Ferry, *The Common Thread: A Story of Science, Politics, Ethics, and the Human Genome* (Washington, DC, Joseph Henry Press, 2002).

than any existing today and that more closely resemble industrial production lines than conventional university research laboratories'.[21] New partnerships were built between industry and the project to develop 'new technologies, new approaches to automation, and new computational strategies'[22] that improved on those developed by Fred Sanger.[23] It can be argued that the eventual success of the HGP depended mostly on the development of technologies for DNA sequencing rather than on the solution of fundamental scientific questions.[24]

The cost of the sequencing at the time of the HGP was expensive and labour intensive. The result is a combined 'reference genome' of a small number of anonymous donors. At that time, the cost of a finished sequence was approximately $10 per base. A well-equipped laboratory could produce about 500 bases of sequence a day, and solid data indicated that more than 90 per cent of the 6 feet of DNA in every cell was 'junk' (ie, repetitive sequences with either no known function or noncoding 'spacer' DNA).[25] It is estimated that the HGP cost US $2.7 billion and involved 20 different laboratories around the world.[26] Since then the costs of sequencing have fallen rapidly. In 2007, James Watson, the co-discoverer of the DNA double-helix—became the first individual to have his genome sequenced and the cost was around $1 million. In 2008, Applied Biosystems of Foster City, California, announced that it had sequenced the genome of a Nigerian man for less than $60,000. In 2009, the company Complete Genomics announced that it could sequence an individual genome for $5000.[27] It is anticipated that these costs will continue to fall. The X Prize is offering the Archon X Prize for Genomics, which will give $10 million for the first team to sequence 100 human genomes in less than 10 days for less than $10,000 each.[28]

While the HGP and new sequencing techniques have resulted in more sequence information, it is also recognised that there is still a need for more detailed genomic sequence data as a well as for the tools to analyse it. A survey conducted on the tenth anniversary of the HGP

> revealed how ill-equipped many researchers feel to handle the exponentially increasing amounts of sequence data. The top concern—named by almost half of respondents—was the lack of adequate software or algorithms to analyse genomic

[21] Watson, 'The Human Genome Project Past', above n 9, 46.

[22] Colins, Morgan and Patrinos, 'The Human Genome Project', above n 6, 286.

[23] M Garcia-Sancho, 'Mapping and Sequencing Information: The Social Context for the Genomics Revolution' (2007) 31 *Endeavour* 18–23.

[24] J Esparza and T Yamada 'The Discovery Value of "Big Science"' (2007) 204 *Journal of Experimental Medicine* 701–04.

[25] CP Lorentz, ED Wieben, A Tefferi, DAH Whiteman and GW Dewald, 'Primer on Medical Genomics Part I: History of Genetics and Sequencing of the Human Genome' (2002) 77 *Mayo Clinic Proceedings* 773–82.

[26] www.genome.gov/11006943.

[27] P Aldous, 'Genome Sequencing Falls to $5000' (2009) *New Scientist* (6 February).

[28] http://genomics.xprize.org/.

data, followed closely by a shortage of qualified bioinformaticians and to a lesser extent raw computing power. Other concerns include data storage, the quality of sequencing data and the accuracy of genome assembly.[29]

These significant informatic and bioinformatic challenges to the proper management and interpretation of large datasets of sequence information need to be addressed.[30] While sequencing to date has been carried out in specialist centres there is the possibility that the development of next generation sequencing instruments could mean that new ways of carrying out science evolve.

## NEW SCIENTIFIC QUESTIONS

The HGP created large reference datasets of new information, developed new sequencing technology and computational power that provided the basis for new scientific questions to be addressed.[31] The success of the HGP as a basis for other studies is indisputable as 'tens of thousands of research programmes, many focused on identifying and characterizing specific genes, have benefited enormously from the creation and study of this database'.[32] The HGP provided the sequence detail but the next step has been to understand the function of genes in regulating proteins (functional genomics) and the effect that these have on the development of disease. The HGP marked the transition from genetics or 'the more established approach of studying DNA structure and function of individual genes'[33] to genomics which involved '"high throughput" data—and technology-intensive approaches to studying DNA structure and function'.[34] The studies that have followed the HGP have 'revealed the complexity of the genome, indeed almost every aspect of human biology, is far greater than was previously thought'.[35] In the ten years following the HGP, new types of scientific questions can be asked, as 'vast numbers of polymorphisms can be studied simultaneously, rather than focusing attention on a small number of genes' and secondly, 'very many more individuals can be genotyped in a single study' to genomics which involved 'high throughput' data—and technology-intensive approaches to studying DNA structure and function' to genomics which

[29] D Butler, 'Human Genome at Ten: Science after the Sequence' (2010) 465 *Nature* 1000–01, 1001.
[30] ER Mardis,'A Decade's Perspective on DNA Sequencing Technology' (2011) 470 *Nature* 19–203.
[31] LJ Palmer and LR Cardon, 'Shaking The Tree: Mapping Complex Disease Genes with Linkage Disequilibrium' (2005) 366 *The Lancet* 1223–34.
[32] R Weinberg, 'Point: Hypotheses First' (2010) 464 *Nature* 678.
[33] J Reineke Pohlhaus and RM Cook-Deegan, 'Genomics Research: World Survey of Public Funding' (2008) 9 *BMC Genomics* 472.
[34] *Ibid.*
[35] D Butler, 'Human Genome at Ten', above n 29, 1000.

involved 'high throughput' data—and technology-intensive approaches to studying DNA structure and function'.[36]

Since the HGP, the scientific agenda has become more orientated to understanding the aetiology of the more common diseases, such as cardiovascular conditions and cancer. By 2020, it is expected that common chronic diseases will account for almost three-quarters of deaths worldwide.[37] Many of the genetic biobanks that were established over the past ten years were designed as research tools to support studies into these conditions. 'With the revolution in genotyping came the realization that much smaller relative risks could be unambiguously identified, rather than focusing on the single gene effects.'[38] There has been a transition from the early genetic studies which focused on the identification of single gene, highly-penetrant disorders to bigger studies investigating the more common, polygenic diseases, such as cancer, cardiovascular conditions and diabetes. These affect more people but it is also more difficult to understand the genetic basis of these diseases as the genetic influences may be small, complex and difficult to isolate accurately.[39] 'Such diseases can exhibit familial clustering, but there is no clear inheritance pattern because of the polygenic aetiology and the substantial contribution from the environment.'[40] The techniques that were focused on families, such as family-based linkage analysis, are not suited to common diseases. This is because such studies 'can rarely identify genetic mutations associated with these disorders because people are typically over 50 years before clinical presentation; by this time their parents may be dead and susceptible children too young to manifest the disorder'.[41]

The investigations into common, polygenic conditions are assisted by the large sequencing projects that followed the HGP, which sought to map the human genome in greater detail. The SNP (single nucleotide polymorphisms) database[42] was developed by the SNP Consortium to identify the most common single nucleotide variations in the genome whereas the Human HapMap consortium[43] sought 'to quantify the association between SNPs in the genomes of human populations with differing ancestry (linkage

[36] N Day, 'Commentary: How Small is Small? (2009) 38 *International Journal of Epidemiology* 274–75, 274.

[37] S O'Rahilly and NJ Wareham, 'Genetic Variants and Common Diseases: Better Late Than Never' (2006) 355 *New England Journal of Medicine* 306–08.

[38] Day, 'Commentary', above n 36, 274.

[39] PR Burton, AI Hansell, I Fortier *et al*, 'Size Matters: Just How Big is BIG?: Quantifying Realistic Sample Size Requirements for Human Genome Epidemiology' (2009) 38 *International Journal of Epidemiology* 263–73.

[40] AD Hingorani, T Shah, M Kumari, R Sofat, L Smeeth, 'Science, Medicine, and the Future: Translating Genomics into Improved Healthcare' (2010) 341 *British Medical Journal* 1037–42, 1038.

[41] *Ibid*.

[42] www.ncbi.nlm.nih.gov/SNP/.

[43] http://hapmap.ncbi.nlm.nih.gov/index.html.en.

disequilibrium)'.[44] These catalogues provide a basis for the Genome-wide Association Studies (GWAS)[45] that have been used to compare 'the frequency of typed (and inferred) SNPs in large numbers of unrelated people affected by a disease and unaffected unrelated controls'.[46] Dense genotyping chips are now able to identify large areas of an individual's genome at a reasonable cost, typically analysing a million SNPs, out of a possible approximately three billion in the human genome, so that it is possible to examine hundreds of thousands of variations at the same time. Whole genome research requires very large sample sizes of patients, healthy controls, and individuals who vary in respect of the trait in question, in order to detect and analyse genetic risk variants, and to produce results of satisfactory levels of statistical significance. The size of the sample is a key determinant of quality.[47] In particular, biobanks, longitudinal studies and cohorts have been useful for providing the large sample sizes of both controls and affected individuals that are needed to carry out GWAS.

## Genome-wide Association Mapping

The hope for GWAS is that it will enable important epidemiological questions to be answered in powerful and rigorous ways.[48] At the outset, these studies have enabled the exploration of the genetics underlying the conditions of interest even in cases where the underlying biology is as yet unknown, working on the assumption that common diseases would be associated with common genetic variants.[49] A major aim of GWAS is to advance the understanding of biological function, including gene–gene and gene–environment interactions, with a further end of developing treatments, including the targeted treatments of pharmacogenomics, and diagnostics.[50] It is recognised that to carry out a good GWAS requires great care, attention to detail, and coordinated resources.[51] These are extremely powerful,

[44] Hingorani, Shah, Kumari, Sofat, Smeeth, 'Science, Medicine, and the Future', above n 40, 1038.

[45] WTCCC, 'Genome-wide Association Study of 14,000 Cases of Seven Common Diseases and 3,000 Shared Controls' (2007) 447 *Nature* 661–78.

[46] See http://hapmap.ncbi.nlm.nih.gov/index.html.en.

[47] WYS Wang, BJ Barratt, DG Clayton and JA Todd, 'Genome-wide Association Studies: Theoretical and Practical Concerns' (2005) 6 *Nature Reviews Genetics* 109–18; KM Channon and H Watkins, 'Coronary Artery Disease Genetics: Bigger is Better' (2004) 25 *European Heart Journal* 900–01.

[48] TA Pearson and TA Manolio, 'How to Interpret a Genome-wide Association Study' (2008) 299 *Journal of the American Medical Association* 1335–44.

[49] *Ibid.*

[50] JP Ioannidis, 'Personalized Genetic Prediction: Too Limited, Too Expensive, Or Too Soon?' (2009) 150 *Annals of Internal Medicine* 139–41.

[51] See Wang, Barratt, Clayton and Todd, 'Genome-wide Association Studies', above n 47 and Channon and Watkins, 'Coronary Artery Disease Genetics', above n 47.

highly methodologically and technological complex techniques. It is necessary to look in some detail at these studies in order fully to understand the potential challenges that they pose for governance and regulation.

The need for large, diverse, well-characterised samples has led to an increased emphasis on cooperation at both a national and at international levels.[52] It has also led to a desire to make use of already existing archived samples.[53] Large genomic reference libraries, such as HAPMAP[54] and the soon to be completed 1000 Genomes Project,[55] provide data to researchers on the web which are essential to help identify areas of the genomes to target and in the design of studies. Furthermore, the validity of results needs to be confirmed in replication studies before they can be relied upon, providing further reason for sharing data.[56] GWAS require efforts to be centralised and well organised.[57] Such efforts include statements aimed at standardising models for performing studies and reporting results.[58] GWAS are also fuelling efforts to develop new methodologies to handle the vast amounts of data created.[59] New models of collaboration and data sharing, with the express aim of maximising public benefits, have been created by GWAS research.[60]

There is no doubt that GWAS are producing an unprecedented amount of data and that this data is a powerful tool for the identification of individuals and also of their genetic relatives. It has been demonstrated that the transition from data considered private to data which is identifiable is happening very swiftly. In 2004 it was shown that between 30 to 80 SNPs could uniquely identify an individual out of a population of 10 billion;[61] currently, most GWAS are sampling a million SNPs for each individual.[62] Previous assumptions about anonymity for aggregated samples have been overturned with the finding that, if an individual's DNA data can be accessed, then it is possible to identify that individual out of a large

52 AT Hattersley and MI McCarthy, 'What Makes a Good Genetic Association Study?' (2005) 366 *The Lancet* 1315–23.

53 DR Karp et al, 'Ethical And Practical Issues Associated With Aggregating Databases' (2008) 5 *PLoS Medicine.*

54 HapMap Consortium, 'A Second Generation Human Haplotype Map of over 3.1 Million SNPs' (2007) 449 *Nature* 851–61.

55 J Kaiser, 'A Plan to Capture Human Diversity in 1000 Genomes' (2008) 319 *Science* 395.

56 JPA Ioannidis et al, 'Assessment of Cumulative Evidence on Genetic Associations: Interim Guidelines' (2007) 37 *International Journal of Epidemiology* 120–32.

57 R Gibbs, 'Deeper into the Genome' (2005) 437 *Nature* 1233–34.

58 J Little et al, 'Strenghthening the Reporting of Genetic Association Studies (STREGA): An Extension of the STROBE Statement' (2009) *PLos Medicine* e100022.

59 M Pop and SL Salzberg, 'Bioinformatics Challenges of Sequencing Technology' (2008) 24 *Trends in Genetics* 142–49.

60 GAIN Collaborative Research Group, 'New Models of Collaboration in Genome-Wide Association Studies: The Genetic Association Information Network' (2007) 39 *Nature Genetics* 1045–51.

61 Z Lin, AB Owen and RB Altman, 'Genomic Research and Human Subject Privacy' (2004) 305 *Science* 183.

62 Pearson and Manolio, 'How to Interpret a Genome-wide Association Study', above n 48.

aggregated sample.[63] Genomic data can also potentially lead to inferences about relatives; for example, it has recently been shown that siblings may be identified with considerable accuracy from an individual's genomic data.[64] In some instances this technology can be used in combination with other sources of information to piece together identifying information about individuals in 'anonymised' research data, including even accurate guesses of surnames.[65]

There are varying accounts of how successful GWAS have been, which to some extent reflects emphasis on different aspects of these studies and their aims. Methodological problems which have slowed progress have been addressed, and since 2005, nearly one hundred sites of genetic variation for forty common diseases and traits have been found and replicated in GWAS, many variants in genes not previously suspected of having a role in disease, and some in genomic regions containing no known genes.[66] There have been reports of variants for range of disorders including Inflammatory Bowel Syndrome, age-related macular degeneration (ARMD), and skin pigmentation.[67] However, most of the variants found so far account for only a small degree of the relative risk of developing a disease or of having a trait, and the findings collectively explain only a very small proportion of the underlying genetic component of most diseases, ARMD being a rare exception to this.[68]

Others are more sceptical about the role of GWAS, considering that although useful for illuminating the genetic architecture of a disease (its distribution amongst a population or species), it will be less useful for understanding disease aetiology.[69] Although technical progress has meant that GWAS have produced results faster than expected, initial hopes for the proportion of heritability (a measure of the proportion of variation in a given population that is attributable to genetic variation) that could be discovered with GWAS have turned out to be optimistic.[70] However, in assessing the

[63] N Homer *et al*, 'Resolving Individuals Contributing Trace Amounts of DNA to Highly Complex Mixtures Using High-Density SNP Genotyping Microarrays' (2008) 4 *PLoS Genetics* e1000167.

[64] CA Cassa, B Schmidt, IS Kohane and KD Mandl, 'My Sister's Keeper?: Genomic Research and the Identifiability of Siblings' (2008) 1 *BMC Medical Genomics* 32.

[65] J Gitschier, 'Inferential Genotyping of Y chromosomes in Latter-Day Saints Founders and Comparison to Utah Samples in the HapMap Project' (2009) 84 *American Journal of Human Genetics* 251–58.

[66] Pearson and Manolio, above n 47.

[67] O'Rahilly and Wareham, 'Genetic Variants', above n 37.

[68] P Kraft and DJ Hunter, 'Genetic Risk Prediction: Are We There Yet?' (2009) 360 *The New England Journal of Medicine* 1701–03.

[69] W Bodmer and C Bonilla, 'Common and Rare Variants in Multi-Factorial Susceptibility to Common Diseases' (2008) 40 *Nature Genetics* 695–701.

[70] DB Goldstein, 'Common Genetic Variation and Human Traits' (2009) 360 *The New England Journal of Medicine* 1696–98.

success of GWAS it should be noted that small effect sizes do not rule out biological, clinical or pharmacological importance.[71]

Ascertaining the validity of GWAS results is a highly complex task.[72] There are many possible shortcomings and findings are often not replicated[73] yet there are also many reasons why it is difficult to replicate GWAS.[74] There is a suspected publication bias that may lead to an overestimation of the significance of GWAS,[75] and there exists a well-established 'winner's curse' whereby initial findings are often either not replicated, or found to be of lower significance than first thought.[76] Clinical applications of GWAS are also still uncertain, with some diversity of opinion on their likely applications. Possible applications include improved tailoring of drug use to individuals,[77] and hopes for tests that may give more accurate indications of lifelong exposure than biochemical tests.[78] However, it is recognised that the use of such findings in clinical care is complex and currently limited.[79]

The majority opinion warns that risk assessment is not a likely outcome of GWAS in many areas, and that the studies are a long way yet from clinical use.[80] The very complexity of the causal pathways in common diseases is thought to limit the chances of accurate prediction.[81] The scientific community has been warned to be careful in overstating the possibilities of personalised medicine from the findings of GWAS.[82] In combination with molecular cell based approaches, researchers hope that the results will be useful for ascertaining the pathogenesis of common disease.[83] The current

[71] JN Hirschhorn, 'Genomewide Association Studies: Illuminating Biologic Pathways' (2009) 360 *The New England Journal of Medicine* 1699–701.

[72] J Attia *et al*, 'How to Use an Article about Genetic Association B: Are the Results of the Study Valid?' (2009) 301 *Journal of the American Medical Association* 191–97.

[73] Hattersley and McCarthy, 'What Makes a Good Genetic Association Study?', above n 52.

[74] A Hamsten and P Eriksson, 'Identifying the Susceptibility Genes for Coronary Artery Disease: From Hyperbole through Doubt to Cautious Optimism' (2008) 263 *Journal of Internal Medicine* 538–52.

[75] Ioannidis, 'Personalized Genetic Prediction', above n 50.

[76] JP Zollner and JK Pritchard, 'Overcoming the Winner's Curse: Estimating Penetrance Parameters from Case-Control Data' (2007) 80 *American Journal of Human Genetics* 605–15.

[77] SEARCH Collaborative Group, 'SLCO1B1 Variants and Statin-Induced Myopathy: a Genomewide Study' (2008) 359 *The New England Journal of Medicine* 789–99.

[78] RA Hegele, 'Plasma Lipoproteins: Genetic Influences and Clinical Implications' (2009) 10 *Nature Reviews Genetics* 109–21.

[79] J Attia *et al*, 'How to Use an Article about Genetic Association C', above n 72.

[80] Kraft and Hunter, 'Genetic Risk Prediction', above n 68.

[81] ACJW Janssens and CM van Duijn, 'Genome-based Prediction of Common Diseases: Advances and Prospects' (2008) 17 *Human Molecular Genetics* R166–R172.

[82] J Jakobsdottir *et al*, 'Interpretation of Genetic Association Studies: Markers with Replicated Highly Significant Odds Ratios May be Poor Classifiers' (2009) 5 *PLoS Genetics* e1000337.

[83] K Ozake and T Tanaka, 'Genome-Wide Association Study to Identify SNPS Conferring Risk of Myocardial Infarction and Their Functional Analyses' (2005) 62 *Cellular and Molecular Life Sciences* 1804–13.

risk estimates for genetic variants in common diseases is currently very unstable.[84] In various studies, it has been found that adding genetic results to current methods of disease prediction adds very little power.[85] Moreover, the widespread heterogeneity from population to population gives additional reason to exercise caution in the adoption of predictive tests.[86] Nonetheless, such results are often drawn upon by commercial companies offering direct-to-consumer genetic testing.[87]

## The Development of Infrastructures

Following the HGP there has been a further commitment by funding bodies to invest in infrastructure and platforms to facilitate research. During 2003–2006 it has been estimated that aggregate spending on genomics research from 34 funding bodies around the world averaged at around $2.9 billion annually.[88] The United States was the leader in funding genomics research during this period, spending 35 per cent of all worldwide spending on genomics research. However, Ireland and the UK spent more per head of capita than any other country on genomics research, as measured both by genomics research expenditure per capita and per Gross Domestic Product.[89] Within the UK, the key government funders in this area are the Medical Research Council (MRC) and the Biotechnology and Biological Sciences Research Council (BBSRC). The charitable bodies that make a significant contribution in this area are Cancer Research UK, the Wellcome Trust and the British Heart Foundation. National funding bodies have also contributed to international projects by funding aspects of larger projects that have been carried out in their own countries.

Following the HGP there have been a number of projects that have been established as reference projects for the scientific community. This project provided the basis, in terms of know-how in establishing and co-ordinating international partnerships of this kind for the projects that followed such as the HapMap project (2002–2005) and more recently the 1000 Genomes Project (2007–). The purpose of the HapMap Project 'was to identify chromosome regions with sets of strongly associated SNPs, the haplotypes in

[84] Janssens and van Duijn, 'Genome-based Prediction', above n 81.
[85] Janssens and van Duijn, 'Genome-based Prediction', above n 81.
[86] RS Cooper, B Tayo and X Zhu, 'Genome-wide Association Studies: Implications for Multiethnic Samples' (2008) 17. *Human Molecular Genetics* R151–R155.
[87] J Kaye, 'The Regulation of Direct-to-Consumer Genetic Tests' (2008) 17 *Human Molecular Genetics* R1–R4.
[88] J Reineke Pohlhaus and RM Cook-Deegan, 'Genomics Research', above n 33.
[89] D Overbye, 'Human DNA: The Ultimate Spot for Secret Messages (Are Some There Now?)' (2007) *The New York Times*.

those regions, and the SNPs that tag them'.[90] The 1000 Genomes Project will develop a new map of the human genome that will provide a view of biomedically relevant DNA variations at currently unmatched resolution.[91] The purpose of all of these projects is to establish reference libraries of sequence information that was, or will be, accessible on the web to everyone with the aim of advancing scientific enquiry.

In addition, funding bodies, such as the Wellcome Trust and the National Institutes of Health (NIH) in the USA, have funded projects that seek to establish repositories of information generated by the use of GWAS methodology, which have a more restricted access.

> The Genetic Association Information Network (GAIN) (2006–2008) completed an ambitious program to genotype existing research studies in six major common diseases, and combine the results with clinical data to create a significant new resource for genetic researchers.[92]

The resulting data are being deposited in a database within the National Library of Medicine at NIH, funded by GAIN, for the broad use of the research community. Originators of the initial studies received additional grants to make their own analyses, but access is restricted to bone fide researchers who have been approved by a Data Access Committee (DAC) and have obtained research ethics committee approval in their own jurisdiction. Similar systems exist for the WTCCC (UK),[93] European Genotyping Archive[94] and the dbGaP[95] in the USA. These new forms of governance structures are in addition to the oversight procedures that usually apply for research.

There are also plans to network biobanks to create a research infrastructure within Europe through the Biobanking and Biomolecular Resources Research Infrastructure (BBMRI).[96] The BBMRI aims to 'improve the accessibility and interoperability of the existing comprehensive collections, either population-based or disease-orientated, of biological samples' and their associated information to build 'a globally unmatched, Europe-wide platform for translational medical research'.[97] Over the past two years,

[90] HapMap Project: www.hapmap.org/abouthapmap.html.

[91] 1000 Genome Project: www.1000genomes.org/page.php.

[92] GAIN, http://fnih.org/index.php?option=com_contentandtask=viewandid=338andItemid=454.

[93] WTCCC, 'Genome-wide Association Study', above n 45. The WTCCC (2007–) is a collaboration of 24 leading human geneticists based in the UK that aims to analyse thousands of DNA samples from patients suffering from different diseases to identify common genetic variations for each condition. Aggregated data are placed on the internet, but access to the more detailed genotypic and phenotypic data is obtained only through the Data Access Committee or the principal investigator, who can also decide on further collaboration.

[94] European Genotyping Archive: www.ebi.ac.uk/ega/page.php.

[95] dbGaP: www.ncbi.nlm.nih.gov/projects/gap/cgi-bin/about.html.

[96] BBMRI: http://www.bbmri.eu/.

[97] BBMRI Brochure October 2010.

these plans have gained momentum as 9 countries across Europe (the UK is not one of these) have made a total investment of €97.8m to fund national BBMRI infrastructures to co-ordinate biobanking at a national level. In addition to the development of infrastructure there have been organisations that have sought to develop tools—software, protocols, procedures and governance mechanisms—required to establish and to run biobanks. P3G[98] is an international organisation that has been funded by Genome Canada and Genome Quebec, to develop standards for population biobanks, in order to facilitate the sharing of information and samples between biobanks.

As well as infrastructure development, many funders now partially fund large international consortia or encourage collaborations on a grand scale that is in accordance with the scientific aims. European funding projects stipulate that collaborative projects must have a number of institutions from different member states of the European Union and often these must include new member states. Funders, such as the National Institutes of Health in the USA now require that large projects should demonstrate a plan for how data and results will be shared. These policies are also being applied to some ongoing longitudinal projects and already completed projects. Researchers do have the opportunity to argue why their data is unsuitable for sharing, such as the fact that the data may involve vulnerable groups or be identifiable, or be contrary to the basic tenets of the consent form. However, the trend is towards deposition for use by others rather than keeping data for one specific research project.

## THE EFFECT ON PRACTICE

These trends in combination have had a marked effect on scientific practice enabling global collaborations to be established and maintained, and for data to be generated and shared with ease. We are moving to a situation where it is possible to search different databases in a seamless way and to move with the click of a mouse between datasets held in different institutions, crossing national jurisdictional boundaries in virtual space. Open-access policies are changing the way that data is generated and distributed, as well as enabling new ways to mine data. Whereas in the past, data sharing

> primarily been done with known colleagues, based on mutual respect, trust and a common interest. The conditions of access would be negotiated on an individual basis and would vary according to particular circumstances. Funders now require that data sharing be considered in every newly funded research project, unless there are justifiable reasons why this should not be so. With these policies, the

[98] P$^3$G Consortium: www.p3gconsortium.org/.

> question for many researchers has become *how* to share data, whereas previously it was *whether* data should be shared at all.[99]

There is now a distinction between data generators and data users. The development of datasets with large numbers of participants is still desirable but access to these datasets is now not held solely by the person who established it. This means that a collection is no longer exclusive to one research group or laboratory, with access being determined by the collector. Instead it is now available to the wider research community and access is determined by a committee or some other third party. This means that the 'secondary users of the data are far removed from the researchers who carried out the collection of the samples and data, as well as from the research participants'.[100]

These technological developments have in turn had an effect on scientific practice, which

> has become increasingly interdisciplinary, with the rapid formation of flexible and dynamic research collaborations around the world. For example, the use of new methodology such as GWAS requires large numbers of clinically well-characterised samples to be collected from patients; laboratory staff and researchers to manage the genotyping pipeline; bioinformaticians, statisticians and other data analysts to interpret the data; and leadership from principal investigators.[101]

These advances also challenge our legal and ethical frameworks as data sharing practices give a new twist to the old questions of informed consent, protection of privacy and the governance of medical research. However, as we shall see in Chapter Nine it is also raises a number of issues about reward, attribution and acknowledgement for scientific endeavour.

## THE CHALLENGES FOR REGULATION

The changes that have occurred within the scientific domain raise a number of challenges for the current legal frameworks and regulatory structures. This is because they are nationally based, while research activity is increasingly global. Although it is easy to cross national borders with the click of a mouse, there are different legal instruments that govern research, which apply in different countries. Within Europe, we are moving towards more uniformity in terms of data protection with common legal instruments to protect privacy such as the Directive 45/46/EC and case law on the Article 8 right to life from the European Court of Human Rights. These would

[99] J Kaye, C Heeney, N Hawkins, J de Vries, P Boddington, 'Data Sharing in Genomics: Re-shaping Scientific Practice' (2009) 10(5) *Nature Review Genetics* 331–35, 332.
[100] Kaye *et al*, 'Data Sharing in Genomics', above n 99, 331.
[101] Kaye *et al*, 'Data Sharing in Genomics', above n 99, 331.

have general application to data sharing and biobanking within the medical research context. There are now a number of legal instruments within Europe that deal specifically with biobanking concerns. These are the Additional Protocol to the Convention on Human Rights and Biomedicine 1977[102] and the OECD guidelines[103] that may be adopted as the basis for a binding legal instrument. As well as our legal framework being nationally orientated, the regulatory bodies that govern research are also nationally based. This means that researchers have to ensure that they conform to the requirements of research ethics committees and for medical research in their own countries, as well as those of the countries of their collaborators. To find out the legal requirements in each country in a large consortium can be time consuming and require legal expertise that may not be accessible to many groups.

Within the current governance framework, the characteristics and requirements of the national research governance systems can have a direct effect on the ability of scientists to operate at an international level. With a changing scientific agenda and the technological abilities to share and generate data, scientists who are at the cutting edge of genomic science engage routinely in networks that share data and samples through specific projects, as well as in large consortium that seek to build research resources. Local regulatory changes that are restrictive and out of tune with international trends can have a significant effect on the ability of scientists to collaborate and plan international initiatives. Therefore, it is advantageous to have stable national governance structures that are in accordance with the requirements of other jurisdictions, as this facilitates collaboration and the sharing of data and samples. Contractual agreements such as Material Transfer Agreements and Data Sharing Agreements are often the legal mechanisms that are used to enable such collaborations.

Stability and certainty is not a characteristic of the research governance system for England and Wales. During the period 2005–2009 there were a number of significant events that lead to substantial changes in the way that research was governed. There were three significant pieces of statute that pertained to medical research that were introduced in the UK in 2004. These were the Human Tissue Act 2004, the Medicines for Human Use (Clinical Trials) Regulations 2004 and the Mental Capacity Act 2005.[104] The most significant for those working with biobanks—and that which was often commented on by our respondents—was the implementation of the Human Tissue Act 2004 (HTAct). This Act was introduced in response to the

[102] Council of Europe—Committee of Ministers, *Recommendation Rec (2006) 4F of the Committee of Ministers to Member States on Research on Biological Materials of Human Origin*, Strasbourg, 15 March 2006.

[103] OECD guidelines on Human Biobanks and Genetic Research Databases (HBGRDs) 2009.

[104] A Alonzi and M Pringle, 'Mental Capacity Act 2005' (2007) 35 *British Medical Journal* 335.

discovery in 1999 that a number of hospitals—starting with the Bristol Royal Infirmary and Alder Hey in Liverpool—had routinely retained organs and tissue from children for research purposes without consent.[105] The intent of Parliament in implementing the legislation was that the HTAct would put consent on a firm footing. It would also ensure greater accountability of the collection and use of tissue for research purposes by the establishment of the Human Tissue Authority. Also at this time, the European Clinical Trial Directive[106] was also implemented into UK law through the Medicines for Human Use (Clinical Trials) Regulations 2004. This introduced a number of changes in the way that clinical trials were governed in the UK. There has been considerable discontent about the burdens and effectiveness of this regulatory change.[107]

During this period there were also a number of events that had an impact on public perceptions and trust in the health sector. The first was the Harold Shipman affair,[108] where a practicing doctor murdered a number of his patients over a period of years. As a way of regaining public trust and to prevent this happening again, the General Medical Council introduced a system of monitoring and enforcing additional training requirements for doctors.[109] The concern over the introduction of electronic medical records and who would be able to access them also had a profound effect on public confidence.[110] Outside of medical research but in a related field, the National DNA database established for crime prevention also gained notoriety as two individuals took a claim against the UK for keeping their DNA samples unlawfully. This resulted in a series of legal cases that finally went to the European Court on Human Rights.[111] These events in combination

[105] D Hall, 'Reflecting on Redfern: What Can We Learn from the Alder Hey story?' (2001) 84 *Archive of Disease in Childhood* 455–56; M Hunter, 'Alder Hey Report Condemns Doctors, Management, and Coroner' (2001) 322 (7281) *British Medical Journal* 255; D Price 'The Human Tissue Act 2004' (2005) 68 *Modern Law Review* 798.

[106] Directive 2001/20/EC of the European Parliament and of the Council of 4 April 2001 on the approximation of the laws, regulations and administrative provisions of the Member States relating to the implementation of good clinical practice in the conduct of clinical trials on medicinal products for human use [2001] OJ L/121/ 34.

[107] J D Neaton *et al*, 'Regulatory Impediments Jeopardizing the Conduct of Clinical Trials in Europe Funded by the National Institutes of Health' (2010) 7 *Clinical Trials* 705–18; J Heam and R Sullivan, 'The Impact of the "Clinical Trials" Directive on the Cost and Conduct of Non-Commercial Cancer Trials in the UK' (2007) 43(1) *European Journal of Cancer* 8–13; S Yusuf, 'Damage to Important Clinical Trials by Over-Regulation' (2010) 7 *Clinical Trials* 622.

[108] *The Shipman Inquiry. Safeguarding Patients: Lessons from the Past, Proposals for the Future* [chairman Dame Janet Smith] (London, Stationery Office, 2004).

[109] General Medical Council, *A Licence to Practise and Revalidation* (London, GMC, 2003).

[110] J Hendy et al, 'Implementing the Nhs Information Technology Programme: Qualitative Study of Progress in Acute Trusts' (2007) 334 *British Medical Journal* 1360.

[111] *S and Marper v The United Kingdom* (App no 30562/04) [2008] ECHR 1581 (4 December 2008).

have a bearing on the way that biobanks have been developed in the UK and the way that policy has been viewed.

## IN CONCLUSION

The purpose of this chapter has been to provide an understanding of the changing developments within science that have provided the rationale, methodology and technical tools that have encouraged the development of biobanks. The way that genomic research is currently carried out and has evolved is due to the Human Genome Project that established a new way of 'doing' and conceptualising the way that science could be done. It galvanised resources and people across the globe to focus on one project that would provide a repository of sequence data that could be used by everyone in the research community. In doing so it had the effect of changing scientific practice and policy irrevocably. Inevitability, this has raised a number of challenges for our regulatory frameworks both within England and Wales and at an international level. The global nature of genomics research has created a number of challenges for these regulatory frameworks which have had to adapt and respond to the changing scientific agenda and methodology. The next chapter describes and analyses the current governance framework for medical research and biobanks in the UK. It demonstrates that a stable and coherent governance structure that is needed to facilitate this new way of doing research is currently not evident in England and Wales.

# Part II

# Key Legal and Sociological Findings

# 3

# Mapping the Regulatory Space

*Susan MC Gibbons*

TO LAY A proper foundation for understanding and critically evaluating the current governance framework applicable to biobanks in England and Wales, we have taken a three-stage approach. The first stage involved a two-pronged investigation into the legal documentation and regulatory bodies that apply to biobanks. The first task in this investigation was to identify and examine the national, regional and international documents or instruments—including legislative materials, common law authorities, codes of practice, guidance materials, electronic resources and so forth—that apply to biobanks, biobanking-related activities, or those who work with biobanks. The second, closely interrelated task was to identify the key public authorities, organisations and other actors that govern (or, at least, that appear or seek to govern) biobanks, biobanking-related activities; and the different stakeholders involved, whether through formal or informal means; and the nature and sources of their power, authority and legitimacy.[1]

Taken together, these two, largely empirical, steps provide the basis for a third, more analytical step—one that leads to a much richer, more profound and more accurate understanding of the current governance framework and how it operates. In essence, this third step involves integrating and analysing the findings from the first two steps in such a way as to uncover how the various bodies interrelate and how that interrelationship influences the governance of biobanks in England and Wales. This analysis provides a basis for comparison with the empirical findings (documented in Chapters Five to Eight below) so that we can obtain some understanding of how the documentary sources of guidance and the various actors and stakeholders, through their powers, day-to-day practices and complex interactions, influence the operation and regulation of biobanks, affect the behaviour, perceptions and attitudes of others, and either constrain or promote the establishment, management, use and organisation of biobanks and related activities.

[1] For empirical evidence and further analysis as to how documentary sources of guidance and apparent regulatory actors in fact influence biobanking practices and professionals, see chs 5, 6, 7, 8 and 9 below.

Therefore, the main purposes of this chapter are twofold. The first is to describe the legal research methodology that we have developed and used in the Governing Genetic Databases project—and, in particular, our third-stage 'regulatory space' approach. Then, through applying that approach, the chapter's second main purpose is to present our key findings, preliminary analyses, and some initial conclusions about important lessons that may be drawn from undertaking this crucial threefold legal 'mapping' exercise. The findings, analyses and conclusions presented in this chapter provide the essential legal foundation—theoretical, descriptive and empirical—for the further critical analyses, reflections and recommendations to be developed in subsequent chapters.

## THE EXISTING GOVERNANCE BACKDROP—LAYERS OF COMPLEXITY

As will become evident as this chapter unfolds, generating a complete 'map' or picture of the existing governance structures applicable to biobanks in England and Wales—let alone to medical research more generally—is a profoundly challenging, difficult and complex endeavour. Before embarking on the threefold mapping analysis, then, it is important to understand some of the key factors that, in combination, render the existing biobanking governance framework so complex and unwieldy.

Perhaps the most significant among them is the fact that, to date, no purpose-designed, specific legal framework or governance structure has existed for biobanks, whether at a national, regional (including European) or international level. There is no single, tailor-made, comprehensive, dedicated statute or other legal instrument pertaining to biobanks. Similarly, no particular public authority or other regulatory body is charged with overseeing the biobanking sector. Consequently, what we are left with is a frankly bewildering array of general statutes, legislative provisions, regulations, directives, common law doctrines, codes of practice, guidelines, ethical statements, conventions, declarations, opinions, recommendations, statements, policies, resolutions, guidance notes, fact sheets, circulars, reports and so forth.[2] The range of bodies that have promulgated these materials, or that have played at least some role in governing biobanks (formally or informally), is correspondingly vast. As we shall see, it includes domestic, supranational and international legislative, policy-making, executive, administrative, judicial, professional, industrial, advisory, financial,

[2] For an earlier analysis and discussion, and for further references on point, see: J Kaye and SMC Gibbons, 'Mapping the Regulatory Space for Genetic Databases and Biobanks in England and Wales' (2008) 9(2) *Medical Law International* 111; SMC Gibbons, 'Are UK Genetic Databases Governed Adequately? A Comparative Legal Analysis' (2007) 27(2) *Legal Studies* 312; SMC Gibbons, J Kaye, A Smart, C Heeney and M Parker, 'Governing Genetic Databases: Challenges Facing Research Regulation and Practice' (2007) 34(2) *Journal of Law and Society* 163.

charitable, special interest and ethical oversight agencies. It is also constantly changing (and growing), as new actors and organisations emerge, existing bodies are merged, abolished, divided up, expanded, refocused or otherwise reformed, and new networks, alliances and consortia are forged. This is an ongoing feature of the governance of biomedical research in the UK.

As if this smorgasbord of governance, examined in greater detail below, were not challenging enough, the various documents and regulatory actors populating the biobanking regulatory universe are essentially unco-ordinated. Their provisions and activities frequently overlap. Yet, at times, they are mutually inconsistent—often in quite important respects.[3] It is unclear how the different instruments, provisions and regulatory bodies should be ranked or prioritised where choices, conflicts or discrepancies arise. Obviously, this poses particular problems for practitioners and professionals when they try to ascertain what the law, relevant rules or applicable guidelines require of them. Moreover, despite the plethora of documentary sources and regulatory actors actively engaged in the field, some conspicuous gaps still remain in the regulatory matrix.[4]

Compounding all of this complexity, at least three additional factors hamper any straightforward mapping of the current biobanking governance framework in England and Wales. First, as was noted in Chapter Two, scientific progress and research funding pressures increasingly demand that scientists collaborate. But, when we move beyond the purely domestic context—for example, when biosamples or data are transferred between collaborators, biobanks or researchers based in different countries—the lack of any harmonised regional or global framework, coupled with similar complexities, confusions and inconsistencies in the biobanking regulations within many other jurisdictions, and variations in how they interpret and implement multinational instruments,[5] all present barriers, to researchers and legal commentators alike.

Secondly, determining precisely what is a 'biobank' remains unclear and unsettled in the legal documentation, academic discourse and in practice.

[3] The slew of rules and guidelines pertaining to consent (eg, its permissible forms; how it is taken from biobank or medical research participants) is one glaring example. Other examples include marked divergences over the meanings and requirements for anonymisation, pseudonymisation, coding and so forth of participants' biosamples and data.

[4] Notable lacunae include topics around intellectual property rights; commercialisation; benefit sharing; open or public access to databases and research results; credit and career rewards for those who create biobanks that are subsequently used by others, particularly on an open-access basis; harmonisation of ethical norms, technical standards and terminology; and the differing forms of oversight (including ethical) that may be required or most appropriate for different types of biobank collections.

[5] A prime example is the notoriously inconsistent interpretation and implementation of the Data Protection Directive 95/46/EC (n 35 below) by different EU Member States' legislatures and courts. For examples and a recent discussion of outstanding issues in the UK context, see R Wong, 'Assessing the Status of Medical Information in the Light of the UK Data Protection Act 1998' (2008) 5 *Web Journal of Current legal Issues*. Inconsistent implementation of the EU Clinical Trials Directive 2001/20/EC (n 37 below) has also been blamed for increasing bureaucracy and reducing the number of global clinical trials done in the UK by two-thirds: A Gulland, 'EU Directive Threatens UK's "Spectacular" Record in Clinical Trials' (2009) 338(7696) *British Medical Journal* 678.

This radically contested concept defies any easy definition. In practice, its contours are constantly shifting and evolving, as the relevant biosciences advance and new generations of technology emerge. As Heeney has observed, biobanks are a classic example of what some sociologists would term a 'boundary object'.[6] Different stakeholders or commentators may well hold widely contrasting views about what should constitute, or count, as a 'biobank'. This quality of being a 'boundary object' carries certain benefits. It enables those from disparate academic disciplines, professional backgrounds or other intellectual, social or political domains to communicate meaningfully by finding common ground. However, for policymakers or lawmakers attempting to fashion some universally acceptable, pragmatically workable definition of a 'biobank', the very fluidity, imprecision and multifarious nature of the concept presents considerable challenges.

Despite some valiant standardisation and harmonisation initiatives instigated over recent years, terminological inconsistencies still abound.[7] At last count, our own research work had uncovered no fewer than 59 distinct generic terms, used in the literature and leading governance-related documents, to denote human tissue, biosample, DNA, other genetic material, and/or data collections that, broadly speaking, may be categorised as 'biobanks'. Internationally, the term 'biobank' seems to be emerging as the preferred generic label—hence its use in this book. But other highly popular terms abound. Leading the way are variations on the words 'bank',[8] 'biobank' (or 'bio-bank'),[9] 'database'[10] and 'collection'.[11] Terms

[6] The concept of 'boundary objects', originally coined by sociologists Star and Griesemer, refers to entities that inhabit several intersecting communities or social worlds, but that are interpreted, viewed or used differently by the members of each community. Boundary objects are thus 'plastic' enough to satisfy the differing informational requirements of each community, yet robust enough to maintain their integrity and a 'common identity' across multiple sites: SL Star and JR Griesemer, 'Institutional Ecology, "Translations" and Boundary Objects: Amateurs and Professionals in Berkeley's Museum of Vertebrate Zoology, 1907–39' (1989) 19(3) *Social Studies of Science* 387, 393. See further: GC Bowker and SL Star, *Sorting Things Out: Classification and its Consequences* (Cambridge, MA, MIT Press, 2000); KR Fleischmann, 'Boundary Objects with Agency: A Method for Studying the Design–Use Interface' (2006) 22(2) *Information Society* 77.

[7] See further Gibbons, 'Are UK Genetic Databases Governed Adequately?', above n 2, especially 324–28.

[8] Specific examples encountered include: biological bank, biological sample bank, biological specimen bank, bank of human biological materials, DNA bank, DNA sample bank, gene bank, genetic databank, sample bank, storage bank and tissue bank.

[9] Examples here include: human biobank, human tissue biobank, population biobank, DNA biobank and virtual biobank.

[10] Leading examples include: DNA database, genetic(s) database, (human) genomic(s) database, human genetic(s) research database, population database, population(al) genetic database, research database and virtual genetic database.

[11] Among the wealth of examples are: (human) tissue collection, (bio)sample collection, collection of genetic samples, collection of human tissues, collection of DNA, (large-scale) DNA collection, collection of tissue samples, population collection, collection of biological materials, genetic collection and repository collection.

such as 'resource'[12] and 'repository'[13] are relatively commonplace, too. Their prevalence appears to be increasing—perhaps reflecting shifting patterns in scientific practice, international collaborations, funding policies, research priorities and technological capacity. Terms we have encountered less frequently include 'biolibrary' (or 'bio-library'), 'stored' biological materials or samples, tissue 'establishment' and (research) 'platform'.

Thirdly—and of particular concern to policymakers—political, social, cultural, moral and ethical sensitivities cast a long, dark shadow across much of the modern biomedical field. Public trust and confidence have been shaken severely over recent years. This has been fuelled by the mass media, particularly following a catalogue of major scandals. These have included gross misconduct by certain biomedical professionals;[14] repeated data protection lapses, within both the public and private sectors; perceived threats to privacy (including genetic privacy); the advent of controversial new forms of research (for example, involving stem cells, genetically modified organisms and transgenic or hybrid embryos); the trend towards governmental and intergovernmental aggregation of information about individuals into increasingly massive, databases to be used and shared for multiple purposes; various expensive IT debacles, notably within the National Health Service (NHS) over the electronic patient care records scheme; and the ever-burgeoning, ever-controversial police National DNA Database. Given the acute public sensitivities and suspicions involved, it is perhaps unsurprising that politicians have been reluctant to grasp the nettle by embarking on a thoroughgoing, comprehensive programme for biobanking sector reform—however much it may be beneficial or urgently needed.

For all of these reasons, identifying and mapping out the principal components of the existing governance structure in England and Wales is far from straightforward. However, at least attempting to do so is both vital and pressing. Not least, it is essential for understanding the current situation and state of the law; for detecting whether (and where) any improvements may be warranted; and for formulating properly informed, empirically grounded and theoretically justifiable recommendations for change.

Helpfully, a number of concepts and tools are available that can assist us in charting a clearer, more meaningful path through the morass. As well as helping with our immediate analysis here, these concepts and tools also may prove useful to UK policymakers, lawmakers and regulatory bodies—both in designing and assessing biobanking reforms, and in keeping the system

[12] Variations here include: biological resource centre, biology resource, community resource project, human biological resource and research resource.

[13] Examples here include: bio-repository, biorepository and (human) tissue repository.

[14] These include the Harold Shipman affair, and the nationwide hospital organ retention scandals that ultimately led to the Human Tissue Act 2004. See, eg: D Price, 'The Human Tissue Act 2004' (2005) 68 *Modern Law Review* 798; K Liddell and A Hall, 'Beyond Bristol and Alder Hey: The Future Regulation of Human Tissue' (2005) 13 *Medical Law Review* 170.

under future review. We have already mentioned the notion of 'boundary objects'. Further key concepts and tools will be introduced and outlined in the next section. Then, the remainder of this chapter will present our primary research findings, and some preliminary conclusions that can be drawn from undertaking our threefold legal analysis. Drawing on the concepts and tools introduced below, the chapter will map out the 'regulatory space' for biobanks in England and Wales—and, in so doing, reflect critically on its key components, their strengths and weaknesses, and some implications and lessons that may be seen to follow.

## LEGAL AND REGULATORY THEORY: HELPFUL CONCEPTS AND TOOLS

### 'Regulatory Space'—Metaphor and Methodology

The 'regulatory space' metaphor, originally propounded by Hancher and Moran,[15] offers one very useful conceptual tool and framing device for understanding and mapping out the complex dynamics and patterns of interaction between the actors who populate existing regulatory regimes. Within the growing body of scholarship on law and regulation, a spectrum of theoretical accounts has emerged. Broadly speaking, three families may be identified: (1) public interest theories; (2) private interest theories; and (3) institutionalist theories.[16] The first two tend to be actor-centred, and to have a normative orientation. Institutionalist approaches, by contrast—into which category the various 'regulatory space' approaches fall—tend to be neutral as to the value, goals or proper role of regulation.[17] Looking beyond actors—and, in particular, decentring the State as the traditional, presumed locus of regulatory power and activity—institutionalist approaches seek to unravel the complex dynamics of institutions, networks or systems, in order to give an account of the wider interactions and interdependencies that shape regulatory processes and behaviours. Often empirical and descriptive in orientation, many institutionalist approaches employ socio-legal concepts and methodologies. Leading subcategories include network theories,[18] systems theory[19] and 'regulatory space' approaches.[20]

[15] See, eg, L Hancher and M Moran, 'Organizing Regulatory Space' in R Baldwin, C Scott and C Hood, *A Reader on Regulation* (Oxford, Oxford University Press, 1998) esp 148–72.

[16] This categorisation derives from B Morgan and K Yeung, *An Introduction to Law and Regulation* (Cambridge, Cambridge University Press, 2007).

[17] Although, note that their findings may well have normative implications.

[18] See, eg, I Ayers and J Braithwaite, *Responsive Regulation: Transcending the Deregulation Debate* (Oxford, Oxford University Press, 1992) ch 3 (tripartite approach).

[19] See, eg, N Luhmann, *A Sociological Theory of Law*, trans E King and M Albrow (London, Routledge & Kegan Paul, 1985) ch 5 (autopoetic theory of social systems within organisations).

[20] For good overviews and leading accounts of the regulatory space metaphor, see: Hancher and Moran, 'Organizing Regulatory Space' above n 15; C Scott, 'Analysing Regulatory

From closely observing the patterns of interactions between actors who populate specific sectors, regulatory space analyses generate accounts of the complex dynamics at play within the regulatory matrix. Accordingly, the 'regulatory space' metaphor offers a highly useful framing device and analytical tool for interrogating, mapping, interpreting and diagnosing problems with existing regulatory regimes. As we shall see in subsequent chapters, it also serves as an enlightening precursor to applying normative theories or principles to formulate specific recommendations for regulatory design or reform.

In summary, regulatory space analysis involves mapping four principal elements: (1) the various occupants of a regulatory domain; (2) their competitive struggles to assert power and influence; (3) the key resources deployed in those struggles; and (4) how those resources are distributed. As Scott has observed, the approach is 'holistic': it recognises that there may be plural systems of authority and complex networks of interests, actions and interactions.[21]

Regulatory space analysis rests on a number of chief ideas. For present purposes, four are apposite. First, 'formal' legal authority is not the only source of regulatory power. Laws, legal rules and legal actors are merely one, relatively small, part of the regulatory space. Secondly, and following on from this, *formal* authority may be tempered significantly by *informal* authority. Indeed, regulatory authority and responsibility tend to be dispersed between many different public (state) and private (non-state) actors. Thirdly, regulatory power is (unevenly) dispersed because the key resources that confer it are fragmented. Which occupants enjoy power—and how much—depends on how those resources are distributed. According to Scott's classic model, four key resources are: (1) formal legal authority; (2) possession and control of information; (3) possession of wealth; and (4) organisational capacities.[22] Fourthly, because resources, authority and power all are distributed, a diverse range of players interacts through various formal and informal networks or linkages. Their interactions typically are complex, dynamic and horizontal, and involve negotiated interdependence and bargaining.

Thus, the factors determining the shape of any given regulatory domain, and the relative positions of its occupants, are manifold. As we shall see in this chapter, this certainly holds true for the biomedical and biobanking field. Yet, as Black has noted,[23] such a bundling together of so many variables somewhat undermines the utility of the regulatory space concept, risking

Space: Fragmented Resources and Institutional Design' [2001] *Public Law* 329. These papers provide the foundation for the summary of the regulatory space approach presented in this section.

[21] Scott, 'Analysing Regulatory Space', above n 20, 331.

[22] Scott, 'Analysing Regulatory Space', above n 20, 336. Note that, in the course of this chapter, we shall suggest that two additional key resources feature within the biobanking regulatory space. See section subection on 'Codes of Practice and Other Formal or Informal Documentary Guidance Sources', pp 64–66 below and subsection Professional Associations, pp 82–84 below.

[23] JM Black, 'Decentring Regulation: Understanding the Role of Regulation and Self-Regulation in a "Post-Regulatory" World' (2001) 54 *Current Legal Problems* 103, 109.

obfuscation rather than illumination. There are other dangers too. Rigorous regulatory space analysis requires systematic identification of *all* relevant players (or 'actants'[24]) and resources (including documentary instruments that convey power, authority or influence). This demands an extremely thorough knowledge of the field. However, some players may well be 'hidden' from view, or may not be obvious. For example, they may lack formal powers which otherwise would mark them out as being 'regulators' or bodies that can exercise regulatory power. Similarly (and rightly or wrongly), in practice, some occupants may be shut out from power-sharing altogether—for example, because they lack access to key resources. They, too, might well be overlooked. Such errors could have significant implications when it comes to formulating regulatory policies or designing reforms, especially if those bodies properly *should* enjoy some degree of power or voice.

One way to minimise such dangers is to set a more modest target in deploying regulatory space theory as an analytical tool. This is the approach that we have chosen to take in the Governing Genetic Databases project. We have confined our reliance on regulatory space analysis to sketching the *main* influences, as *one* step in our analysis, drawing upon tailored empirical research. We have treated the concept as a helpful analytical construct in mapping out, and thereby understanding better, the principal actors (or actants) and occupants populating the space surrounding biobanks that are used for medical research purposes in England and Wales, and the nature of their roles, interactions and the kinds of resources, power and authority that they wield. Our aim has been to get as good a sense as possible of where medical research regulatory activities principally are focused at the present time, and the implications of this for biobanking governance.

## 'Good Governance' Principles

A second helpful conceptual tool is to apply certain 'good governance' or 'better regulation' principles. As we outlined in Chapter One, two politically favoured sets of principles (being those enunciated by the Hampton Review and the Better Regulation Commission respectively) have been enshrined in the Legislative and Regulatory Reform Act 2006 (LRRA)[25] and the

[24] A central concept within actor–network theory, 'actants' are nonhumans that form part of a social network. According to some sociologists, actants may be 'actants with agency' if they function as independent actors within the network. In the biobanking context, conceivably one might describe certain biobanks, fundamental documentary sources of guidance and/or internet databases of guidance resources (eg, HumGen, PopGen, GEObs—see Appendix 4) as actants with agency. See generally: Fleischmann, 'Boundary Objects with Agency', above n 6, 79; B Latour, *Science in Action: How to Follow Scientists and Engineers Through Society* (Cambridge, MA, Harvard University Press, 1987).

[25] See especially ss 2 and 21.

accompanying statutory Regulators' Compliance Code.[26] For our immediate purposes, applying such principles is useful both for assessing the adequacy of the current biobanking regulatory space (especially from a politico-legal perspective), and for formulating defensible, pragmatically acceptable recommendations for future reform. However, it is also an approach or 'test' that policymakers, lawmakers and regulators themselves may well wish to—or indeed, in some cases, have to—apply.[27] Alongside the foundational norms, values and bioethical principles for governance that we shall elucidate further in Chapter Eleven, so-called sets of 'good governance' principles offer a useful normative and analytical prism. For, in essence, they purport to specify the principal qualities or attributes that 'good' regulatory systems, policies and practices should (or must) exhibit.

Obviously, this immediately begs the question as to what '*good*' governance actually means—and what it *should* mean. Certainly, the two favoured sets of principles outlined above are by no means the only—nor, indeed, necessarily the 'best'—regulatory principles to choose. Numerous different sets of principles have been propounded around the globe, including within the European Union (EU)[28] and the Organisation for Economic Co-operation and Development (OECD).[29] Some variations between them are semantic or terminological; or stem from slightly different emphases being placed on different aspects of broad principles by different bodies. However, some variations are more substantive—perhaps reflecting differing views or underlying philosophies as to what 'good' governance actually constitutes.

Extrapolating from the many lists on offer, one can identify some potentially significant principles that seem to be omitted from (or, at least, not explicitly embodied within) the current UK lists. These include: consultation and public/stakeholder participation; coherence, co-ordination and integration with other laws, regulations and policies; predictability; fairness and non-discrimination; ongoing review and reform; and having a sound legal and empirical basis. Given their adoption by other countries and international bodies—not to mention their potentially significant normative weight—these principles, too, may well be relevant and important to bear

[26] Department for Business, Enterprise and Regulatory Reform, *Regulators' Compliance Code: Statutory Code of Practice for Regulators* (17 December 2007).

[27] As discussed below, many UK regulators must comply with the five principles of good governance, as mandated by the Legislative and Regulatory Reform Act 2006.

[28] See, eg: D Mandelkern, 'Final Report: Mandelkern Group on Better Regulation' (Brussels: European Commission, 13 November 2002) (seven core principles of better regulation); European Commission White Paper, 'European Governance: A White Paper' COM (2001) 428 Final (Brussels, 25 July 2001) at 10 (five principles underpinning good governance).

[29] See, eg: OECD, Recommendation of the Council of the OECD on Improving the Quality of Government Regulation (OCDE/GD(95) 95, adopted 9 March 1995) aka 'Policy Recommendations on Regulatory Reform' (seven principles); OECD, OECD Guiding Principles for Regulatory Quality and Performance (Paris, OECD, 2005).

in mind for designing appropriate UK biobanking regulatory structures, especially as we move beyond the purely domestic frontier.

Interestingly, while both the Hampton and LRRA good governance principles include accountability, one factor not explicitly mentioned in either is *legitimacy*. Insofar as the 'accountability' principle is framed and constructed within the UK, it focuses very much on instrumental goals, especially efficiency. But ensuring *legitimacy* (including democratic legitimacy) is also vitally important when it comes to designing, implementing and enforcing regulatory structures. Alongside its obvious normative significance, instrumentally speaking, legitimacy is essential for earning and maintaining public trust and confidence—both in the biobanking governance system and, through it, biobanking itself. As we shall see from the legal findings and analyses to which we now turn, evidence strongly suggests that the current biobanking governance framework in England and Wales suffers from a marked legitimacy deficit, in several respects.

## MAPPING THE REGULATORY SPACE: DOCUMENTARY GOVERNANCE SOURCES

As noted above, a primary goal of our legal research has been to obtain as clear a descriptive picture or 'snapshot' as possible of the contemporary legal framework and regulatory structures that apply to biobanks in England and Wales.[30] For the reasons outlined above, this has proven to be a Herculean task. Accordingly, to make the empirical legal research stage as effective, comprehensive and manageable as possible, we broke it down into two principal steps. First, we began by identifying all of the legal, ethical, general guidance and other *documents*, instruments and other text-based materials that apply—or potentially could be applied—to biobanks. Our list includes both formal and informal 'paper', documentary and electronic materials. Secondly, and in parallel with the first step, we sought to identify the full array of *actors* who are involved with governing biobanks, biobanking activities, or those who work with biobanks in England and Wales, in any way. Again, our list encompasses both those who wield 'formal' and those who wield 'informal' power, authority and/or influence (or, in some cases, both).

Having completed these two tasks, we then employed our modified 'regulatory space' methodology (described above) to integrate and analyse the two streams of legal research findings, so as to generate a richer, deeper, truer, and more revealing account of the current governance framework

[30] The 'snapshot' presented in this book captures the biobanking regulatory space as it looked when we completed our legal research in late 2008. Some updating work was done during the writing of this book in early 2009. However, as already noted, the biobanking regulatory landscape is constantly evolving. This is fuelled, to a considerable extent, by an ongoing programme of general regulatory, NHS, (bio)technology, and other health sector reforms.

surrounding biobanks. This analysis, which we present below, also drew on our own knowledge of the field, and the burgeoning international legal and bioethical literature on biobanking, population genetics, public health and biomedical research. Set out below are our major findings and conclusions. Overall, our analysis identifies a wide array of documentary instruments and actors who wield many different types of authority and influence. It highlights the haphazard, highly complex character of the existing governance structure, and exposes some interrelationships between the principal stakeholders that we did not anticipate at the beginning of this enquiry.

For clarity and convenience, it is helpful to split our findings into two rough categories: what we have loosely termed 'formal' and 'informal' instruments and actors. Broadly speaking, by 'formal', we mean those documentary sources and actors that enjoy either direct formal legal authority (including delegated), or at least some 'official' recognition as agents or conduits of governance. By 'informal' authority or influence in the field, we mean those instruments and actors that are extra-legal, unofficial or non-binding in nature. We stress that these designations are intended as loose descriptions only. They are certainly not hard-and-fast or precise categorisations.

The documentary governance framework applicable to biobanks in England and Wales consists of three broad classes of instruments and other text-based materials. Selected examples drawn from each group appear in Appendices 1, 2 and 3 at the end of this book.[31] Major instruments are noted individually below.

## Formal Documentary Governance Sources

Primarily, the subsisting framework comprises a substantial and highly diverse range of formal statutes, regulations, statutory provisions, EU directives, other legislative instruments and statutory codes of practice (see Appendix 1 for key examples). Most of these materials are applicable to medical research, medical professionals, public health or medical research in general, rather than to biobanks specifically. The extensive catalogue includes both domestic and supranational instruments. Secondly, the documentary framework incorporates various statutory and non-statutory codes of practice, guidelines and other guidance-related documents, webpages and other internet-based resources or databases, that seek to define, elaborate, disseminate or explain the law, or how it should operate in particular circumstances

[31] Despite the considerable length of the Appendices, they by no means provide an exhaustive account or statistical survey of the wealth of materials available. Providing fully comprehensive lists would be entirely impracticable—not least, because the governance landscape is constantly changing and evolving. Thus, Appendices 1–4 simply aim to present a selective, but broadly representative, 'snapshot' sample.

(see Appendix 2). Thirdly, it comprises a number of common law doctrines on various topics, coupled with an ever-growing body of judicial rulings on the proper status, interpretation and application of relevant laws and professional best practice norms (Appendix 3).

Turning to consider the first class more closely, within the multiplicity of general and narrower domestic laws governing medical research—and, hence, biobanks, to a greater or lesser extent—major formal legislative instruments include the Human Tissue Act 2004 (HTAct), Data Protection Act 1998 (DPA), Human Rights Act 1998,[32] Human Fertilisation and Embryology Act 1990 (as amended), Human Fertilisation and Embryology Act 2008, and Patent Act 1964 (as amended), together with a host of supporting regulations,[33] statutory codes of practice and official guidance materials.[34] Many domestic instruments implement binding EU Directives—notably, the Data Protection Directive 95/46/EC,[35] Tissues and Cells Directive 2004/23/EC,[36] Clinical Trials Directive 2001/20/EC[37] and Databases Directive 96/9/EC.[38]

Narrower relevant provisions include, for example, section 251 of the National Health Service Act 2006.[39] This allows the supply of personally identifiable NHS patient information without consent for certain purposes, including medical research, with the permission of the National Information Governance Board for Health and Social Care (NIGB) Ethics and Confidentiality Committee.[40] Also relevant is the Regulatory Enforcement and Sanctions Act 2008 (RESA). Under RESA, a very wide range of public authorities may be permitted to wield various 'civil sanctions' as an alternative to flexing their prosecutorial muscles.[41] As we elaborate further below, the bodies affected by RESA include leading 'formal' biobanking regulatory space actors—most notably, the Human Tissue Authority and Information Commissioner.

In violation of numerous 'good governance' principles, this wealth of legislative enactments creates a complex, overlapping, yet unco-ordinated and (paradoxically) incomplete statutory system. For biobank managers,

[32] See especially Art 8 of the European Convention on Human Rights (the right to respect for private and home life), which is enshrined in the Human Rights Act 1998, sch I.

[33] See, eg, the Data Protection (Processing of Sensitive Personal Data) Order 2000 (SI 2000/417), especially art 9, which permits the processing of sensitive personal data (including health data) for research purposes in certain circumstances.

[34] See, eg, the statutory Codes of Practice issued by the Human Tissue Authority under the HTAct, discussed in subsection 'Human Tissue Authority (HTA)', pp 70–73, below.

[35] [1995] OJ L/281/31.

[36] [2004] OJ L/102/48. Two technical annexes (Commission Directives 2006/17/EC and 2006/86/EC) provide detailed rules.

[37] [2001] OJ L/121/34.

[38] [1996] OJ L/77/20.

[39] These sections repealed and replaced ss 60–61 of the Health and Social Care Act 2001.

[40] Until its abolition on 31 December 2008, this function was performed by the Patient Information Advisory Group (PIAG).

[41] Regulatory Enforcement and Sanctions Act 2008, pt 3.

users, regulators and commentators alike, this raises serious problems for accessing and comprehending the legal requirements. In part, difficulties arise because biobanks typically hold both physical biomaterials (tissues, biofluids and their derivatives) and information (data)—or, as Parry has helpfully classified them, the 'corporeal' and the 'informational'.[42] Thus, they straddle the data–tissue divide currently enshrined within our law, as is reflected in the separate, largely parallel, statutory regimes governing data and tissues respectively, principally under the DPA and HTAct.[43]

Additional problems compound the inadequacy of the existing biobanking governance framework. First, the formal rules governing medical research keep changing—seemingly in an ever-accelerating process of evolution and flux. This perpetuates uncertainty and complexity. Over recent years, the domestic legislative framework has altered significantly and repeatedly. Many new statutes have been passed or come into force. Major recent legislation affecting medical research and biobanks includes the HTAct, Mental Health Act 2007, Mental Capacity Act 2005, the Medicines for Human Use (Clinical Trials) Regulations 2004,[44] and (to a limited extent) the Human Fertilisation and Embryology Act 2008. Meanwhile, at time of writing, the Coroners and Justice Bill 2008–09, Health Bill 2008–09, National Health Service Reform Bill 2008, and draft regulations under the Human Fertilisation and Embryology Act 2008 all are before the Westminster Parliament.

Secondly, each law must be read in conjunction with all others to develop a true understanding of the legal regime. However, given their lack of co-ordination—not to mention their sometimes conflicting provisions, procedures and standards—this is extremely difficult.[45] Thirdly, both existing and draft laws address medical research, but usually only from the perspective of the particular topic(s) or area(s) that they seek to regulate. For example, in regulating the storage and use of 'relevant material', the HTAct addresses how its consent requirements should (or should not) apply for research purposes. The Mental Capacity Act 2005, which sets protections for people who lack decision-making capacity, also details how these should apply in relation to medical research consent. Overall, however, the law does not cover *all* dimensions of medical research governance in any comprehensive way—much less the extra challenges posed by biobanks. Significant issues left unaddressed include such things as benefit-sharing; ownership and property rights (including vis-a-vis the many different kinds of biosamples and

[42] B Parry, *The Fate of Collections: Exploring the Dynamics of Trade in Bio-Information* (New York, Columbia University Press, 2004).

[43] See further Gibbons, 'Are UK Genetic Databases Governed Adequately?', above n 2, esp 314, 320.

[44] SI 2004/1031. These Regulations implemented the EU Clinical Trials Directive (above n 37).

[45] See further Gibbons, 'Are UK Genetic Databases Governed Adequately?', above n 2.

bodily materials); and feedback of individually relevant research results to research participants.[46]

## Codes of Practice and Other Formal or Informal Documentary Guidance Sources

Turning to the second class of materials—codes of practice and other formal or informal guidance-related documents—the frankly overwhelming volume of statutory and non-statutory (but nevertheless still influential) codes of practice, directions, guidelines, best practice guidance documents, recommendations, statements of ethical principles, fact sheets, toolkits, opinions, circulars, letters, reports, position statements, policies, webpages dedicated to providing guidance, internet databases of documentary materials, and so forth is simply mindboggling. Again, this situation clearly violates the good governance diktats—particularly transparency, proportionality and the giving of 'authoritative, accessible advice easily and cheaply'.

Appendix 2 attempts to convey an impression of the plethora of materials that exists. However, by necessity, it is highly selective and incomplete. It easily could have been twice the length, had it included a fuller range of significant consultation papers and follow-up reports; parliamentary select committee reports; other leading reports, reviews,[47] inquiries[48] and task force publications; recommendations and progress reports; green and white papers; House of Commons Library, public perception,[49] and other research papers; and so forth, that have been published on issues pertaining to genetics and biobanking. Were it to enumerate all of the additional reports, guidance materials and publications put out by the raft of informal governance actors currently active in the field—a sample of which bodies appears in Appendix 4—the list simply would have dwarfed the version presented many times over.

As we shall see in more detail shortly, this glut of guidance materials has been issued both by formal regulatory authorities and by other informal stakeholders, many of them acting, effectively, as self-appointed 'regulators'.

[46] Feedback is addressed in guidelines issued by the Medical Research Council. But these do not apply to researchers who are funded by other bodies.

[47] Prominent chaired reviews covering relevant, governance-related topics include: Warnock (human fertilisation and embryology legislation); Hampton (regulatory reform); Cooksey (publicly funded UK healthcare research); Warner (NHS research ethics review system); Darzi (comprehensive NHS reforms and Next Stage Review); Wanless (long-term NHS resource requirements to meet future technological, demographic and medical trends); and Sainsbury (UK Government's science and innovation policy).

[48] See, eg: Harold Shipman Inquiry; nationwide organ retention scandal inquiries (Kennedy; Redfern; Retained Organs Commission); Modernising Medical Careers Inquiry (Tooke).

[49] See, eg, V Armstrong *et al*, 'Public Perspectives on the Governance of Biomedical Research: A Qualitative Study in a Deliberative Context' (London, Wellcome Trust, 2007).

Prolific formal regulatory actors include the Department of Health, Human Tissue Authority (HTA), Information Commissioner and Information Commissioner's Office (ICO), General Medical Council (GMC), and the National Research Ethics Service (NRES).[50] The huge range of stakeholders producing what we have loosely termed 'informal' materials includes, most prominently, the British Medical Association (BMA) and other professional associations; key advisory bodies such as the Nuffield Council on Bioethics; leading domestic research funders such as the Medical Research Council (MRC), Research Councils UK, and large-scale private sector charities such as the Wellcome Trust; the various Royal Colleges; industry bodies; accreditation and technical standards organisations; international consortia, especially those seeking to promote harmonisation of standards and practices; and assorted interest groups. Several leading European and international bodies also have been active. Unlike the general domestic trend, many of their documents do pertain specifically to biobanks, or to specific aspects of biobanking policy or practice. Here, the main actors include certain EU agencies, the Council of Europe, the Human Genome Organisation (HUGO), the Council for International Organisations of Medical Sciences (CIOMS),[51] the OECD and the World Health Organisation.

Unfortunately, all of these domestic and international materials tend to embody competing visions of appropriate norms, policies, standards and day-to-day practices.[52] Their sheer number and diversity suggest both deep uncertainty over, and pronounced gaps within, the 'official' governance framework. Such deficiencies have allowed those stakeholders who possess key regulatory space resources to step into the void, or power vacuum, left by the lack of any specific, tailored governance framework for biobanks, and to compete at will for their own views, values, preferences, priorities and interests to assume prominence. By issuing and publicising norm-setting guidance documents, good practice or best practice guides, reports, recommendations and so forth, such actors can materially influence—if not control—the shape and content of the governance framework. This is not necessarily the most desirable regulatory strategy. (We return to this theme again below.)

As noted above, Scott has identified four classic key 'regulatory space' resources: formal legal authority; possession and control of information; possession of wealth; and organisational capacities.[53] We have interpreted

[50] Formerly the Central Office for Research Ethics Committees (COREC).

[51] CIOMS was created under the auspices of the World Health Organisation and United Nations Educational, Scientific and Cultural Organisation (UNESCO) in 1949.

[52] See also: A Cambon-Thomsen, E Rial-Sebbag and BM Knoppers, 'Trends in Ethical and Legal Frameworks for the Use of Human Biobanks' (2007) 30 *European Respiratory Journal* 373, especially 374–75 and table 2; S Eriksson, AT Höglund and G Helgesson, 'Do Ethical Guidelines Give Guidance? A Critical Examination of Eight Ethics Regulations' (2008) 17 *Cambridge Quarterly of Healthcare Ethics* 15.

[53] Scott, 'Analysing Regulatory Space', above n 20, 336.

Scott's model as including within the 'control of information' resource the production of (influential) documentary sources of governance. However, from our own application of the regulatory space approach, we believe that two further key resources also play a vitally important role within the biobanking sphere.[54] The fifth is the ability to *publish and disseminate* materials effectively—especially among scientists, medical researchers, biobank mangers and policymakers—coupled with the ability to *persuade* such parties to accept those materials as being authoritative or persuasive. From our research findings, in practice, the most effective dissemination methods seem to be distribution via professional association membership networks, controlling the purse-strings (a power exercised by medical research funding bodies to compel grant recipients to comply with the funders' preferred policies, as discussed below), engaging in advocacy work, such as by issuing press releases that attract mass media coverage, and having a strong, sophisticated internet presence.

## Common Law Doctrines, Case Law and Judicial Rulings

Turning finally to the third class of documentary governance sources, one noteworthy feature of the traditional English legal landscape is the marked prominence of judicial rulings, and the elevated position and power that this gives to judges. In many areas relevant to biobanking, crucial matters are governed—either solely or substantially—by case law rulings, whether from the English courts (including tribunals), the European Court of Human Rights, or the European Court of Justice. As Appendix 3 illustrates, such areas include the taking of consent (especially from living participants); the doctrines of negligence, battery and capacity; safeguarding confidentiality and privacy;[55] property rights (albeit, to a limited extent); intellectual property; and the interpretation and application of criminal offences and sanctions.

Another interesting zone of interaction worth highlighting here concerns the relative status of guidelines and professional 'best practice' norms, especially via-a-vis the courts. English judges generally recognise these as being persuasive. But they have explicitly reserved to themselves the role of being final arbiters of the correct legal standards.[56] This is another notable area, then, where regulatory space analysis highlights competition

[54] On the sixth key resource, see subsection 'Professional Associations', pp 82–84 below.

[55] Especially the equitable duty of confidence, as expanded under the right to respect for private and home life contained in the European Convention on Human Rights, Art 8 (now enshrined in the Human Rights Act 1998, sch I).

[56] See, eg: *Bolitho v City and Hackney Health Authority* [1998] AC 232 (HL); *AB v Leeds Teaching Hospital NHS Trust* [2004] EWHC 644 (QB), [2004] 2 FLR 365.

for dominance between actors. Here, Parliament, medical professionals, professional associations, other code of practice or guideline producers (both formal and informal), and judges all are in the fray.

Having completed our map of the broad patchwork of documentary governance sources potentially applicable to biobanks in England and Wales, one important insight rapidly emerges. The regulatory framework for biobanking cannot be detached from the general framework governing medical research, medical professionals and public health. Equally, it cannot be understood without unpacking the true complexity of that general framework. For, as we have seen, no dedicated, specifically tailored legal framework exists: biobanking simply falls within the broader governance regime. Moreover, in marked contrast to most other leading European nations, no overarching legislation governs medical research involving human participants in the UK. This is reflected in the fact that the UK has not implemented the Council of Europe's Convention on Human Rights and Biomedicine 1997,[57] or its additional Protocols. Conspicuously, it also contrasts with research involving animal experimentation. In this, the UK is among the most heavily regulated jurisdictions on the planet.

## MAPPING THE REGULATORY SPACE—FORMAL GOVERNANCE ACTORS

Moving on from mapping the documentary governance framework, it is time now to look in greater detail at the key governance actors populating the regulatory space. Once again, for analytical purposes, it is convenient to divide them, roughly speaking, into two broad categories: (1) those actors that are 'formally' constituted, empowered or 'officially' recognised in some way within the law; and (2) those whose status, authority or influence are essentially 'informal' only. In both cases, the aim is the same: to establish and explore the power, status, roles and interrelationships of all leading regulatory actors. An illustrative (but by no means complete) cross-sectional sample of relevant governance-related actors appears in Appendix 4.

It is important to note that our primary focus here is on identifying those regulatory space actors that enjoy—at least, according to our *legal* research findings—some discernable governance function, authority or power to influence others. Many additional stakeholders inhabit the biobanking regulatory space. However, when viewed through a legal or regulatory prism, such stakeholders do not appear to possess any material capacity to shape the governance framework, or to influence the attitudes or behaviour of others. On the whole, individual patients, participants and their families, as well as the majority of special interest groups, individual biobanks, researchers,

[57] Oviedo, 4 April 1997, ETS No 164.

research teams, scientists, other professionals, smaller charities, and private sector bodies (including pharmaceutical companies) are not obvious contenders. Fascinatingly, though—as we shall see further in subsequent chapters—a somewhat different picture emerges when we compare and integrate our *legal* and *sociological* empirical research findings. Epitomising one of the dangers of relying exclusively on a regulatory space methodology (as identified above[58]), such a comparison shows that, indeed, there are some significant, ostensibly 'hidden' governance actors in the biobanking sphere.[59]

## Formally Constituted, Empowered or Recognised Governance Actors

Turning, then, to consider 'formal' governance bodies, three obvious, high-level players are legislators,[60] the judiciary[61] and government.[62] We have already discussed the raft of legislative provisions applicable to biobanks; and similarly noted the wide range of matters that are governed, substantially or entirely, by common law doctrines and judicial case law rulings. Much biobanking regulation in England and Wales (as in most countries) is carried out at the national level. However, several European and international bodies also are influential. Some European institutions enjoy formal law-making power via legislative instruments, case law and treaties. Some also enjoy informal authority, through issuing various non-binding recommendations, reports and so forth.[63] As Appendices 2 and 4 illustrate, key formal EU regulatory bodies are the European Parliament, European Commission, Council of Ministers and European Court of Justice. Within the Council of Europe, see the Committee of Ministers, Parliamentary Assembly and European Court of Human Rights.

[58] See subsection '"Regulatory Space" – Metaphor and Methodology', pp 56–58 above.

[59] Notably, while the prominence of research ethics committees and funders is reinforced by both research streams, some additional 'hidden' key actors emerge. Principal among these are eminent individuals and officeholders; researchers' own personal contacts (notably trusted colleagues, collaborators, and their own institutional, biobank, or project-specific steering or advisory committees); technical accreditation bodies; and pre-existing, similarly situated entities (including large-scale, path-breaking biobanks and pharmaceutical companies).

[60] Including the Westminster Parliament, Welsh Assembly, European Parliament, European Commission, and various parliamentary committees (see Appendix 4 for examples).

[61] Including domestic courts and tribunals (eg, the Information Tribunal), the European Court of Human Rights, and the European Court of Justice.

[62] Including Government ministers; ministerial and non-ministerial departments; executive agencies; executive, advisory and tribunal non-departmental public bodies; task forces; ad hoc advisory groups; and other reviews.

[63] See especially Recommendation Rec (2006)4 on research on biological materials of human origin, and numerous other non-binding but authoritative Recommendations covering, inter alia, medical research, human tissue banks, and the protection of medical data and data used in research.

Domestically, within government, by far the pre-eminent and most dominant institution is the Department of Health.[64] Both the Secretary of State for Health and the Department of Health (which has authority to act in many areas under powers delegated by the Secretary of State) enjoy considerable influence over medical practice and research. In particular, they may issue binding regulations, ministerial orders, directions and formal guidelines. Given that the Department of Health's remit covers all NHS employees, in practice it can—and, indeed, does—exercise considerable influence over the internal governance mechanisms applicable to medical practice, biobanks and biomedical research generally.

In addition to these three top-level players—parliamentary, judicial and executive/governmental—our research has identified a small handful of lower-level governance actors who also possess some formal authority to regulate core aspects of biobanking in England and Wales at the day-to-day level. This includes through issuing directions, publishing codes of practice and other forms of official guidance, setting standards, operating licensing schemes, monitoring compliance, and implementing and enforcing the law. However, one of our most surprising findings has been how few of the vast army of actors who populate the regulatory space actually hold any 'formal' governance powers or responsibilities. Aside from the legislative, judicial and executive triumvirate, during the period when we conducted our research only four bodies possessed enforcement or sanctioning powers directly relevant to biobanking—the GMC,[65] the HTA, the ICO, and (to a limited extent) the Human Fertilisation and Embryology Authority (HFEA).[66]

Since that time, the goalposts have shifted. Comprehensive reforms have begun to be rolled out, notably under the Health and Social Care Act 2008. Among other things, these curtail the GMC's regulatory functions. While the GMC retains authority to register doctors, and to set and advise on standards of professional conduct, performance and medical ethics, its adjudicative functions in fitness-to-practice cases are being removed and transferred to a new Office of the Health Professions Adjudicator.[67] This is one area

[64] Other active, relevant departments include the Ministry of Justice (which replaced the former Lord Chancellor's Department and Department for Constitutional Affairs), the Department for Business, Innovation and Skills (formed from the merger of the Department for Innovation, Universities and Skills (aka DIUS) and the Department for Business, Enterprise and Regulatory Reform (aka BERR) in June 2009), and the Home Office.

[65] Under the Medical Act 1983 (as amended), the GMC registers doctors to practise medicine in the UK. Its role includes fostering good medical practice, issuing guidance, and promoting high standards of medical education. Many biobank professionals and researchers are registered medical practitioners.

[66] The HFEA is a statutory body with regulatory powers. However, its relevance to biobanking is limited. Its primary remit focuses on licensing and monitoring UK clinics that offer certain fertility treatments. But it also oversees all UK-based research into human embryos. While it supervises certain research-related activities, then, as well as the storage of gametes and embryos, it does not regulate biobanks or biobanking per se.

[67] Health and Social Care Act 2008, pt 2 and sch 6.

where, one might say, perceived failures of the GMC system to meet the good governance principles (notably accountability and transparency)—coupled with growing public concern over a perceived underlying legitimacy deficit (lack of independence through doctors enjoying power to set and police their own benchmark standards, and to regulate and exonerate 'their own')—have prompted UK lawmakers to adopt a radical new approach.

Also fairly recently, certain research ethics committees (RECs) have been accorded greater official recognition and authority. Today, many RECs effectively straddle the formal–informal divide, with elements of their remits, status and authority now deriving from both camps. The role of RECs within the biobanking regulatory space is a particularly fascinating one. It is also one which, as we shall see, provided us with our second major surprise finding. It is illuminating, then, to look briefly in more detail at the three leading formal (or quasi-formal), day-to-day biobanking regulatory bodies—the HTA, the ICO, and RECs—and to compare and contrast their roles, status, powers and governance styles.

*The Human Tissue Authority (HTA)*

The HTA is an independent, executive non-departmental public body sponsored by the Department of Health. Established under the HTAct, it regulates the removal, storage, use and disposal of human bodies, organs and tissue for various 'scheduled purposes' listed in the HTAct, including medical research.[68] Its remit is very broad. Its core statutory functions are issuing codes of practice, and licensing and inspecting a wide range of activities, including the storage of tissues for research. As indicated in Appendix 2, for biobanking, key HTA codes of practice are Code 1 (Consent), Code 5 (Disposal of human tissue), Code 8 (Import and export of human bodies, body parts and tissue) and Code 9 (Research).

As a regulatory body, the HTA is responsible for ensuring that licensed establishments comply with the HTAct, its codes of practice, and the Human Tissue (Quality and Safety for Human Application) Regulations 2007.[69] The HTAct creates several criminal offences. Penalties range up to an unlimited fine, three years' imprisonment, or both. Criminal offences include removing, storing or using human tissue for any scheduled purpose without appropriate consent; storing or using human tissue donated for one scheduled purpose for some other purpose; falsely representing to others that appropriate consent is in place when it is not; and carrying out licensable activities without a licence.[70] Failing to observe the codes of practice is not a criminal offence per se.

[68] Sch 1.

[69] SI 2007/1523. These Regulations implement the Tissues and Cells Directive 2004/23/EC, above n 36.

[70] HTAct, s 5.

Intriguingly, however, the HTAct contains the somewhat cryptic, veiled threat that the HTA may 'take into account' any non-observation when carrying out its responsibilities in respect of licensing.[71]

One criminal offence directly relevant to many biobanks is that concerning non-consensual DNA analysis (sometimes dubbed 'DNA theft').[72] However, in order to be prosecuted under this provision, a person must intend that DNA be analysed without 'qualifying consent', and that the results be used for some purpose other than a listed 'excepted purpose'. Similarly, while the consent and licensing requirements for storing relevant material for research purposes[73] appear to give the HTA considerable control over how human tissue is used and stored, in reality numerous exceptions apply. Acellular, subcellular and genetic materials—such as extracted DNA—all fall outside the definition of 'relevant material'. This effectively excludes many biobanks from the HTA's reach altogether. Those who store relevant material intending to use it themselves for 'qualifying research' are exempt from the licensing requirement.[74] Moreover, no consent is needed for health-related research using material or DNA analyses taken from living donors, where researchers cannot identify the donors and the research has been ethically approved under regulations.[75] Thus, the HTA's powers are circumscribed, and the HTAct has surprisingly little relevance for regulating biobanks directly.

Under the HTAct, the HTA has several powers at its disposal to superintend compliance with, or to redress breaches of, the HTAct or its codes of practice. These include power to issue, suspend and withdraw licences; inspect premises; and stop the use or transfer of human tissues for purposes, or in ways, that are not permitted. These powers largely determine the scope and focus of the HTA's activities. Thus, in relation to research, its main focus is on ensuring that correct procedures are in place for seeking consent (where required[76]), and ensuring that relevant institutions—including certain biobanks—hold the necessary licences. In practice, however, it appears that the HTA in fact leaves much of the oversight of consent procedures to RECs. Yet, as discussed below, RECs—unlike the HTA—lack any enforcement powers. Moreover, by no means all tissue users or biobanks require licences or participants' consent. Hence, again, in reality, the HTA only regulates a small range of biobanking activities.

[71] HTAct, s 28(2).

[72] HTAct, s 45.

[73] HTAct, s 16(2)(e)(ii).

[74] The Human Tissue Act 2004 (Ethical Approval, Exceptions from Licensing and Supply of Information about Transplants) Regulations 2006 (SI 2006/1260). 'Qualifying research' means research that is ethically approved in accordance with the Regulations.

[75] HTAct ss 1(7)–(9) and *ibid*; HTAct, s 45 and sch 4, para 6.

[76] Research exceptions to consent apply in specified circumstances. See n 75 above.

Where regulatory breaches are identified, the HTA will take regulatory action in accordance with its published enforcement policy.[77] However, even insofar as the HTA does command authority over biobanking activities and professionals, its stated regulatory attitude is not to be remote or authoritarian. Instead, 'collaboration' is one of its 'founding principles'.[78] It seeks to engage closely with professionals and public stakeholders in developing its policies and regulatory arrangements, so as to create and foster 'a robust and proportionate regulatory system' under which 'professionals support and understand the regulatory framework they must follow'.[79]

As is elaborated elsewhere in this book, our own sociological research shows that practitioners *do* regard the HTA as a key authority responsible for regulating biobanking and biomedical research in England and Wales. But, while its preferred co-operative strategy may enable the HTA to govern in accordance with practical realities, it also runs the many well-rehearsed risks associated with regulatory 'capture'.[80] Moreover, given the largely invisible nature of biobanking in England and Wales, there is clear potential for dominant stakeholders to exercise undue influence. Well-resourced, well-organised institutions, industry groups, professional associations or funding bodies in particular—especially those that possess and can exploit key regulatory space resources—may well succeed in shaping the HTA's decision-making, standard-setting, guidance-drafting or policy-making decisions so as to reflect their own particular values, agendas, priorities or interests. Yet, the HTA faces a difficult dilemma. Co-operation arguably is essential. For, precious little independent information is available—either to the HTA or, indeed, to any other regulatory actor—as to what biobanks or other tissue collections actually exist in England and Wales, where they are located, and how, when, why or by whom they are used. Biobanks, biobanking practitioners, funders and industry associations largely enjoy monopoly control over the key regulatory space 'information' resource. The HTA's budget and financial resources, too, are tightly constrained.

As noted above, under the Regulatory Enforcement and Sanctions Act 2008 (RESA) a very wide and diverse range of public authorities may now be permitted to wield various civil sanctions, as an alternative to invoking criminal measures, where criminal offences already exist.[81] This applies both to the HTAct and the DPA. Under RESA, four new kinds of civil sanctioning powers have been introduced. These include fixed and variable

[77] Human Tissue Authority, 'Regulatory Enforcement Policy' (version 0.5, 1 October 2008).

[78] Human Tissue Authority, 'How We Work', available at www.hta.gov.uk/about_hta/how_we_work.cfm.

[79] *Ibid.*

[80] See generally: A Ogus, *Regulation: Legal Form and Economic Theory* (Oxford, Hart Publishing, 2004) 57–58; G Wilson, 'Social Regulation and Explanations of Regulatory Failure' (1984) 32 *Political Studies* 203.

[81] Regulatory Enforcement and Sanctions Act 2008 (RESA), pt 3.

monetary penalties, stop notices, compliance requirements and restoration requirements.[82] By allowing for this wider range of alternative sanctions, the stated aim of RESA is to give formal regulators more options and greater flexibility to enforce the law, using more appropriate and efficient alternatives to criminal prosecution. The approach positively encourages both co-operation and negotiation over conditions and sanctions between regulators and regulatees. Under the RESA regime, three categories of regulators may be given civil sanctioning powers.[83] One is various 'designated regulators' listed in RESA itself.[84] That list includes the HTA, Information Commissioner and HFEA.[85] Ministers also have power to grant additional specific regulators civil sanctioning powers via-a-vis certain regulatory offences.

As yet, it is too soon to say how RESA's expansion of the HTA's sanctioning portfolio will affect its conduct—or that of its regulatees. However, it is reasonable to anticipate a further erosion of the HTA's willingness to seek prosecutions. A combination of factors is relevant. There remains widespread hostility within the scientific community to the HTAct's criminal sanctions. So, dominant interest groups opposed to criminalisation may well seek to 'capture' or strongly influence the HTA away from criminal measures. The HTA's own avowed preference for co-operation and negotiated regulatory action also is likely to favour RESA's civil alternatives. Together with the HTA's limited resources, and its efficiency-related duties under the LRRA, the *Regulators' Compliance Code*, the five good governance principles, and the Hampton Principles, the balance of pressure on the HTA (and ICO too) certainly leans heavily towards minimal, civil interventions. Potentially, all of these factors could result in the HTA favouring civil options, even in circumstances where criminal prosecutions may be more desirable—perhaps in order to serve some public policy value(s) other than efficiency and economy.

### *The Information Commissioner's Office (ICO)*

In respect of the personal data concerning living individuals stored by biobanks, the principal formal regulatory authority is the ICO. The ICO is an independent public authority responsible, among other things, for overseeing and ensuring the protection of personal information in the UK under the DPA. Its main functions here include promoting good practice, providing information and advice, resolving complaints, and enforcing the legislation—including through applying legal sanctions.

[82] RESA, s 42.
[83] ss 37–38.
[84] Sch 5.
[85] Those who enforce offences under (inter alia) the Medicines Act 1968 also may be so empowered.

The ICO has a wide range of tools available to it for taking regulatory action under the DPA. Formal powers include bringing criminal prosecutions and issuing cautions where offences are committed. It also may take various non-criminal enforcement measures—including, now, the RESA civil sanctions (discussed above)—and conduct audits. The ICO may assess organisations to check that their data processing complies with the DPA, and serve enforcement notices or 'stop now' orders where the DPA has been breached. Failure to comply with any enforcement notice itself constitutes a criminal offence. Under the DPA, the ICO may issue notices requiring organisations to furnish information, and execute search warrants (albeit, only with a court order). In 2010, the powers of the ICO were increased so that the ICO may now enforce fines of up to £500,000.[86] However, there is still considerable concern that the ICO—despite having such expanded powers—may be reticent in using them.

This seemingly formidable 'paper' armoury of enforcement powers does not, however, paint a completely accurate picture—not least, in the biomedical context. Several factors combine to limit the practical impact or 'bite' of the data protection regime. One major factor is the law itself. The relevant laws—notably, the DPA, Data Protection Directive 95/46/EC, and certain judicial decisions[87]—are notoriously confusing, complex, incomprehensible and impenetrable. They have been roundly condemned for many years, even by judges. In *Campbell v MGN Ltd*,[88] for example, Lord Phillips MR pointedly observed that the DPA 'is certainly a cumbersome and inelegant piece of legislation'.[89] As it stands, the DPA regime applies only to 'personal data' about currently living individuals. Once people are deceased, its application in respect of them ends. Given the familial nature of much genetic data, genealogical information, and medical and lifestyle histories (all forms of data commonly held by biobanks), the DPA's definition of, and focus on, 'personal data', and its underlying paradigm of the atomised, individual 'data subject', are of questionable merit.

Guidance on the law, too, is problematic. As Appendix 2 illustrates (but can only really hint at), a glut of supposed guidance materials exists, addressing various aspects of data protection laws and practice. Yet, even the 'official' (albeit currently non-binding) guidelines from the ICO, and materials from the European Data Protection Supervisor and EU Article 29

[86] 'News—ICO Gains Powers to Fine £500,000 for Data Security Breaches' (9 April 2010) *The British Journal of Healthcare Computing & Information Management*, www.bj-hc.co.uk/archive/news/2010/n1004008.htm (last accessed 19 Sep 2011).

[87] See, eg: *R v Department of Health, ex p Source Informatics Ltd* (No 1) [2001] QB 424 (CA); *Durant v Financial Services Authority* [2003] EWCA Civ 1746, [2004] FSR 28 (CA).

[88] *Campbell v MGN Ltd* [2002] EWCA Civ 1373, [2003] QB 633 (CA), at [72].

[89] *Campbell v MGN Ltd*, above n 88, at [72]. See also *Totalise plc v The Motley Fool Ltd* [2001] EWCA Civ 1897, [2002] 1 WLR 1233 (CA) where the Court of Appeal labelled the DPA 'difficult to construe'.

Data Protection Working Party, are often convoluted, uncertain and difficult to reconcile. How the DPA interacts with other strands of law too is unclear. Overall, this situation clearly undermines the good governance principles that regulation should be transparent, consistent, readily accessible and accountable.

Like the HTA, within the biomedical sphere the ICO also depends heavily on data controllers and data processors to adhere voluntarily to non-legally-binding guidance standards developed through consultation. Here, too, an information deficit, coupled with limited resources, arguably hamper its regulatory effectiveness. Following a torrent of public scandals and fiascos during 2007 (with more occurring since then) over repeated, often entirely avoidable, serious data protection losses, lapses and thefts, the Information Commissioner called publicly for his powers to be extended, and for maximum criminal penalties under the DPA to be increased.[90] As with the scandals that prompted the HTAct, the intense political fallout, embarrassment and media coverage, once again, forced the Government's hand.[91]

The Prime Minister also commissioned the Information Commissioner and Director of the Wellcome Trust (the UK's largest independent medical research charity) to carry out a full-scale independent review. Unsurprisingly, their report called for wide-scale reforms.[92] Most interestingly for present purposes, they advocated moving towards a more beefed-up version of the HTA-style governance model. This would include introducing a formal, statutory code of practice to set the benchmark for guidance standards, and creating a new, corporate 'Information Commission' to replace the Information Commissioner (a single officeholder). Additionally, their report called for increased funding and powers, because, 'for a regulatory system to bite properly, it must have teeth—and it is clear that the ICO's teeth need to be made sharper'.[93]

From a 'regulatory space' perspective, these developments and comments illustrate the significant impact that sufficiently proactive, dynamic, high-profile, eminent or authoritative single individuals, officeholders or experts can have—especially when they are able to deploy key resources such as publicising their policies or preferences effectively, and accessing and influencing other key stakeholders. (We will return to this theme again in subsequent chapters, when we highlight the role played by 'hidden'

[90] S Bridge, 'Give Me More Power, Information Watchdog to Tell MPs', *The Guardian* (4 December 2007).

[91] See DPA, ss 55A–55E, inserted into the DPA by the Criminal Justice and Immigration Act 2008.

[92] R Thomas and M Walport, 'Data Sharing Review Report' (11 July 2008) available <http:// www.justice.gov.uk/docs/data-sharing-review.pdf>.

[93] Para 7.3.

key individuals as biobanking governance agents, brought to light by our empirical research.)

Despite this general push for more rigorous regulation, and the burgeoning armoury of measures available to the ICO, in the medical research arena the ICO prefers to adopt a 'light touch' approach, issuing guidance and relying on systemic measures to ensure that the DPA is followed. In keeping with the good governance transparency principle, the ICO has explicitly articulated and published its strategy for regulatory action.[94] According to this, its overriding imperative is to 'take a practical down to earth approach'. Following a 'carrots and sticks' approach, it aims to 'adopt a targeted, risk-driven approach to regulatory action—not using our legal powers lightly or routinely, but taking a tough and purposeful approach on those occasions where that is necessary'. The ICO explicitly accepts as its guiding principles (as it now must) the LRRA's five principles of good regulation. In addition to its statutory powers, the ICO's strategy document also refers to negotiation as a form of regulatory action that it aims to use widely to bring about compliance. Negotiated resolutions may be backed by formal undertakings given by organisations to the Information Commissioner.

Overall, then, our research findings suggest a gap between the data protection regime's regulation-on-paper and its actual delivery. The ICO has in the past employed its investigative and enforcement powers only in extreme circumstances, where public scandals erupt, or in response to formal complaints by data subjects. The medical research community has largely avoided intrusive oversight as it has been regarded by the ICO as being an area that is relatively unproblematic. For the medical research community and biobanks, then, so long as scandals and complaints are both avoided, the ICO will assume a predominantly information-and-guidance-providing role, rather than expending limited funds on proactive investigations or heavy-handed oversight.

### *Research Ethics Committees (RECs)*

Since the 1970s, RECs within the NHS have safeguarded research participants by conducting prior ethical reviews of proposed projects and studies.[95] RECs are staffed by unpaid expert and lay members. Approximately 150 NHS RECs are distributed geographically around the UK. Many more are located within other organisations, including universities and pharmaceutical companies.

[94] Information Commissioner's Office, 'A Strategy for Data Protection Regulatory Action' (version 1.1, November 2005).

[95] See generally: National Research Ethics Service, *Governance Arrangements for NHS Research Ethics Committees (GAfREC)* (July 2001). For further discussion, see Gibbons, 'Are UK Genetic Databases Governed Adequately?', above n 2, 333–38 and references cited therein.

Until recently, the REC system was voluntary. RECs were entirely non-statutory. Today, however, certain RECs—notably, those legally recognised under the Clinical Trials Regulations[96]—enjoy limited statutory status and formal authority to give ethical opinions on clinical trials involving investigational medicinal products. Over recent years, much work has been done to rationalise, streamline, standardise and improve REC procedures and decision-making. Consequently, RECs increasingly have come to straddle the formal–informal divide. Yet, still in keeping with their historically exclusively 'informal' nature, no RECs possess any formal investigatory or enforcement powers; nor can they veto, investigate or stop projects that violate ethical requirements. Thus, their exact status and position both remain legally ambiguous. As regulatory actors, they can, perhaps, best be described as 'quasi-formal' in nature.

One of our most surprising research findings has been the extent to which RECs—rather than the more obvious, 'formal' regulators—are actually responsible for day-to-day regulatory action. RECs exercise by far the greatest measure of control over biobanks; essentially, through acting as medical research 'gatekeepers'. Their role is pivotal. This is despite their comparative lack of resources, organisational capacity or formal powers as compared with the other agencies, particularly the HTA, ICO, Department of Health and government—each of which either lacks direct responsibility for, or has shied away from exercising, active supervision of biobanks.

In systemic terms, our own regulatory space analysis suggests that, over time, a complex intrinsic interdependency has evolved. The principal UK policymakers, lawmakers, formal regulators and the governance framework itself all implicitly rely upon RECs to keep biobanks and biobanking activities in check. Yet, because RECs lack formal powers, one quid pro quo for this arrangement is that those other agencies must provide the missing 'teeth'. RECs accordingly depend heavily upon the existence of other bodies and mechanisms—both formal and informal—to ensure compliance, enforce their ethical conditions and penalise violators.

As Chapter Seven explores further, in practice, we have found that RECs do enjoy considerable status and influence. Ethical approval effectively is mandatory for biobanking research activities. However, in regulatory theory terms, this is only because the Department of Health, NHS, GMC, professional associations and all leading research funding bodies (notably the MRC) typically insist upon prior ethical approval. For example, all research involving NHS staff, facilities, premises, patients, users (including relatives and carers), patient data, organs or bodily material must be REC-approved. Serious violations of ethical conditions may prompt remedial action by these other regulators. Possibilities include disciplinary action by NHS Trusts,

[96] Above n 44.

professional misconduct or fitness-to-practice proceedings (for example, before the Office of the Health Professions Adjudicator), or the suspension or termination of researchers' or biobanks' funding.

Stepping back to reflect critically on the implications of this model of governance—one which has evolved essentially ad hoc, and through historical accident rather than any deliberate design—is illuminating. For, it reveals at least five significant pitfalls in relying on such an implicit, unarticulated 'systemic symbiosis' to govern biobanks, rather than (say) having a dedicated, formally empowered, clearly designated supervisory authority (such as a national bioethics commission). First, there is a risk that other agencies simply may decline to respond to ethical violations, or deem biobank governance to fall outside their remits and responsibilities. Secondly, while many RECs now ask researchers to submit ongoing progress reports, REC approvals rest on *proposed*, not *actual*, protocols for running specific projects or 'research tissue banks' (RTBs) which now may seek 'generic' approval.[97] Once REC approval is given, RECs do not have any clearly established, proactive monitoring or quality assurance role. Thirdly, despite concerted efforts over recent years to overhaul and streamline the REC process—including via the National Research Ethics Service (NRES), with its revised *Standard Operating Procedures for Research Ethics Committees in the United Kingdom (SOPs)*[98]—there is still a persistent lack of uniformity in REC standards, procedures, practices and decision-making across the nation.

Fourthly, RTBs that are given generic or blanket approval subsequently bear responsibility for assessing and monitoring individual research projects that utilise them as tissue and/or data resources. No individual project REC approval is required. Only limited requirements for RTBs to report back to RECs apply. This system effectively hands RTB managers considerable power and autonomy. Arguably, this could be seen as rendering biobanking activities insufficiently visible, transparent or accountable—thereby undermining both the good governance principles and overall systemic legitimacy. However, certain anecdotal evidence strongly suggests that the vast majority of REC members do not adequately understand genetic research, or its implications.[99] Therefore the effect of passing control over research to RTB managers puts such control into the hands of professionals who may have greater expertise and knowledge to judge the merits of proposed research initiatives than RECs. The key issues then become how accountable those professionals may be for their decisions, and the extent

[97] See National Research Ethics Service, *Standard Operating Procedures for Research Ethics Committees in the United Kingdom ('SOPs')* (version 4.0, April 2009) and Gibbons, 'Are UK Genetic Databases Governed Adequately?', above n 2, 335.

[98] National Research Ethics Service, *Standard Operating Procedures*, above n 97.

[99] R Ashcroft, AJ Newson and PMW Benn, 'Reforming Research Ethics Committees' (2005) 331 *British Medical Journal* 587.

to which this changes the research governance system—in particular, the role of RECs as gatekeepers for research approvals.

Finally, in addition to the official SOPs, other purported guidelines for RECs positively abound. Domestically, the Department of Health, MRC, Association of the British Pharmaceutical Industry, other industry associations, the GMC, several Royal Colleges and various professional groups all have issued their own documents, suggesting how RECs should conduct themselves and organise, assess or determine ethical reviews.[100] Numerous international bodies also have issued guidelines directed at bioethics committees generally, wherever they may be located around the globe. For example, the United Nations Educational, Scientific and Cultural Organisation (UNESCO), through its International Bioethics Committee, has published a series of guidebooks on establishing, educating and running bioethics committees. Regional bodies (such as the European Federation of Pharmaceutical Industries and Associations) and other international organisations (including the World Health Organisation) have done likewise. This overwhelming—and oftentimes conflicting—array of materials again underscores the deficiencies within the current framework. It evidences another key arena of competition between stakeholders to influence RECs, made possible by the lack of clear, comprehensive, formal provision. It also demonstrates the pivotal (yet deficient) role performed by RECs within the current governance framework, given the fact that so many stakeholders have targeted them.

## MAPPING THE REGULATORY SPACE—INFORMAL GOVERNANCE ACTORS

Having mapped out both the documentary governance framework and the formal governance actors, the final step in our three-stage analysis is to identify any other significant 'informal', 'unofficial' or erstwhile 'hidden' actors that perform a relevant governance role in respect of medical research or biobanking in England and Wales. In keeping with our regulatory space methodology, this step has required us to look beyond the formal legal framework, and to delve more deeply into the networks, patterns and methods of behaviour, and complex social interactions between all stakeholders within the field.

Undertaking this analysis delivered a third, major surprise finding—namely, the revelation as to just how densely and extensively populated is the biobanking and biomedical regulatory space, especially with informal, self-appointed, or would-be governance actors, of all manner of descriptions. Appendix 4 seeks to convey a sense of this melting pot. However,

[100] J Saunders, 'More Guidelines on Ethical Research?' (2008) 33 *Journal of Medical Ethics* 683.

as noted already, it is far from exhaustive. It simply seeks to illustrate, by presenting the principal actors plus a cross-sectional, representational sample of the others, the astonishing breadth of the hundreds of actors involved.

As can be seen, prominent informal actors come from all sectors: public (including both State and general public), private and charitable. Their respective power, status, authority and impact vary. Such capacities in turn derive from varying sources—not least, the actors' standing or reputation in the eyes of others, and their possession of (and ability to exploit) key regulatory space resources. Prominent informal actors span domestic, regional, supranational and international agencies. Indeed, they cover virtually the full gamut of institutional types. Notable sub-categories include non-governmental organisations; advisory bodies; review bodies; committees (standing or ad hoc); professional associations; research funders; charities; Royal Colleges; industry groups; accreditation and technical standards organisations; consortia; large-scale biobanks; academic and professional networks, including transnational networks of professionals and experts ('knowledge actors'[101]); informal partnerships and coalitions; and assorted special interest, human rights, and lobbying groups. One could also plausibly include within the list the press and other mass media entities; publishers (especially academic); and leading scientific, medical, legal, regulatory, (bio)technological and similar journals and magazines.

### European and International Informal Governance Actors

As noted already, while much biobanking regulation occurs at the national level in England and Wales, several regional and international bodies enjoy considerable informal influence. Typically, this is by virtue of their prestige, power, sheer size, resource base or organisational capacity—or simply through their having been active in issuing documents, reports or publishing other materials that tackle otherwise un-regulated or under-regulated topics. Noteworthy examples are various pan-European entities. These include professional associations (for example, the European Society of Human Genetics); industry groups (such as the European Association of Tissue Banks); networks (including the European Network of Research Tissue Banks); research funding bodies (notably, the European Science Foundation, European Research Council, and

[101] 'Knowledge actors' are transnational networks of professionals and experts, who meet together in transnational fora and act as major agents and channels of regulatory change. See further: F Gilardi, J Jordana and D Levi-Faur, 'Regulation in the Age of Globalization: The Diffusion of Regulatory Agencies Across Europe and Latin America' in G Hodge (ed), *Privatization and Market Development: Global Movements in Public Policy* Ideas (Cheltenham, Edward Elgar, 2006) 127; A-M Slaughter, *A New World Order* (Princeton, NJ, Princeton University Press, 2004).

European Medical Research Councils), and relevant research institutes (for example, the European Molecular Biology Organisation, or EMBO).

Internationally, no organisation possesses any formal regulatory or enforcement power. But, as Appendices 2 and 4 illustrate, several prominent non-governmental organisations have assumed responsibility for co-ordination and leadership in significant areas of health governance, including biobanking. Principal among such bodies are the United Nations, UNESCO (especially via its International Bioethics Committee),[102] and the World Health Organisation.[103] UNESCO's Declarations in particular, while non-binding and unenforceable, nevertheless constitute strong normative statements of principle and policy that are widely respected. Other important international bodies are the World Medical Association (an influential lobbying group, representing doctors and physicians, and guardian of the Declaration of Helsinki); CIOMS; the OECD; and HUGO. The HUGO Ethics Committee has been particularly active and influential in the field of genomics.[104]

## Leading Informal Governance Actors—The NHS

Turning to scrutinise more closely the actual, day-to-day impact of informal governance actors, not surprisingly, the NHS—especially in its capacity as an employer—features prominently. Most biobanking research and associated activities in England and Wales are undertaken by NHS staff, by other researchers in association with NHS staff, or using NHS resources (patient biosamples and data) collected and stored by NHS staff. We have already noted some of the functions, activities, controls, guidelines, sanctions and so forth applied by NHS organs. In addition, all NHS biobanks and research projects must undergo extensive internal and external scientific and ethical scrutiny. This can entail multiple REC, NHS Research and Development (R&D) Office, Caldicott Guardian,[105] and NHS Data Protection Officer approvals, strict reporting requirements, laboratory accreditation procedures, performance audits, formal technical

[102] See especially UNESCO Declarations on human rights and the human genome, human genetic data, and on bioethics and human rights, as well as numerous reports (Appendix 4).

[103] Note also several major, multinational initiatives underway (eg, via the Public Population Project in Genomics (P3G) Consortium) that seek to achieve agreements between biomedical scientists, ethicists and lawyers to harmonise standards and practices, so as to foster interoperability of biobank databases and the easier international sharing of data and biosamples.

[104] See, eg, its statements on benefit sharing, the principled conduct of genetics research, and human genomic databases (Appendix 4).

[105] Caldicott Guardians are senior NHS and social services staff members, appointed within each institution to protect patient-identifiable information and confidentiality. Their responsibilities include advising NHS overseeing, monitoring, and reviewing institutional information governance protocols and policies, and staff, departments and ethics committees (including RECs).

standards compliance checks, inspections and so forth. In addition, the many guidelines for research practice circulating within the NHS, coupled with ongoing professional training courses, serve to inculcate and shape professional cultures and practices (and, hence, practitioners' attitudes and behaviours) in particular ways.

Looking beyond the NHS, our legal research findings delivered a final startling surprise. Despite the host of informal stakeholders actively engaged in the arena and competing for influence, two kinds of informal actors play an unexpectedly—and disproportionately—powerful and pervasive role. These are professional associations and the leading biomedical research funding bodies. Like RECs—the other principal, quasi-formal, quasi-informal governance actors (discussed in detail above)[106]—both professional associations and funding bodies act as 'gatekeepers'. Using different methods, they each control the means by which medical research can proceed. Here, once again, using the regulatory space methodology has been vital for flushing out these underlying patterns of interactions between stakeholders, and for understanding the complex dynamics leading to such dominance.[107] In particular, it helps us to explain how their authority and influence may be seen to stem from their possession and control over key regulatory resources—not least, information, effective avenues for disseminating their views, policies and preferences, wealth, and organisational capacity. It is illuminating to consider each of these two groups of actors briefly in turn.

## Professional Associations

Traditionally, professional associations within England and Wales have commanded high levels of status and respect. This is particularly so where their membership is restricted to a small professional elite. Between them, the leading UK medical professional associations—especially the GMC, Royal Colleges,[108] Academy of Medical Sciences and BMA—exert considerable influence over medical research (and, hence, biobanking) policy, practice, practitioners and governance. To date, few professional associations have ventured beyond medical research generally, to address the specific areas of genetics or genomics. However, several specialist bodies and industry organisations have sprung up. They too have been active, and so are relevant here. Leading examples include the British Society for Human

[106] See subsection 'Research Ethics Committees (RECs)', pp 76–79 above.

[107] Our legal research findings here are strongly reinforced by our empirical research findings. See further chs 5, 6, 7, 8 and 9.

[108] Notably, the Royal Colleges of Physicians, Pathologists, General Practitioners, and Medicine.

Genetics, Association of the British Pharmaceutical Industry and British Association for Tissue Banking.

Much of the authority of professional associations appears to derive from their expertise, coupled with their carefully cultivated elite status and reputations. Many boast a long and distinguished history. The Royal College of Physicians, for example, was established in 1548 under Royal Charter. New members typically are elected by current members, in recognition of their contributions to the profession. By so restricting and controlling membership to a self-selecting elite, such bodies ensure that their power and prestige are maintained. Consequently, their views, guidelines and normative pronouncements tend to carry considerable weight. Many also possess significant regulatory space resources—notably, expertise, information and wealth—accumulated over many years.

Utilising such resources, professional associations make their impact felt through various strategies and methods. Three principal ones are: (1) issuing, disseminating and overseeing professional 'good practice' and other guidance materials;[109] (2) acting as self-styled guardians of 'professional culture'; and (3) lobbying and advocacy work.

The first two methods are largely directed at professionals and practitioners; the audience for the third is usually policymakers, lawmakers and the general public. Typical activities relating to the first two methods include issuing guidelines (such as to professional associations' general membership and RECs); providing information; setting standards; commissioning research; publishing reports and recommendations; and educating professionals, thereby inculcating certain professional cultural understandings and norms. As we shall see in subsequent chapters,[110] professional cultures played a very significant (albeit usually implicit) governance role for our Governing Genetic Databases project interviewees. Consequently—and particularly in the absence of any 'formal', or universally accepted, standards and procedures—we would add to Scott's list a sixth key biobanking regulatory space resource: *control over professional education, the setting of good practice norms and the inculcation of professional cultures.* As our empirical research suggests, the ability to control such things conveys considerable power to shape individuals' attitudes and day-to-day behaviour, both consciously and subconsciously.

Advocacy work by professional associations entails such activities as courting media attention; issuing press statements; responding to public consultations; giving evidence, such as before parliamentary committees; having members who enjoy dual affiliations (such as sitting on the House of Lords); and actively lobbying policymakers and lawmakers. A prime

[109] See Appendix 2 for examples.
[110] See especially ch 5, and ch 6 below.

illustration of the medical, scientific, and medical charity communities' power to influence governance through lobbying is the HTAct. A host of exceptions to its consent requirements—many of them expansive, especially the research exception[111]—were hastily cobbled into the Human Tissue Bill during its passage through Parliament. This followed intense pressure by professional bodies and charities, particularly targeting members of the House of Lords, and vociferous concerns expressed by them to MPs and the media that a blanket consent requirement could hinder or preclude beneficial medical research.[112] That the exceptions entailed an embarrassing *volte-face* for the Government, whose initial rhetoric repeatedly stressed that consent would be the 'watchword' throughout,[113] and 'the fundamental principle' to which the Government 'would stick',[114] is particularly telling.

## Funding Bodies

Control over the purse-strings is another powerful 'gatekeeping' governance tool, enjoyed by funding bodies across the public, private and charitable sectors. Medical research and biobanks are typically highly expensive enterprises. UK Biobank, for example, was allocated some £61m by the Wellcome Trust, MRC, Department of Health, Scottish Executive, and Northwest Regional Development Agency. Almost invariably, at least some external funding support is essential. By setting priorities and policies, and vetting funding applications, funding bodies effectively control *which* proposals are rejected or allowed to proceed—and, moreover, on what *terms*. While all funding applications are peer-reviewed, funding calls predetermine the types of applications that may be made. Thus, in practice, a small number of funding bodies can wield substantial control over national and (where collaborative projects are involved) international research agendas, strategies and priorities.

Funders are able so to do so because those who receive grants from funders must comply with their funders' policies and conditions. This means that funders can prescribe certain norms and conditions as to how research should be carried out by researchers, biobank operators and (indirectly) the biobanking community generally. One example of a policy driven strongly by funding bodies that has had a significant effect on

[111] See text to n 75 above.

[112] See, eg, Price, 'The Human Tissue Act 2004', above n 14, 800; C Dyer, 'Human Tissue Bill is Modified Because of Research Needs' (2004) 328 *British Medical Journal* 1518; HL Deb vol 664 col GC427, 15 September 2004 (Lord Jenkin).

[113] HC Deb vol 416 col 986, 15 January 2004 (Ms Rosie Winterton, Minister of State for Health and chief sponsor of the Bill).

[114] HC Deb vol 423 cols 88–89, 28 June 2004 (Ms Winterton).

practice is the development of open-access policies for publishing and for the sharing of genomic reference datasets. Any analysis of the regulatory space that focused simply on the formal, legally constituted bodies would not capture the importance of funders—not only in directing policy (such as in relation to open access), but also in working with the scientific community to formulate new strategic directions, and writing guidance and formulating conditions for funding.

Funders also have different mechanisms of oversight for their activities. While national funding agencies ultimately are accountable to the Secretary of State, charitable institutions that fund research are accountable to the Charities Commission and internal boards. As funders can be very influential in directing the scientific agenda there needs to be appropriate recognition that they do not *just* allocate funding, but that they are one of the key actors in this field—despite the fact that they are not instantly recognisable as 'regulators'. Like the quasi-formal RECs, they also play an important role in governing research. However, unlike formal regulators such as the ICO, they do not have formal powers to regulate and therefore are not subject to the same accountability mechanisms as other regulators.[115] It is this almost 'hidden' or 'de facto' effect of their activities that raises issues about the transparency and accountability of their regulatory activities.

Turning to consider the public sector, two key government departments currently responsible for funding research relating to biobanks are the Department of Health and the Department for Innovation, Universities and Skills. Several publicly funded research councils, too, are active—most notably, the MRC (which has invested extensively in biobanks[116]); the Economic and Social Research Council (which funds the Genomics Network); the Biotechnology and Biological Sciences Research Council; and the Engineering, Physical Sciences Research Council. The recently established Office for Strategic Coordination of Health Research acts as a central co-ordinating body for health research, overseeing budgetary decisions and research strategies of the National Institute for Health Research, MRC and Department of Health. Also relevant are key European and international funding agencies. Prominent examples include the European Research Council, and the US National Institutes of Health.

The power of the leading funding bodies to define and direct how research is conducted—and, consequently, to shape opinions, expectations, underlying assumptions and the prevailing wisdom as to how it *should* be conducted—is profound. For example, in addition to stipulating particular grant conditions, the MRC and its Regulatory Support Centre

[115] Such as the *Regulators' Compliance Code: Statutory Code of Practice for Regulators* (17 December 2007).

[116] J Kaye and P Martin, 'Safeguards for Research Using Large Scale DNA Collections' (2000) 321 *British Medical Journal* 1146.

have produced a wealth of operational and ethical guidelines, statements of principle, position statements and good practice guidance. These cover a huge spectrum of topics relevant to biobanks. Specific matters addressed include data and tissue handling, consent, vulnerable participants, investigating misconduct, good research practice, clinical research governance, open access to research findings and data curation.[117] Despite lacking any formal, legal force, in reality all MRC-funded research must comply with these materials—lest funding be withdrawn or disciplinary proceedings ensue. Generally speaking, there is no robust, external, independent or mandatory oversight of funders' decisions, priorities, policies or practices as would be required were they perceived to be a regulator on behalf of the state.[118]

Looking now within the private and charitable sectors, UK charitable funding far outstrips that of any other European country. Indeed, the Association of Medical Research Charities represents over 100 charities that support medical research and genetics within the UK. Three leading charities that, between them, wield substantial de facto power within the domestic regulatory space are the Wellcome Trust, Cancer Research UK and the British Heart Foundation. As privately endowed charities, such organisations are independent from government and industry. To some extent, such bodies are held answerable under general laws governing charities, and to patients and the public who are their donors. They are required to fulfill their mandate to obtain and maintain charitable status, and the possibility of the risk of adverse media publicity leading to a loss of public support, trust and confidence (on which biobanks fundamentally depend) can help to play a major role in maintaining standards. This affects both funders and high-profile biobanks.

UK Biobank offers an illustration of regulatory innovation. On the one hand, given the vacuum left by the absence of any formal, external, specifically tailored biobanking regulation or policy, 'big players' (or 'actants') such as UK Biobank can arrogate to themselves significant informal influence and freedom of choice. Thus, to date, UK Biobank has been able to articulate and follow its own preferred norms, standards, policies and practices for this emerging field. Examples from its Ethics and Governance Framework document[119] include its blanket decision not to feed back research results to individual participants, and its assertion (as yet judicially untested) that UK Biobank 'owns' participants' biosamples. As a high-profile, highly ambitious, and path-breaking initiative, UK Biobank's standards and procedures

[117] See Appendix 2 for relevant examples.

[118] Although note that oversight by the Office for Strategic Coordination of Health Research could have implications for the MRC's independence in terms of its funding policies, research strategy and decision-making.

[119] UK Biobank Ethics and Governance Framework (version 3.0, October 2007).

may well become norm-setting, as other biobank creators and policymakers look to copy its example (thereby avoiding having to reinvent the wheel themselves). Indeed, as Chapter Five will show, such copying is already taking place. Furthermore, given UK Biobank's sheer size and scope, involvement with it is likely to become increasingly important for researchers in England and Wales. Accordingly, how it conducts itself, and its decisions over which projects are granted access, may well materially influence the future shape and direction of UK genetics research.

Yet, on the other hand, the very fact that UK Biobank *has* an Ethics and Governance Framework document—and, moreover, that its funders felt the need to establish the UK Biobank Ethics and Governance Council as a dedicated, independent advisory body and guardian of the Ethics and Governance Framework—indicate a marked concern to be seen to be legitimate, trustworthy and accountable. Apparently as a direct consequence of the lack of any external, overarching, formal governance structures, then, UK Biobank's creators perceived a need to establish some high-profile governance arrangements of their own. UK Biobank has been seen to be developing good practice under the scrutiny of its Ethics and Governance Council; therefore, it has been left relatively free of external oversight from formal regulators such as the ICO and the HTA.

One may justifiably conclude that this situation undesirably excludes other actors—notably, researchers, biobank operators, individual participants, their family members, donor communities, participant groups, citizens and the public(s) alike—from enjoying adequate protection, representation, power or voice within the biobanking regulatory space. The impact and status of patient or participant interest and lobbying groups in genetics is quite variable. Organisations such as the Genetic Alliance UK (formerly called the Genetic Interest Group) and GeneWatch are active. However, there are few bodies that have a legitimate role to play as 'watchdogs' and that are seen in that role as being an essential element of the regulatory framework. Yet, paradoxically giving such actors a greater say within the regulatory framework might well serve to enhance its legitimacy, thereby bolstering public support for biobanking.[120]

## CONCLUSIONS

This chapter has set out in detail our three-stage legal research methodology, and used it to present our major legal research findings. It began by introducing a range of helpful concepts and tools—notably, 'boundary objects', the 'regulatory space' metaphor and methodology, and the various 'good governance' principles. Then, through applying our modi-

[120] Ayres and Braithwaite, *Responsive Regulation*, above n 18.

fied version of the regulatory space approach, this chapter: (1) mapped out the key formal and informal documentary governance sources relevant to biobanking in England and Wales that applied up until 2009; (2) identified the leading formal and informal governance actors that were active during this period; (3) explored the complex dynamics of their sometimes competitive, sometimes symbiotic, interrelationships; and (4) highlighted a number of important conclusions, implications, and lessons that can be learned from our legal research findings. As we have seen, those findings included a number of significant—and sometimes quite worrying—surprises.

First, our analysis of the legal documentation demonstrates that there is no specifically tailored regulatory framework, or comprehensive, purpose-designed legal instrument(s) governing biobanks in England and Wales. Instead, the regulation of biobanks comes under the general framework for biomedical research which is based on a complex patchwork of references to medical research in general legislation, as well as a bewildering array of formal and informal documentary sources of governance. At the time when this research was carried out no one piece of legal documentation applied at a national level either to medical research in general, or biobanks in particular, that was authorative and based on a clear normative orientation, policy direction, or solid theoretical, bioethical, ideological, principled or teleological foundation. Another significant factor that researchers have had to deal with is the fact that the governance documentation in this field is constantly changing with ongoing minor and major reforms to aspects of the law and continuous development of new guidelines by a variety of actors. A number of governance documents overlap and are at times mutually inconsistent. This situation makes it extremely difficult for scientists, researchers, biobank operators and, indeed, anyone else to ascertain what the law, relevant rules or guidelines require. This is especially problematic where conflicts arise between different laws, rules, guidelines, standards or regulatory actors. It is frequently unclear which (if any) should be given precedence when their demands cannot be reconciled.

Biobanks and the nature of their activities raise specific challenges for the regulatory framework as they do not fit neatly within existing parameters. Because biobanks typically store and use both corporeal and non-corporeal materials (tissues; genetic material; data), additional complications flow from the law's current, artificial attempt to bifurcate tissue and data regulation under separate (but overlapping, partially inconsistent and arbitrarily limited) parallel regimes. Biomedical research, clinical trials and therapeutic applications also tend to be subject to different instruments, regimes and rules. Yet, in reality, the boundaries between these too are flexible and imprecise. As boundary objects, 'biobanks' themselves are difficult entities to define with any legalistic precision. They cover a wide and diverse spectrum of types.[121] Nationally, regionally and internationally, there remains a pervasive lack of harmonisation or agreement over correct terminology, definitions, standards, norms and practices. Each of these issues makes it difficult to fit biobanks within existing regulatory and legal frameworks, making coherent governance of them problematic.

In the absence of any specific public authority or other regulatory body being charged with governing and overseeing the biobanking sector, an equally expansive range of formal, informal and quasi-formal/informal actors crowd the regulatory space. A small number of 'formal' regulatory actors have remits and responsibilities that happen to extend to biobanking activities in some way. In addition to Parliament, the judiciary, and Government, primary examples are the Department of Health, Secretary of State for Health, HTA, ICO and GMC. Somewhat surprisingly, these bodies tend to have a 'hands off' approach to the regulation of medical research. Despite having legal enforcement powers, the principal 'formal' bodies, such as the ICO and the HTA, tend to favour co-operative regulatory action strategies and negotiated bargaining. To date, their engagement with the emerging field of biobanks has been limited. Those that are willing to regulate more proactively (notably, the ICO), or that have no choice in the matter (notably, RECs), arguably lack sufficient powers, authority, resources and/or co-ordination to be fully effective. This raises significant questions about accountability, and the ability, or willingness, of these bodies to follow up on the standards that they set or the decisions that they make. By contrast, a highly significant role is played by quasi-formal/informal RECs, and by the vast swathe of 'informal' stakeholders (not least, other professional associations and medical research funding bodies) that inhabit this regulatory space. However, the powers that these bodies have, and the way in which they influence the regulation of the biobanking sector, vary significantly depending upon the resources that they have at their disposal.

Scott identified four classic key 'regulatory space' resources: formal legal authority; possession and control of information; possession of wealth; and organisational capacities.[122] However, in the field of biomedical research there are two additional resources that need to be added to this list to get a true understanding of how regulation and governance are carried out in this area. A fifth resource that is essential to having influence is the ability to publish and disseminate materials effectively—especially among scientists, medical researchers, biobank mangers and policymakers—coupled with the ability to persuade such parties to accept those materials as being authoritative or persuasive. The sixth resource is the ability to exercise some control over professional education, the setting of good practice norms, and the inculcation of professional cultures.

[121] See further ch 4 below, and SMC Gibbons, 'Regulating Biobanks: A Twelve-point Typological Tool' (2009) 17(3) *Medical Law Review* 313–46.
[122] Scott, 'Analysing Regulatory Space', above n 20, 336.

Therefore, in the biomedical regulatory space, there are at least six key resources that confer power, status or authority. These are: (1) formal legal authority; (2) possession and control of information; (3) possession of wealth; (4) organisational capacities; (5) the ability to publish or disseminate one's views and preferences effectively, and to persuade others to accept them as authoritative or persuasive; and (6) control over professional education, the setting of good practice norms, and the inculcation of professional cultures.

These key resources are not distributed evenly among all stakeholders; or even between different categories or groups of stakeholders. The absence of any overarching, external governance framework or dedicated supervisory authority (that otherwise might operate to mitigate such imbalances) consequently permits those actors that possess and can most effectively exercise the key regulatory space resources to come to the fore. Prime examples are professional associations and funding bodies. The lack of clear authority and national guidance means that the majority of 'informal' stakeholders seek to fill the regulatory gaps. Not least, they do so by generating guidelines and other documents (which frequently overlap and are mutually inconsistent), and by utilising the regulatory space resources at their disposal. This system relies heavily on an implicit symbiosis, or intrinsic interdependency, that has evolved, ad hoc, between RECs and other leading formal and informal actors (particularly the HTA, ICO, Department of Health, GMC, and the NHS).

Our analysis reveals that an incomplete and inconsistent regulatory framework can also lead to innovation. A good example is UK Biobank, which has set up its own governance systems to ensure accountability and transparency. In so doing, it has become a norm-setter for other similar high profile projects, both in the UK and internationally. However, while this system can be responsive to new developments, there are also considerable inadequacies with the system. Not least, it can lead to duplication of effort, a lack of certainty over which guidance should be followed, and a lack of clarity over what should constitute 'best practice' in emerging areas.

One of the dangers of having a governance system that allows 'informal' bodies to step in to fill the regulatory gaps is that those constituencies that do not possess the necessary regulatory resources, or may (inadvertently) be excluded from playing a role, or from having an effective voice in how medical research and biobanks operate and are governed. Key stakeholders that appear to be excluded in this way under the present system include participants and their families, participant groups, special interest groups, and the public(s). Potentially, their rights, values, priorities and interests could be marginalised, squeezed out, or put at risk. Paradoxically, this situation could prove to be self-defeating in the long run for the biobanking sector, and for the many informal stakeholders who currently strive for authority, power and influence. It also could undermine the UK's future as a world leader in the biomedical and biotechnology fields. Public trust, confidence, support and participation all are crucial to the success of biobanks. Yet, as widespread research consistently shows, all four of these factors depend heavily upon the existence of reliable, independent and transparent governance frameworks that people feel can be trusted.[123] One solution to this problem is for the state to fund advocacy and special interest groups as essential elements of a robust regulatory system.

The current regulatory system also challenges many of the 'good governance' principles that can be found in the Regulators' Compliance Code[124] and the principles of better regulation—notably, consistency, transparency, accountability, targeting and proportionality.[125] If we consider the system as a whole, rather than analysing the activities of individual government regulators that must adhere to the Regulators' Compliance Code, there are obvious deficiencies in the biomedical research framework. First, the system as a whole is not *consistent*[126] and coherent—there is neither a clear authoritative body for decision-making, nor any clear set of principles that might provide a foundation for the development of best practice for biobanking. Researchers do not know what are the laws, rules and standards, because these are difficult to find out and, when found, vary markedly depending upon the regulatory body. They are not joined up to provide clarity and ease of implementation. The bodies that regulate research do not work together in a transparent way—another requirement for consistency.

The basic principle of *transparency*[127] that 'regulations should be simple and user-friendly' is lacking under this current system, as it is difficult to determine what applies and when, and who will enforce it. It is also difficult to say that the system embodies the principle of *accountability* as not all of the decisions made have any mechanism by which they can be scrutinised or required to be justified. For example, REC decisions are not published, there is a lack of uniform decision-making across the country, and RECs do not have the power to enforce their decision-making. As the current system stands it is very difficult to say that it is *targeted*[128]—that is, that the system is focused on the

[123] See, eg: D Chalmers and D Nicol, 'Human Genetic Research Databases and Biobanks: Towards Uniform Terminology and Australian Best Practice' (2008) 15 *Journal of Law and Medicine* 538, 543; AV Campbell, 'The Ethical Challenges of Genetic Databases: Safeguarding Altruism and Trust' (2007) 18 *King's Law Journal* 237; B Salter and M Jones, 'Biobanks and Bioethics: The Politics of Legitimation' (2005) 12 *Journal of European Public Policy* 710, 713, 717; D Chalmers and D Nicol, 'Commercialisation of Biotechnology: Public Trust and Research' (2004) 6 *International Journal of Biotechnology* 116; M Hansson, 'Building on Relationships of Trust in Biobank Research' (2005) 31 *Journal of Medical Ethics* 415.

[124] *Regulators' Compliance Code: Statutory Code of Practice for Regulators* (17 December 2007).

[125] Better Regulation Task Force, *Principles of Good Regulation* (revised 2007) available at <http://archive.cabinetoffice.gov.uk/brc/publications/principlesentry.html>.

[126] *Ibid.*

[127] *Ibid.*

[128] *Ibid.*

problem of biobanking and minimises side-effects. It is also difficult to say that the system as a whole is *proportionate*[129] given that it lacks basic coherency, clarity, transparency and accountability. Other characteristics of good governance also appear to be lacking, such as: consultation and public/stakeholder participation; predictability; fairness and non-discrimination; systems for overall ongoing review and reform; and a sound legal and empirical basis.

All of these principles are important for ensuring that a governance system has legitimacy and credibility. Ensuring *legitimacy* (including democratic legitimacy) is vitally important when it comes to designing, implementing and enforcing regulatory structures. With many diverse groups having a role in the regulation of biobanks that are not established by Parliament, the current system exhibits a marked legitimacy (including democratic legitimacy) deficit. With many of the better regulation principles also being at issue, the legitimacy of the current governance structure for biobanks is brought into question. Overall, then, our legal research leads to the conclusion that the subsisting governance framework applicable to biobanks in England and Wales is seriously deficient, and that it is in urgent need of reform. In this chapter we have analysed the regulatory space that applied to biobanks up to 2009. The chapters that follow provide an account of the perspectives of those who have first-hand experience in navigating this complex and evolving regulatory space.

[129] *Ibid.*

# 4

# Dynamic Networks of Practice

*Catherine Heeney*

THE GOVERNING GENETIC Databases (GGD) project[1] started from the premise that no satisfactory definition of a 'genetic database,'[2] was available which could serve as a basis for designing a governance framework. Whilst broad notions of types of research being carried out existed alongside knowledge of high-profile projects, a more systematic breakdown of activities associated with types of 'genetic database' in the UK was wanting.[3] This chapter, therefore, addresses one of the key aims of the GGD project, which was to provide a description of the classes and characteristics of 'genetic databases' functioning in the UK. This account is constructed primarily from the data collected during fieldwork involving 49 qualitative interviews with users of 'genetic databases' and is therefore based on respondents' reports of their experiences and interactions with different types of 'genetic databases' currently in existence and the range of practices associated with these entities. However, this chapter does not aim to provide a typology[4] specifically designed for the purposes of guiding the design of future governance, as it does not organise 'genetic databases' according to one 'structural dimension' ideally suited to this purpose.[5] Rather this chapter can be read as the groundwork for such a typology, as it sets out a number of features, or structural dimensions, which

[1] See ch 1 above for more on the work leading up to the Governing Genetic Databases (GGD) project.

[2] Throughout the chapter the entities being studied will be referred to as 'genetic databases'. We began the project and the fieldwork with this definition. However, this term has been placed in inverted commas throughout in order to acknowledge that it both could easily be substituted by other terms and that it is also an umbrella term for entities which often were more than one thing.

[3] SMC Gibbons, J Kaye, A Smart, C Heeney and M Parker, 'Governing Genetic Databases: Challenges Facing Research Regulation and Practice' (2007) 34 *Journal of Law and Society* 163–89.

[4] 'The study of classes with common characteristics; classification, esp of human products, behaviour, characteristics etc., according to type; the comparative analysis of structural or other characteristics; a classification or analysis of this kind.' *Oxford English Dictionary*.

[5] I Chompalov and W Shrum, 'Institutional Collaboration in Science: A Typology of Technological Practice' (1999) 24 *Science, Technology, and Human Values* 338–72.

serve as ways of classifying 'genetic databases' and reports on the range of observed characteristics within these.

The research design, data gathering and data analysis for the empirical work, which informs the chapter, will be outlined first. Then follows a description of the 'structural dimensions',[6] or features of 'genetic databases' encountered, together with an account of associated practices. This account will, therefore, proceed by means of presenting the broad features, which are simultaneously ways of classifying 'genetic databases'. These are: age, size, materials held, participants and recruitment, organisational setting and methodological approach. Under each of these features a description of the range of characteristics found will be presented.[7] Concluding observations will be made about associations between some of the characteristics of the 'genetic databases' that have been encountered and implications of these associations for the task of producing a typology upon which to ground governance models. Any attempt to produce a typological account is in tension with the dynamic nature of the characteristics and the associations between them and with the wider environment. However, this dynamism and interaction is approached as being an important part of an account of the practices associated with 'genetic databases' as they are themselves a type of 'open system'.[8]

## CHALLENGES INHERENT IN PRODUCING A TYPOLOGICAL ACCOUNT

The difficulties of attempting to provide a typological account of 'genetic databases', provide insight into the way these entities behave and for this reason they are outlined in this section. The first barrier to the production of a definitive typology of 'genetic databases' was the fact that, as noted above, a clear definition of these entities did not exist.[9] This had implications for the task of classifying 'genetic databases' as it is helpful to have a clear idea of the boundaries and definitions of the entities, which are to be classified or typed.[10] However, in order to do justice to the exploratory way in which the concept of a 'genetic database' was approached in the GGD project it has been important to hold on to a sense of the variety of features and characteristics 'genetic databases' can have. Simultaneously, however, to produce this account, it has still been necessary to abstract from the detail or the 'thisness' of any one 'genetic database' in order to find general

[6] *Ibid.*

[7] Sample is used to denote physical biological samples.

[8] JC McKinney, 'Typification, Typologies and Sociological Theory' (1969) 48 *Social Forces* 1–12.

[9] Gibbons *et al*, 'Governing Genetic Databases', above n 3.

[10] Chompalov and Shrum, 'Institutional Collaboration in Science', above n 5.

features.[11] Indeed, the fact that types are not 'dictated to observation by some natural arrangement of the phenomena themselves',[12] is part of the reason it requires a social researcher to come along and devise them.[13] A typology must produce a simplified account of the world and pass over the 'the unique, the extraneous, and nonrecurring'.[14] However, in what follows the tension between the general and particular involved in drawing out types will be highlighted rather than ignored. The reason for this approach is the belief that it is helpful in developing an understanding of the practices that constitute 'genetic databases' and in framing governance models to acknowledge the inevitability of 'exceptions' and the difficulties of prediction.[15]

The 'genetic databases' encountered in the fieldwork had a variety of features and not only many comparable examples of the same features. Bowker and Star capture this type of entity within the concept of a 'polythetic class', which they describe as 'a class that is defined by the congruence of multiple characteristics no one of which is essential'.[16] These challenges need to be met and addressed when producing a typological account[17] but this difficulty also reveals the heterogeneity of practices in this area and the multiple ways in which something could function as a 'genetic database'. For example, some genetic databases map onto an IT structure, some have an associated sample collection and others are physically based within one organisation, some may be associated with several sources of data none of which are exclusively associated with a 'genetic database'. It is complicated, therefore, to find features that allow classification of all examples of a 'genetic database'. Furthermore, how a feature is defined is also important. For example, whether a 'genetic database' is based on computerised records depends upon how this is viewed (a centralised system containing data on all aspects of the research, data stored physically on premises where other aspects of the research took place etc).

That these difficulties where experienced as such denotes the template provided throughout the empirical work by 'a pure type' based on abstraction from the classes and characteristics of 'genetic databases' encountered empirically.[18] From the point of view of recruitment, through interviewing and for the purposes of drawing up a typology an *ideal* database would have

[11] H Becker, 'Constructive Typology in the Social Sciences' (1940) 5 *American Sociological Review* 40–55.

[12] McKinney, 'Typification', above n 8, 5.

[13] Becker, 'Constructive Typology', above n 11.

[14] McKinney, 'Typification', above n 8, at 3.

[15] Becker, 'Constructive Typology', above n 11.

[16] GC Bowker and SL Star, *Sorting Things Out: Classification and its Consequences* (Cambridge, MA, MIT Press, 1999) 62.

[17] Gibbons *et al*, 'Governng Genetic Databases', above n 5; McKinney, 'Typification', above n 8; and Becker, 'Constructive Typology', above n 11.

[18] Becker, above n 11.

one Principle Investigator (PI) who would be based within in one institution and who worked exclusively with one 'genetic database' with a dedicated IT structure, which had one clear source of data and samples, which were collected at roughly one point in time and were connected with only one institution. Aside from making classification systems easy to identify, this ideal of neatly nested researchers and databases would mean that respondents could be expected to have complete knowledge of *their* database and the ability to describe all of the characteristics, which might prove useful for governance models, ideally in a uniform way. Therefore, an ideal set of objects would succumb easily to the Aristotelian system of classification functioning, 'according to a set of binary characteristics that the object being classified either presents or does not present'.[19] This binary logic, as already noted, was difficult to apply to the entities, which were described by the GGD respondents; it was equally difficult to apply to the informants themselves, the means, adopted to find out about the 'genetic databases'. It became evident during the process of identifying people to be interviewed that it is as rare to find a scientist who was not involved in multiple relationships as it was to find a 'genetic database' or sample or data collection that was not connected to any combination of collections, research groups, IT systems and organisations. Again, the huge inconvenience this presents to finding an easy classification system is revealing of the sort of systems 'genetic databases' are:

> Their autonomy and isolation are always relative rather than absolute, and *de facto* they are always, to varying degrees and in widely varying ways, a part of a more extensive network of social relationships within which they are nested.[20]

McKinney could easily be describing the 'genetic databases' encountered, except that, as I have already pointed out they were not nested, unlike the ideal type would have been, within laboratories, within premises, within organisations and sectors. The various 'genetic databases', as with the respondents who described them and many of their characteristics, therefore, formed less a neat hierarchical or nested structure and more a 'rhizomatic' set of interconnections.[21] Such a system functions via 'Principles of connections and heterogeneity: any point of a rhizome can be connected to anything other, and must be. This is very different from the tree or root, which plots a point, fixes an order'.[22] In a world that follows the logic of the rhizome, relationships and objects are not nested hierarchically but are rather built upon a variety of objects and relationships, which are in turn connected for a multitude of reasons and in a variety of ways. As we shall

[19] Bowker and Star, *Sorting Things Out*, above n 16, at 98.
[20] McKinney, 'Typification', above n 8, at 10.
[21] G Deleuze and F Guattari *A Thousand Plateaus* (Minneapolis, University of Minnesota, 1987).
[22] *Ibid* 7.

see, the way in which some samples or data come to be included in a piece of biological research, while not random could not always be predicted by looking at an organisational setting, the professional or scientific interests of researchers, or by gaining an insight into what might have been the intention of the individuals who had originally been responsible for instigating a study. Moreover, the experience of people working with these entities is often not that they are one discrete thing but several things, related to other things via diseases, people, technologies and so on.

Respondents were often not comfortable with the terminology 'genetic database' and indeed suggested better alternatives to describe what they had. Furthermore, despite the database having some of the characteristics, which marked it out for inclusion in the GGD project, respondents occasionally resisted the notion that what they worked with was a 'genetic database' at all. One possible explanation for this, which has already been mentioned, is that historically 'genetics' was only a part of the research and, in some cases not the part the interviewee was most interested in. However, it was clear that the issue of definition was not simply about a lack of standardised nomenclature or indeed the non uniform nature of practices; there was also a clear *performative* or political dimension, in other words terminology can be used to shore up a particular stance on the issues being raised.[23] Thus the following interviewee expressed dissatisfaction with the term 'genetic database,' despite there being both biological samples, upon which a research project was based and a focus on traditional genetic themes such as carrier status, both of which fitted criteria for inclusion in the GGD project, 'I think *you* could choose to call it a database, a genetic database if you wish but I'm not too bothered about that' (Int 19). However, this reluctance is illustrative of one type of challenge in drawing an account from the descriptions of the practitioners and users; respondents were not always inclined to say they had a 'genetic database' much less waiting to answer a set of questions about it.

## Methodological Reflections[24]

One of the basic objectives to be achieved by the empirical work was to provide a descriptive account of the characteristics of scientific research associated with, or based on entities that could be defined as 'genetic databases'. The first challenge began even as the empirical work did; for the

[23] See esp, Bowker and Star, *Sorting Things Out*, above n 16 and also A Hedgecoe and P Martin 'The Drugs Don't Work: Expectations and the Shaping of Pharmacogenetics' (2003) 33 *Social Studies of Science* 327–64; SP Wainwright, C Williams, M Michael, B Farsides and A Cribb 'From Bench to Bedside? Biomedical Scientists' Expectations of Stem Cell Science as a Future Therapy for Diabetes' (2006) 63 *Social Science & Medicine* 2052–64.

[24] For a more detailed account of the methodology see the section on 'Sociological research methodology' in ch 1.

purposes of sampling it was necessary to define, at least broadly, what sort of entities we would be investigating. As a satisfactory definition of 'genetic database' was lacking, due to the dynamic nature of this research area and the fact that a key goal of the project was to find out *what was out there*, the aim was to be inclusive. Nevertheless, there was of course a limit to how open it was possible to be, and a number of criteria needed to be specified in order to be able to recruit individuals into the study.[25]

Recruitment was managed using a visual map of types of organisations known to or likely to have 'genetic databases'. These included NHS hospitals, universities, specially designed premises and facilities at the offices of pharmaceutical companies. It was at the recruitment stage, moreover, that it became clear how individual respondents were not nested, within projects or attached to 'genetic databases', which were within organisations, within sectors and so on. Rather the same individuals or projects were appearing in more than one place on the map. With the exception of some (less than five respondents), who had specific roles on the larger projects, respondents generally worked on or were responsible for several projects. These in turn were usually funded by a number of different funding bodies, from different sectors, who had funded the project, parts of the research or the collection, to varying degrees. In order to span the field of biomedical research the organisations or institutions with 'genetic databases' were grouped into the following broad sectors: Public Research, University, Public Health, Charity, and Private. The category of Public Research included organisations or institutions, which were specially constituted for the purposes of creating and maintaining a 'genetic database'.[26] They were not within a university or a hospital though in all cases having strong ties with one or both types of organisation.

Respondents were grouped originally into one of the following seven professional categories: Bioinformatician, Researcher/Clinician, Scientist/ Geneticist, Clinicians, Epidemiologists, IT Specialists and Public Health Professionals.[27] By far the biggest group were the clinicians who were research active, these individuals made up 22 of the 49 interviews. Those respondents who were biologists interested in genetics with no health role or training, were less than half this number, while those who were epidemiologists, clinicians without any research interests, public health professionals or IT specialists were nine in total. The same caveat applies here, as elsewhere in this chapter, therefore: there was not complete mutual

[25] The data was gathered in 49 qualitative interviews with individuals who work with 'genetic databases' in England and Wales. The criteria which informed recruitment is described in ch 1, see n 24 above.

[26] Following analysis these 'genetic databases' are captured under the term 'repository' in chs 5–9.

[27] Professional categories were changed in some cases during the analysis process and presentation of results in chs 5–9.

exclusivity between professions. Indeed the cases of respondents who were both clinicians and researchers lead to the creation of the hybrid category clinician/researcher.[28]

## CLASSIFICATION OPTIONS FOR 'GENETIC DATABASES'

In the sections below, the following features of 'the genetic databases' will be presented: age of study, size,[29] materials held, study participants and modes of recruitment, organisational setting (which will include funding source), and methodology employed. In presenting the data in this descriptive way the intention is to demonstrate the range of observed characteristics associated with these common or comparable features of 'genetic databases'. The characteristics presented are drawn from the descriptions given by the respondents themselves, using interview data, from notes taken during and after the interview and with reference to project and personal websites. Data will be presented in support of the range of observable characteristics of that were encountered. Quotations are used where they illustrate the respondents' experiences of features or characteristics of the 'genetic databases' being discussed in the interview. The data are qualitative and, therefore, are not being used as a basis for claims of representativeness or universality.

### Age

Given that the majority of 'genetic databases' that the respondents were involved with drew on several data sources and sample collections, and that in some cases different parts of the 'genetic database' came into existence at different times, attributing an age to these entities is not straight forward. Here a Clinical Scientist talks about samples, which are still being used in current research despite having been collected for past projects:

> [W]e have what we call legacy samples, samples that we've had here for a long time (R21: Pharma).

Nevertheless, the aim in this section is to present this complexity while at the same time presenting the range of ages within and across the databases we looked at. The studies[30] range between 50 years to pre-operational at the time of fieldwork. The majority of the studies that respondents were involved in were less than 20 years old. Whilst there were some notable

[28] Research Scientist in chs 5–9.
[29] See table explaining size categories in appendices.
[30] Respondents, often when prompted referred to a distinct study that could be named, which did not always coincide with all parts of the genetic database.

exceptions of well-established longitudinal studies having evolved so that they now included elements that brought them under the term 'genetic database' the study had not originally been envisaged as such. In one case the study had been running for over 30 years before a biomedical sweep, which would gather blood to enable the extraction of DNA, was included. Another example, which also demonstrates the influence of factors, especially scientific advances, on not only age but also materials held, was provided by this Research Scientist who refers to the introduction of DNA, RNA and other biological samples into the collection at different times, which in turn has affected the types of biological sample needed to produce these:

> DNA we've routinely been taking for I guess the last 6 or 7 years, we've routinely incorporated a DNA sample. We're now also incorporating more routinely samples for RNA analysis and samples for proteomic and samples for metabolonomics (R23: Pharma).

Almost half of our respondents were involved in studies that they described as having begun in the last 10 years. Twelve respondents (representing eleven distinct studies) said the studies they had been or were still involved with had begun between 10 and 20 years previously. In other words, the 1990s and 2000s had marked the beginning of genetic research or the genetic component of the research that most of our respondents were involved in, according to their own reports, at the time of fieldwork. This accords with the considerable changes going on within the field of genetics and genomics due to projects such as the Human Genome Project as described in Chapter Two.

Determining the age of a 'genetic database' caused some respondents to reflect on the various stages in the work they were involved in, parts of which constituted the 'genetic database':

> The actual MREC and opening it up as a project which we could recruit people to and actually start to use was in *month and year*. (R11: Clinical Scientist, Uni/Hos)

Here the age of this web-based data-sharing platform could arguably date from the discussion of the idea, securing funding or from the Multi-site Research Ethics Committee (MREC) permission to begin the recruitment of research subjects. Individuals had several strategies, which included making a distinction between stages or reflecting on the age of some of the 'legacy' samples, which were, in one case 30 years old and still used for research. Two 'genetic databases', which were repositories and which were being set up during the fieldwork of the GGD project proposed to source samples, data or both from the National Health Service (NHS) in both cases results from the genetic research carried out using the repository was to be fed back into it over time. However, in these cases whilst the sources of data and samples were multiple the specially designed infrastructure, which would manage samples and data as they came in, gave them a unified identity, as a 'genetic database'. This was despite the fact that they would become 'genetic', when results from genetic research was done on the samples and data, and were not starting out as such, although genetic research was given as one of the reasons for creating both these repositories. In the majority of cases, research drew on data and samples from various sources and/or samples from clinical work, which were collected incrementally. The existence of family histories and whether or not they were considered part of the 'genetic database' was a further interesting complication in determining age, the fact that family history records had been collected over a long time and where not always stored in electronic form placed them at the limits of what could constitute part of a genetic database. One interviewee, for example, indicated boxes in the corner of the office where family history data was kept, thereby demonstrating the idiosyncratic and difficult to capture status this form of information had in relation to discussion of 'genetic databases'.

## Size

Respondents used a number of different units of measurement when discussing the size of the 'genetic database'. The following were all mentioned by one or more respondents: the number of samples (including blood, tissue and urine) number of DNA or RNA samples, the number of data points, arrays, cell lines, family history, genomes and mutations and the number of individual research participants. One Bioinformatician at a newly-established Repository said that in their collection they had '50 samples per patient' (R20). A Clinical Scientist involved in small-scale clinical studies talked about having, '220 family histories' (R16: Uni), while a Bioinformatician at a repository holding genomic data talked of having '450 material genomes' (R38) and a second (who originated a web-based data-sharing platform) spoke of '100 tests of positive results, positive mutations' (R1: Uni). These answers show the variety of relationships between our respondents and the data to which they had contributed in some way. It also illustrates the difficulty of determining which of these things is 'bigger' or 'smaller'. A concept as seemingly straightforward as size is complicated by the variety of units by which this can be measured. The following Research Scientist at a repository lodging genomic data captures this problem of the complexity of the concept of quantity due to the possibility of deriving different amounts of data from samples:

> Now in terms of data I don't know how I can describe that. You know for all these 60 to 70,000 samples we are holding for example genetic information,

which ranges from 600,000 datapoints to for example 100 datapoints per sample. (R37)

One unit, which was used to measure size, was that of the individual research participant. Taking individual participants as a unit, therefore, the range for our respondents was from under 100 to over 1,000,000. However, as suggested above, the relationship between individual people represented in the database and other units was variable and complex. Moreover, our respondents were commonly involved in several studies of different sizes (in terms of individual participants). One interviewee mentioned having access to around 4,000,000 individuals. However, the explanation for this very large number was that the research participants had been recruited through clinics or their samples sourced from laboratories within the NHS, so the research depended on access to mixture of samples and data from individuals recruited originally as part of a research programme and reuse of sample collected for other purposes within the NHS. A sizable group (13 respondents representing at least 11 distinct studies) were involved in research projects where participants numbered between 10,000 and 20,000. Only five respondents were involved in projects wherein the number of research participants exceeded this number. Seven respondents had between 5,000 and 10,000 individual participants involved in research, while three had between 1,000 and 5,000.[31]

The focus of the research was related both to methodological and size requirements. The two respondents who had very small numbers of individual study participants were looking in the first case at a family with a particular genetic mutation and in the other at a particular chromosomal translocation. One of these, a Clinical Scientist involved in small-scale clinical studies, made the following methodological point in relation to the small numbers in the study:

> We could get the data we want from quite small numbers of samples. So if I found a family where it was clear there were 10 or 15 or 20 affected individuals you could imagine that you could study that family in-depth for some years and you'd discover a new gene. (R12: Uni/Hos)

The methodological approach of the research, which made use of the 'genetic database,' however it is constituted, was also related to the units that individuals tended to focus on. Here the same interviewee talks about the use made of larger collections, thus illustrating the interconnection of disease type, methodological approach and the size of the database:

> [C]omplex trait genetics where we have very large studies looking at susceptibility to common forms of heart disease, particularly at risk of having a heart

[31] In later chapters the size of the databases will be referred to in terms of ranges—see Appendices.

> attack. And there we have huge data collections and huge samples of DNA. (R12: Clinical Scientist, Uni/Hos)

Epidemiological studies also require much greater numbers than small studies on rarer diseases, one study contained 200,000 participants. Therefore, the type of work being done had affected the size, for example, research looking for the causes of common diseases where a wide variety of genes play a part had a larger number of individual research participants than research on a disease wherein a small number of genes with high penetrance were involved.

The following quotation from an Epidemiologist who worked on a well-established longitudinal study demonstrates how the database can in some sense be different sizes depending on what the focus of the work is:

> [I]t started as 14,000 pregnant mums, in reality it's, you have data of between something like 6 and 10,000 depending on what you want to do at any time (R6: Uni).

In some cases the size or expected size was given as a projection and not as a fixed and unchanging fact. For example, there was the case in which patients with a particular disease were to be asked to donate samples across 100 different hospital sites. These samples were then to be collected and transported daily to a repository. The disease is a common complex disease, which suggests that the number of samples in the repository would grow at a very rapid rate. One IT Specialist (R35) who was associated with the same newly established repository, suggested the number would be in the region of 4,000 patients donating samples in any one year. The dynamic and evolving nature of the growth of some 'genetic databases' was evident in the response of this Bioinformatician who originated a web-based data-sharing platform:

> Yes about 100 tests of positive results, positive mutations that can be reported. So typically you know we're seeing legacies of about 100, 100 referrals from each laboratory we look at. So in theory if we went to every service and laboratory 20 odd labs and how many services each month, how many genes each one has perhaps 6 or 7, something like that, you know and there's a few thousand. (R1: Uni)

Evolution both in terms of questions to be asked and the number and type of samples needed to answer them was another factor that ensured that size was a dynamic variable. Variability in numbers within studies was common as well as between them. For example, the studies had different amounts of data and samples for all individual participants. The 'genetic database' evolves according to the questions to be asked, which affects size and other characteristics. However, it likely that size be it measured by data points, genomes, samples or participants is unlikely to diminish but rather to grow at a rapid rate for most of the 'genetic databases' our informants talked to us about.

## Materials Held

In terms of the materials being held in the 'genetic databases' the GGD respondents worked with, this was made up of two main things: data and biological samples. All of the work that our respondents were involved in used digitised data in some form—be it health records, the digitised results from genetic tests, or even whole genome sequences. Not all of the respondents were working directly with, or had access directly to biological samples. For example, two respondents talked of projects where the 'genetic database' was designed specifically to facilitate the exchange of information about patients with genetic 'abnormalities' but samples were not shared or pooled in connection with the database. One of these databases, a 'web-based data-sharing platform' was in some respects a sort of introduction service for geneticists, in that it facilitated access to data about the work being done with collections of samples by clinicians/researchers in geographically dispersed locations. The other was described by a Clinical Scientist who worked on it as a system that helped 'a clinician interpret a result' gained from running a genetic test on their patient in relation 'with known syndromes' (R11: Uni/Hos). Other respondents were involved in databases, which housed sequencing data. This was the case for 'the big genome datasets' (R10: Bioinformatician). The samples which gave rise to these collections were not held by our respondents, who worked for a repository lodging genomic data.

The types of data used in 'genetic databases' varied greatly. The following categories of data, mentioned by at least one interviewee, were as follows: healthcare records, genomes, data on behavioural problems, clinical data, baseline characteristics and phenotype data, epidemiological, demographics, phenotypes, longitudinal history, age, sex, primary disease, pathological stage and grade, histopathology reports, education and social and economic background, details of the delivery of children and antenatal complications, data on birth defects and physical measurements. Ten respondents, all of whom had a clinical role, referred to 'family history' information but it is likely that all respondents with a clinical role had this sort of information. The fact that it was not computerised, as mentioned above, may have been one reason more respondents did not see this as being part of a 'genetic database'. At least 40 of the 49 respondents had access to biological samples. Respondents described having a wide variety of biological materials, these included: blood, urine, faeces, hair, tumours and healthy tissue, and saliva, while as this interviewee suggests, a range of biological samples could be extracted from these raw materials. The types of sample collected were limited by the type of biological material available, which was related to (but not necessarily limited by) other factors such as the disease being studied as in the following quotation. This Research Scientist is describing this relationship:

> [S]o that tends to be plasma and urine and RNA it's predominantly been whole bloods or white cells but obviously in oncology we do tissue as well. (R23: Pharma)

## Study Participants and Modes of Recruitment[32]

The individual participants who supply the research materials, discussed in the previous section, were recruited via a variety of channels, for numerous purposes and have a wide range of characteristics, which made them eligible members of the research population, with which one of our respondents was involved. The following list is a summary of the reported routes by which a person's samples or data were included in one of the 'genetic databases' encountered in the fieldwork. The individuals whose data and samples were being used by at least one of the respondents for the GGD project had at least one of the following characteristics: membership of an affected population such as those with a phenotype associated with a mutation; membership of a group with a genetic mutation predisposing them to a condition, a carrier of such a mutation, individuals with a disease within the expertise of a particular clinician/researcher; those with a family history of a condition (in all of the preceding cases the families of these individuals may also become participants in the research); cohorts of individuals born in a particular year and/or born in an area with geographical proximity to the location of study; those who are part of multiple births, ie twins; people who have been involved in clinical trials and controls, and the general population. The characteristics that individuals had were, therefore, important in bringing them into contact with the researchers we spoke to but the nature of these characteristics varied very widely.

About three quarters of the clinician/researchers in our study recruited individuals into studies via their clinical interaction with them. Another route in was via clinical trials, which produced data and samples. Around ten people talked about having been involved in conducting clinical trials directly and four respondents said they were secondary users of data and samples collected in clinical trials. Three people said that the trial had not originally had genetic research as a primary goal but rather that the opportunity was seized to collect samples for these purposes. Such examples of

[32] In terms of classifying databases according to a schema that would provide a basis for governance models it would perhaps be better to focus only participants or recruitment. However, the characteristics of individuals who found their way into studies are intimately connected with the active practices employed by researchers to involve them, for this reason I have chosen to present them together.

opportunistic collecting of genetic materials also demonstrate how the primary purpose of the 'genetic database' or a given study does not necessarily allow one to predict which materials will be held. For example, this Clinical Scientist, involved in small-scale clinical research and collaborations, said, '[i]n the trials the genetic collections are actually if you like a subsidiary questions, the main, those trials are funded and their purpose is to work out whether a particular treatment is going to prevent [name of disease][33] disease' (R9: Uni/Hos).

Direct recruitment via the clinic was the major route by which this Clinical Scientist who worked on small-scale clinical studies gathered the majority of the research materials. 'I mean the starting point for a lot of the research we do comes from the patients I see in the clinic' (R14: Uni/Hos). Another Clinical Scientist, involved in small-scale clinical studies, describes the course one family took into a research study:

> [T]hen what usually happens in families where there's something unusual or that don't [sic] fit the mould or that don't have the mutation and the genes we know about we ask if they are interested in taking part in a research programme and they sometimes move from the more clinical set of questions into the research questions. (R12: Uni/Hos)

Here the dual role of this interviewee as a clinician/researcher was a crucial component to the eventual inclusion of this group of patients in the 'genetic database'.

Two individuals involved in clinical work described how they either alerted or were alerted to patients with particular traits suitable for inclusion in a research project. As in the following case:

> I'll see a patient with something rare, you might have a look on the internet to see who last published something, who discovered the gene for this rare condition and then you find out their email address and you send a kind of 'You don't know me but I've just seen a patient with your disease, are you interested in their DNA?' (R13: Clinician, NHS)

These latter cases provide an interesting contrast to those systems, which three of the respondents were involved in, wherein the aim was specifically to provide a platform to share this sort of information, so that such exchanges were not left to chance. Other means of finding participants appropriate to a piece of research included searching hospital databases, via special research clinics, through an organised network of clinics with a specific structure for collecting and transferring samples to a central repository, via pre-natal clinics, through a programme of recruitment inviting individuals to attend specially set up 'assessment centres,' through other health activities, such as screening programmes and, as discussed, via clinical trials.

[33] Omitted to protect confidentiality.

Recruitment practices, therefore, range from the serendipitous to the systematic. People become research subjects as a result of exchanges in networks of like-minded colleagues with similar interests, through clinics, through specially set up IT platforms for encouraging data-sharing and national or local campaigns of data collection, reuse of existing health records or combinations thereof. For some of the databases our respondents were involved in, there was a long-term plan to source materials from research participants over time and from existing sources, meaning that participants would not be asked to provide all of the samples or data at the outset, which might eventually be included in the database. So for example, at least two 'genetic databases,' which were also repositories, explicitly planned to source NHS records to follow up individuals of interest at a later stage. Recruitment via one's clinician was only one route, therefore, to being included in a 'genetic database'. 'People' can also enter the database already reduced, as it were, to a particular type of sample. As this Clinical Scientist points out:

> About probably 60% come from outside this building, outside this hospital and I've no exact proportions but another substantial proportion come from outside of this region. So they would come in as DNA samples rather than blood samples and the remainder is taken from within this hospital. (R2: Uni/Hos)

There are, therefore, participants whose representation in a 'genetic database' could be best described as partial i.e. it was limited to a sample of DNA or a number data items. Some of these samples donated by participants find their way into other databases due to sharing and reuse of existing records or samples, as with the case of legacy samples, mentioned above. In the following case this Clinical Scientist involved in small-scale clinical research and collaborations within one large organisation describes the origins of some of his data and samples:

> [T]hrough the epidemiologists in the clinical trials building in the [name of department][34] and they have a fairly standard way of ascertaining their patients and families which works for common illnesses, they're rather clearly identified like xx and that works on hospital discharge databases. (R12: Uni/Hos)

Here the process involved the samples and data being procured by one group of clinical professionals and passed on to researchers, working within the same organisation.

This is also true in the following case of a Bioinformatician analysing data as part of a large international collaboration:

> And that's been organised by the Department of Paediatrics here, so it's organised by a different department and the way that's done is that they hold the names, addresses and so forth, we don't see them, they just have an identifier link but

[34] Omitted to protect confidentiality.

> we don't see the detailed phenotype, we just see the minimal stuff that we need to get to do our work and you know they've recruited basically through NHS so... (R30: Uni)

The data which is passed on is limited, this interviewee thought, in an effort to protect privacy. About a quarter of the respondents mentioned that they were holding data, which had originated in other countries. At least one of the databases, a repository lodging genomic data, housed data that was 'generated by different research groups worldwide and they are deposited into these databases' (R10: Bioinformatician, Repository). For the majority of 'genetic databases', including those held in the private sector, the NHS in the shape of people or computer systems played a large part in the recruitment process. This IT Specialist (at a Repository involved in a programme of sample and data collection) talks about the important role of organized records held by the NHS in building this repository:

> [W]e're getting data in, we're managing that data, we're storing that data but then the next phase will be retrieving healthcare records for follow up purposes and so on. (R40)

## Organisational Setting

As, discussed above, respondents were seldom to be found fitting comfortably within one organisation, or if they were based in one sector or within one organisation they had funding from another. A number of our respondents, due to their particular skill mix were to be found working across organisations and sectors, as well as on several different projects, which had 'genetic databases'. Individuals, therefore, moved between sectors and organisations applying their experience and expertise in new projects and different contexts. Twenty two of the respondents were based in the Public Health system, although with one or two exceptions respondents in this group had joint positions with the academic sector, indeed the threshold between Public Health and the university sector appeared to be very fluid in the case of these respondents. Ten respondents were in the university sector but six of these individuals had had some training or role in the Public Health sector at some time in their careers. Seven respondents were involved in repositories constituted through a dedicated programme of sample and data collection with funding coming from multiple sources and six were associated directly or indirectly with a charity-funded repository lodging genomic data. Thirty of the respondents had worked extensively in the NHS. One of these individuals was now involved with the creation of a repository, while another worked in the private sector. The rest had remained in the public health sector or were in the University sector. Four of the respondents were employed directly by pharmaceutical companies and none were directly employed by a charity. At least three other respondents, however, explicitly said that they were indirectly or partially funded by the private sector. More than half of the respondents were involved in work that was at least partially funded by charities.

The prominence of the universities and the NHS in the research world continues even for those who are outside of the public sector. The following quotation from a person working in the private sector demonstrates this:

> And the way we approach our tissue acquisition is that we look to develop collaborations because we've got expertise, and equipment and the NHS and academic centres have got expertise and equipment and we want to learn off each other and we're, we are looking at, we wish to collaborate. And in the end the NHS has got a monopoly on the patients pretty much as we speak, that may change, or it will change but at the moment it has. (R21: Clinical Scientist, Pharma)

As well as providing an illustration on these cross-sectoral connections, this quotation demonstrates at least one motivation for spanning organisations, sectors, projects and biobanks. As an organisation, the NHS was a prominent actor connected in a variety of ways to the work of our respondents, playing roles, which ranged from a source of research materials, via professional training ground to employer. However, as the last interviewee hinted this may or will change. In cases where projects were contained within one organisation, the practicalities of funding needed to maintain resources of samples and data still ensured that in practice this was only one part of the picture.

Moreover, the pursuit of answers to scientific questions may expand collaborative networks bringing in new samples and data, which in turn change the constitution of the database. The following quotation, from a Clinical Scientist involved in small-scale clinical studies shows how various types of collaboration are involved in constituting the 'genetic database' in ways that transcend organisational boundaries:

> And then the secondary for those who, a last group of people who come along to the clinic are referred by colleagues or general practitioners or themselves sometimes looking for a genetic answer to their problem. And in those families we've tended to become involved in large international collaborations where we've shared DNA with colleagues and we screen genes that are good candidates. And through that a number of important genes are identified which actually cause mutations in our families. (R14: Uni/Hos)

This description of practice, moreover, exemplifies how the interaction of a variety of factors, including the motivations of 'of people who come along to the clinic' and 'colleagues', contributes to the mode in which a piece of research develops and how new links are made and existing boundaries spanned. In a rhizomatic manner,[35] therefore, the disease, the genetic findings

[35] Deleuze and Guattari, *A Thousand Plateaus*, above n 21.

or the social and professional network of the given researcher can lead to new connections, which will take the purpose and the contents of the 'genetic database' in different directions, which, while not random, are also not easy to predict in advance.

### Methodology Employed

One further way of distinguishing the types of study (or genetic database') is according to the methods which were used. Broadly the types of methodology used by the respondents could be grouped as follows: epidemiological, cohort, clinical, longitudinal, cross-sectional, hypothesis driven, comparative and prospective. As ever, these methodological characteristics were not mutually exclusive in practice; for example one medium-sized study was a longitudinal, prospective, epidemiological cohort study. However, these issues about types and methodology are typical of the complexity of defining and delimiting types of 'genetic database' because the rule rather than the exception is that the database is constituted in multiple ways at multiple times by multiple individuals or networks and has any number of parts. For example, data and samples collected in clinical trials can be used for epidemiological purposes or sequencing. The methods associated with a type of research both influence and are influenced by the type of professionals involved, what is possible in relation to the 'genetic database', which in turn depends on other characteristics such as size, materials held and the study population. Where 'genetic databases' are constituted to address specific genetic questions for particular diseases, these promote particular methodological choices, such as, for example, a comparison between two individuals in the same family.

## DISCUSSION

One of the key issues that came out of respondents' descriptions of features and characteristics of 'genetic databases' was the fact that they were anything but static and were evolving constantly in response to a variety of interacting factors. Examples of these factors include the incorporation of newly available biological samples or data relating to DNA into an established project and the general growth in the type and quantity of data for all of the 'genetic databases' encountered via the GGD study. The interrelation between the different characteristics of the database meant that change in, for example, materials held affected how 'age' was described by the respondents, as well as meaning that different parts of the database were different ages and sizes. Continuing changes in scientific knowledge and the availability of new technologies were both factors influencing changes in the 'genetic databases' described by respondents. Despite these complex and shifting relationships between the respondents and the various parts of a 'genetic database,' it has still been possible to identify common 'structural dimensions'.[36] These serve two purposes: first, they provide a means of organising the range of associated observable characteristics and second, they offer options for the classification of 'genetic databases'.

In this chapter the aim has been to present a picture of common features or 'structural dimensions' and the characteristics of 'genetic databases' encountered in the fieldwork, without obscuring the difficulties inherent in the task. The fact that the project necessarily began with an open mind as to what would constitute a 'genetic database' meant that the task of organising these entities into types was carried out in tandem with identifying them. The lack of nested organisational structures, either for respondents or 'genetic databases' meanwhile, presented challenges for determining the boundaries of any given 'genetic database' and defining it as one rather than multiple things. Some of the researchers were skilful entrepreneurs who pieced together research from various sources and collected and acquired data sometimes opportunistically and/or due to the pursuit of research interests. Other respondents who may have had established 'genetic databases', with an identifiable infrastructure, often brought in funding from a variety of sources. The lack of clear existing boundaries and the *multiple* nature of the 'genetic databases' had implications at every stage of the field work from recruitment through to analysis. However, it was often in the interview itself that this was most challenging as respondents often discussed practices in relation to a project or study rather than a 'genetic database' or multiple studies or focused on the particular part of the 'genetic database' that they were involved in and were less able to offer information about other parts. There were 'genetic databases' that were constituted from a dedicated programme of sample and data collection, which were also 'repositories'. However, while they had a comparatively clear organisational infrastructure, the division of labour meant no one informant was *ideal* in terms of having exhaustive knowledge of every aspect of the database. Furthermore, none of these 'repositories' were self-sufficient in terms of procuring samples and data and relied on agreements and/or collaborations with other organisations or individuals working within those organisations. Indeed, the very term we use for these entities 'repositories' suggests this dependence on other sources.

'Genetic databases', like any open system, are engaged in shifting and dynamic interaction with the wider scientific, political and policy environment,[37] which affect their constitution, governance and use. Moreover, the availability of new knowledge, new technologies and new techniques, for

[36] Chompalov and Shrum, 'Institutional Collaboration in Science', above n 5.
[37] McKinney, 'Typification', above n 8.

example, Genome-wide Association Studies (GWAS) and new sources of materials or data also prompt changes in practice, which may affect the constitution of a genetic database. As the fieldwork progressed so did the sense that the reasons for this evolution were many and included, personal and professional interests, the latest knowledge about the disease and its relationship with genetic factors, the entry of new people into the collaboration as well as developments in science and technology and wider political forces. Aside from gleaning descriptive information, the aim of the interviews was to explore with the respondents their attitudes to and experiences of some of these wider influences on governance of human genetic research, which will be further discussed in Chapters Five to Nine.

The 'genetic databases' described during the GGD interviews are not only diverse in their constitution but are also evolving constantly. To return to Deleuze and Guattari's ontological framework defining and delimiting a 'genetic database' would be simple if it was 'treelike' that is to say isolated from other similar entities and developing in a logical and easily explicable way. The description of this phenomena as treelike, with nested and hierarchical relationships, while it would make for tidier descriptions and better predictions of future developments, is prevented by practice which is more often found to resemble a rhizome with multiple and apparently unsystematic connections or *ways in*. 'Perhaps one of the most important characteristics of the rhizome is that it always has multiple entryways'.[38] Moreover, depending on a number of factors such as the interviewee's role in relation to a 'genetic database', the type of research they were interested in doing with it, the origins of the samples and data and the point in the life of the 'database' it could be described in a number of ways. For example, a 'genetic database' might be perceived by a person working on it as a cohort study, which has a genetic component or 'bolt on', which came later, whereas somebody involved in characterising genetic samples might be primarily interested in it as a source of genetic data. As with a disease, which is enacted by multiple actors so the 'genetic database' was sometimes visible as one thing and sometimes it was not.[39] Throughout the fieldwork the ubiquitous concept of the 'boundary object',[40] was useful in envisaging the multiple groups, artefacts and individuals whose interests and endeavors constituted and maintained 'genetic databases' as *a thing*.

[38] McKinney, 'Typification', above n 8, at 21.

[39] AM Mol, *The Body Multiple: Ontology in Medical Practice* (Durham, NC, Duke University Press 2002).

[40] SL Star and JR Griesemer, 'Institutional Ecology, "Translations" and Boundary Objects: Amateurs and Professionals in Berkeley's Museum of Vertebrate Zoology, 1907–39' (1989) 19 *Social Studies of Science* 387–420.

## CONCLUSION

From the outset of the empirical work—that is to say drawing up a list of respondents—it was important to have some definition of what these people would be working on ie a 'genetic database'. However, such a definition was far from straightforward as there were no standardised terms for the clusters of artefacts or working practices we were interested in.[41] What they actually were *in themselves* was impossible to define in a fixed way, especially because, as has been stated above they were often more than one thing, depending on who was describing them. Due to being constituted by many parts and by numerous actors 'genetic databases' are not one but thing but multiple things.[42] However, that they existed in some sense was clear in that they provided the focus of the work of our respondents and of our research, which in turn made them what they *were*. Where it was visible as one thing the behaviour of any given 'genetic database' was that of a 'boundary object' in that its use was not necessarily a settled thing but was constantly open to renegotiation and required maintenance.[43] When capturing the difficulties of drawing up a typology of 'genetic databases' in the UK, the concept of the rhizome was useful in explaining how these difficulties were not simply the result of the data being messy. The data *was* messy but this was because practices were not 'treelike' in their origins, uses, organisation or development. What was encountered, therefore, was a 'rhizomatic' system, wherein 'genetic databases' were constituted from all sorts of connections between collections, existing records and samples held by trusted colleagues, or large organisations or as part of collaborations.

As Star and Greisemer[44] predicted these focal points are fragile and are under constant threat of being replaced or at least redefined. This was the case for the 'genetic databases' encountered in this project, which are always in the process of responding to institutional, biological, social and political requirements. The act of defining something as 'genetic' or indeed 'bio' as in the term 'biobank', which largely replaced the term 'genetic database' in the lifetime of the project, was not free from political significance for some of our respondents. Relying on the reports of the respondents it is possible to say that 'genetic databases' were constituted as a result of opportunism, scientific interest, professional ambition, a will to improve human health either at the micro or the macro level and so on, but ultimately these entities do not exist in a political, economic or social vacuum. The constitution and use of 'genetic databases' often will be beyond the knowledge or control of any

[41] Gibbons *et al*, 'Governing Genetic Databases', above n 3.
[42] Mol, *The Body Multiple*, above n 39.
[43] Star and Griesemer, 'Institutional Ecology, "Translations" and Boundary Objects', above n 40.
[44] *Ibid.*

one actor[45] the wider political, economic and social context plays a crucial part. However, it has been noted that attempts at providing a prediction of future behaviour, which is a reason for producing a typological account of phenomena,[46] may be only partially successful. This is due to the dynamic interactions described above and the fact that as has been shown to be the case in relation to other artefacts in this field '[p]roblems and obstacles, solutions and opportunities, are likely to arise from numerous directions in a relatively unpredictable manner, and no doubt they will interact in complicated ways'.[47]

The findings presented in this chapter are intended to ensure that the challenges in terms of defining a governance framework, which aim to capture the shifting constitution of the 'genetic database,' are acknowledged and addressed. Given the complexity or rhizomatic structure of practice in relation to these changing and contested entities, it is perhaps overly optimistic to suppose that one law or set of regulations could adequately capture this multiplicity once and for all.

[45] See especially ch 9.

[46] Chompalov and Shrum, 'Institutional Collaboration in Science', above n 5.

[47] S Hilgartner 'The Human Genome Project', in S Jasanoff, GE Markle, JC Petersen and T Pinch (eds), *Handbook of Science and Technology Studies* (Thousand Oaks, Calif, Sage Publications, 1995) 302–19, 305.

# 5

# Governance in Practice

## Patterns of Awareness and Engagement

*Susan MC Gibbons and Andrew Smart*

TWO DISTINCT KINDS of 'maps' of the biobanking landscape in England and Wales were presented in the previous two chapters. Drawing on legal research work, and using a modified 'regulatory space' methodology, the principal laws, guidance documents, and many of the key actors involved in governing biobanks during the research period were mapped out by Susan Gibbons in Chapter Three. In Chapter Four, drawing on the sociological research work, the diverse range of activities and entities that were discussed by interviewees were drawn together by Catherine Heeney to show what a 'typology of practice' might include. One striking finding—which, significantly, was common to both 'mapping' exercises—is just how complex, messy, unco-ordinated, and unsystematic are both the current legal framework and contours of everyday practice when looked at in overview.

In this chapter, we begin to integrate our legal and sociological research streams. Specifically, we turn to examine practitioners' *awareness* of the multifarious sources of governance and guidance currently on offer, and the nature, circumstances, and extent of their *engagement* with them. This chapter should be read in conjunction with Chapters Six to Eight—a trilogy of chapters in which, building on the foundations laid in this chapter, we examine the complex array of practitioners' *attitudes* towards governance and their implications. It should also be read in conjunction with Chapter Nine, where Catherine Heeney and Andrew Smart explore 'access' as a case study of enactment of governance in practice.[1]

[1] 'Open-access' (or sometimes 'public-access') policies typically impose mandatory requirements that third parties be allowed to access data, biosamples and research findings at the earliest possible opportunity. In ch 9, access is explored in greater depth to reveal how governance is enacted in certain working practices, what underlying motivations are at play, and what impact they have.

As noted in Chapter Three, even the most careful and rigorous of legal analyses can give, at best, only a partial impression of how biobanking governance actually plays out in practice. For a fully informed, more accurate account, coupling a legal analysis with tailored empirical research is essential. Accordingly, in this chapter we report key data and findings drawn from the sociological research, looking into the sources of governance and guidance actually identified by respondents as being relevant to their practice. The sociological research design has been described in detail elsewhere.[2] However, to reiterate its key aspects pertinent to this chapter, the interviews (conducted between May 2006 and November 2007 by Catherine Heeney (CH)) included a number of questions designed to collect descriptive information, including specific questions about the law and actors involved with governance. The overarching aim was to explore with respondents their awareness, attitudes and experiences of governance in the area of biobanking and genetic research in England and Wales. The transcribed interviews were coded (by CH) and the data analysis presented in this chapter was developed, refined and discussed by all members of the team (but particularly by Susan Gibbons (SG), Andrew Smart (AS) and CH). To mirror the data analysis process, whereby data was thematically coded as relating to 'Governance' and subdivided into the various sub-codes, this chapter is divided into four substantive sections. Each section presents our major findings and data on one specific sub-code, together with analysis and discussion. We compare and contrast these findings with our legal mapping of the regulatory space. Based on our findings, we identify a number of key lessons and conclusions, carrying important implications for any governance reforms.

The first substantive section reports on respondents' awareness of the various 'laws' and other governance or guidance instruments that currently exist. The next section similarly explores their awareness of 'official' governance bodies and guidance sources—although, the data in fact encompass not just 'official' regulatory bodies but actors of all descriptions (formal, informal and quasi-formal), for the reasons explained therein. The following section then turns to look at respondents' engagement with governance and guidance. It identifies the external and internal 'oversight' bodies actually encountered by respondents, and considered by them to be relevant to their practice. The section after this describes the forms and methods of engagement proactively used by respondents to seek guidance, information or advice. A concluding section then summarises our principal findings, observations and key lessons, and begins to reflect on their implications for future biobanking governance and regulation.

[2] See chs 1 and 4 above.

Analysing these data helps to reveal respondents' knowledge and understanding of, and their perspectives on, the current composition of the regulatory space for biobanking activities. As we shall see as the chapter progresses, it also helps to uncover some significant gaps. In particular, these include notable mismatches between, on the one hand, the 'law-on-the-books' and apparent composition of the regulatory space, and, on the other hand, the reality of how biobanking governance operates and is delivered—not least, by whom—in practice.

## INSTRUMENTS: FORMAL 'LAWS' AND OTHER DOCUMENTARY GOVERNANCE SOURCES

### Key Findings

Turning now to the substantive presentation of the major findings in relation to awareness and experiences of governance, respondents were asked what they thought were the main 'laws', codes, or guidelines governing their practice. When asked to reveal the 'laws' relevant to their work, the 49 respondents offered almost as many distinctively different answers. These covered a broad spectrum of instruments and institutions. Indeed, the answers ranged from identifying specific UK statutes and EU instruments, through various guidelines and websites (notably those of the Department of Health, Medical Research Council (MRC), Wellcome Trust, and some professional associations), down to more general legal forms and processes and finally to what we have termed vague legal 'reference points'. Some respondents also mentioned actors or institutions in this context.

The responses occurred with very varying frequency. While 21 people referred to the Human Tissue Act 2004 (HTAct) and 20 mentioned the Data Protection Act 1998 (DPA) by name—the two most common answers given—nearly 30 other 'laws' or documentary governance sources were mentioned just by a single interviewee. Also, while some responses were quite specific, many were opaque and open to interpretation. Thus, in addition to direct references to the HTAct and DPA, several other, vaguer, statements might have been referring one of these two Acts—and/or to associated EU Directives. For instance, one Clinical Scientist alluded to 'regulations about [data] security' (R49: Uni); while a Research Scientist spoke of both 'general guidelines [...] of data protection and the EU Directive on that' (R43: Repository). Two respondents, a Bioinformatician and an IT Specialist (both working at Repositories) specifically noted the relevance of 'the Database Directive' (R29)[3] and 'Article 29' (R40).[4] Similarly, with respect

[3] Databases Directive 96/9/EC [1996] OJ L/77/20.
[4] Data Protection Directive 95/46/EC [1995] OJ L/281/31, Art 29.

to biosamples, one Research Scientist may have been thinking about the HTAct or European regulation when noting the 'regulations [...] regarding [...] retaining tissues and DNA' (R5: Uni). The only respondent who made specific reference to EU tissue regulation was a Repository-based Research Scientist who mentioned it in tandem with the UK law: 'I was aware of HTA, Tissues and Cells Directive, that type of thing' (R37).[5]

Limited recognition of non-UK law was evident elsewhere, too. The third most frequently cited legal instrument (although trailing some way behind the HTAct and DPA, with six mentions) was the EU Clinical Trials Directive 2001/20/EC.[6] Two respondents alluded more generally to 'EU Directives' (R32: Public Health, NHS) or 'European Directives' (R41: Research Scientist, Uni/Hos). With respect to other countries, our respondents from the private sector referred to 'the laws' in Sweden, Denmark and China (R34: Research Scientist, Pharma) and mentioned the 'Swedish Biobank Law' (R21: Clinical Scientist, Pharma; R23: Research Scientist, Pharma), a reflection of the national jurisdictions that were pertinent to their work. Five respondents noted various US federal enactments, regulations or courts.

Eleven respondents—largely Research or Clinical Scientists—described 'good practice' or 'best practice' principles as key governance sources. A few identified specific documents and authorities here. Thus, two followed 'best practice' of the US National Institutes of Health (NIH) (R17: Uni/Hos; R34: Pharma); while one referred to complying with the 'GLP [...] good laboratory practice [...] governed by the, I think it's the GLPMA, the Good Lab Practice Monitoring Authority' (R22: Pharma).[7] More commonly, these respondents made less precise references to 'good laboratory practice' (R3: Uni), or 'Good Clinical Practice' (R49: Uni; R22: Pharma; R47: Uni/Hos). As one put it:

> Yes good clinical practice is right, written down, you can find those documents about good clinical practice on the web and that's a way one conducts a clinician trial. We just adapted that to work for us. (R31: Uni/Hos)

On other occasions, respondents simply alluded vaguely to 'best practice' (R1: Uni; R3: Uni), or 'best practices' (R34: Pharma).

Interestingly, three respondents described 'good practice' standards as having greater significance than 'law' or 'rules'. For two Bioinformaticians

[5] Tissues and Cells Directive 2004/23/EC [2004] OJ L/102/48.

[6] [2001] OJ L/121/34.

[7] The UK Good Laboratory Practice Monitoring Authority (GLPMA) is responsible for monitoring test facilities that must comply with the principles of Good Laboratory Practice (GLP). Those principles are embodied in the Good Laboratory Practice Regulations 1999 (SI 2106/1999), as amended by the Good Laboratory Practice (Codification Amendments Etc.) Regulations 2004 (SI 2004/994). These Regulations implemented European Commission Directives 99/11/EEC and 99/12/EEC, which themselves were based on the OECD's revised Principles of GLP and Compliance Monitoring, issued in 1998.

who worked at different Repositories, good or best practice was all that they had to guide them, since no legal instruments govern either the technical design of databases or IT security:

> So I mean I think, so er I mean there aren't any kind of rules there are just good practice. (R10)

> What you have in this domain is just a few best practices which are recommended to you by IT consultants. However there is absolutely no law and there is no guidance and there is nothing available so it all depends, it's really left to the individual organisation to determine how the level of security they, how they manage the security aspect. (R20)

Meanwhile, the only Genetic Counsellor in our sample referred to 'good practice guidance like confidentiality which depending on what professionals you are can be a bigger issue in some ways than perhaps the more abstract statutory law' (R33: NHS).

Further specific 'laws' or governance documents mentioned by two or three interviewees apiece were the Freedom of Information Act 2000, matters relating to intellectual property, and material transfer agreements. Items mentioned by a single respondent included the Health and Social Care Act 2001; 'the Medical Records Act or whatever it used to be called, Healthcare Records Act' (R33: Genetic Counsellor, NHS); 'Information Commissioner guidance [...] I can't remember the title of the document to be honest' (R20: Bioinformatician, Repository); and—even less precisely—'freedom of access to medical records' (R27: Research Scientist, Uni/Hos), 'UK Biobank law' (R22: Research Scientist, Pharma), and 'abstract statutory law' (R33: Genetic Counsellor, NHS).

Several respondents approached the topic of 'laws' by drawing attention to substantive issues—principally consent and/or confidentiality. However, those who did so rarely mentioned specific laws, or even informal guidance materials. The only exception was one Clinical Scientist who referred to consent and confidentiality as being 'more in the area of common law' (R15: Uni/Hos). Meanwhile, several respondents expressed their awareness of the law by using vague legal 'reference points'. Examples here included 'research governance' (R43: Research Scientist, Repository), 'the research governance framework' (R42: Clinical Scientist, Uni/Hos), 'all those working parties that exist at government level' (R28: Clinical Scientist, Uni/Hos), 'government directives and things' (R6: Epidemiologist, Uni), 'a whole lot of ethical guidelines for science' (R29: Bioinformatician, Repository), 'standard professional ethics' (R25: Research Scientist, Repository), and 'European Union things' (R29: Bioinformatician, Repository). Finally, around a dozen respondents referred to almost as many different institutions or actors to frame their answers about relevant 'laws'. The actors most commonly mentioned in this context were research ethics committees (RECs), Caldicott

Guardians,[8] the Information Commissioner's Office (ICO), the MRC, the Wellcome Trust and the Department of Health.

**Analysis and Discussion**

Reflecting on these findings, five aspects stand out. Foremost among them is the remarkable range and differential frequency of responses. This can be interpreted in various ways. In part, it simply may reflect the fact that different laws genuinely applied (or the same laws applied in differing ways, or to varying degrees) to respondents within our sample. Given that our sample encompassed bioinformaticians, clinical scientists, research scientists, epidemiologists, IT Specialists and other health professionals—all of whom were involved with biobanks of diverse types—to some extent this is no doubt true. However, from a strictly legal perspective, this explanation alone cannot possibly account either for the huge range and variability of responses or the discrepancies that we encountered. Alternatively or additionally, respondents simply may have diverged over how far they had engaged with, understood, or knew about (or, indeed, needed to know in detail about) the 'laws' and other governance materials applicable to their circumstances.

A second notable, interrelated, aspect of these findings is the tremendous range in clarity of expression. This, too, suggests marked variability between respondents, in terms of their engagement with, understanding of, and knowledge of, the relevant 'laws' and rules covering their practice. In combination, these first two aspects of our findings raise issues that are notable for anyone considering how governance is enacted in practice, especially if other professionals and practitioners undertaking biobanking activities are similarly (un)aware over precisely what is required of them. In turn, this raises questions over how far day-to-day biobanking and related activities in England and Wales actually comport with all of the legal, ethical and professional requirements governing them. Insofar as people are not fully aware of (and, therefore, not guided by) such requirements, the answer to this will depend on how far 'informal' or 'implicit' sources of governance or guidance (as discussed in the following chapters[9]) operate, essentially on a de facto basis, to 'plug the gaps'.

Thirdly, although the HTAct, DPA and EU Clinical Trials Directive were cited frequently, suggesting that these three laws were seen as having a particular bearing, this pattern of response, too, may be interpreted in several ways. In part,

[8] As noted in ch 3, Caldicott Guardians are senior NHS and social services staff members, appointed within each institution to protect patient-identifiable information and confidentiality. Their responsibilities include overseeing, monitoring and reviewing institutional information governance protocols and policies, and advising NHS staff, departments and ethics committees (including RECs).

[9] See esp ch 6 below.

it could be attributable to the common importance of these three instruments to biobanking practice generally; or, in the case of the HTAct especially, to its heightened profile around the interview period. Mere awareness alone of the existence of a 'law' does not necessarily imply either a correct understanding of it, or actual compliance. Indeed, where such references were made, they tended to lack detail. Similarly, respondents were not always accurate about what these laws actually required. For example, one Clinical Scientist spoke of being 'aware that we need to be licensed' under the DPA—a factually incorrect statement—but 'that the licence resides at Trust level' (R2: Uni/Hos). It is unclear whether this respondent really did have in mind the DPA registration and notification requirements on data controllers, or was perhaps confusing or conflating these with the HTAct licensing requirements.

Fourthly, by contrast, the low frequency and relative paucity of references to other 'formal' governance documents—including many leading Acts, legislative provisions, regulations, directives, statutory codes of practice, common law doctrines, official guidelines and so forth identified in Chapter Three—was marked. This is another important finding. Once again, it suggests a potential mismatch or gap between legalistic regulation 'on-the-books' and governance as enacted in practice.

Finally, it is noticeable that all of the answers were grounded very firmly in the context of UK and, to a lesser extent, European laws. References to other foreign national laws (and actors, as noted below) were relatively uncommon, and particularly so from respondents who did not work in the private sector. This spectrum may be expected from a sample of respondents recruited from a single jurisdiction, many of whom did not explicitly talk about working across national borders. Nevertheless, the existence of some examples in the dataset does show that it is important to some practitioners in certain contexts—typically, respondents who were engaged in multi-jurisdictional collaborations, received funding from overseas funding bodies and/or worked for multinational organisations. Moreover, given the burgeoning of multinational collaborations, familiarity with foreign legal systems and their requirements is likely to become more of an issue (and a challenge) for UK-based practitioners and biobank operators. This raises questions about the best ways to ensure consistent drafting and implementation of legal, ethical and technical norms, standards, practices and procedures across multiple jurisdictions.

## ACTORS: 'OFFICIAL' GOVERNANCE BODIES AND GUIDANCE SOURCES

### Key Findings

Having asked been asked about specific 'laws', interviewees were also asked which 'official' bodies they thought were influential or relevant to their practice, and what sources of guidance from such bodies they had actually used. Once again, the responses are illuminating, and somewhat surprising.

When the data was analysed to identify specific actors identified as sources of 'official' governance or guidance, it was found that respondents mentioned over 40 different 'actors'. This number does not include actors simply noted as having produced laws, rules or governance requirements in the abstract. It encompasses only those actors specifically identified—either as governance bodies, or as sources of guidance about the law, rules or requirements—that respondents had actually engaged with or attempted to use (whether or not successfully).[10] Also, out of necessity, the list embraces formal, informal and quasi-formal regulatory space actors (as delineated under the schema in Chapter Three). This is because respondents themselves did not draw any hard-and-fast, legally accurate distinctions between 'official' bodies and other external actors.

Broadly speaking, the raft of actors identified as being 'official' sources of governance or guidance fell into nine categories. They covered a remarkably wide spectrum. The nine categories were: (1) leading research funders; (2) RECs; (3) NHS organs and actors; (4) UK government departments and agencies; (5) formal regulators; (6) professional bodies (including the Royal Colleges); (7) technical standards or accreditation bodies; (8) industry associations; and (9) advisory bodies. In keeping with their responses to the 'laws' question, several respondents identified non-UK actors, particularly regulators and major funding bodies from the EU and USA.

Once again, there was marked variability in the frequency of responses. By far, the actors most commonly mentioned were leading biomedical research funders—especially the MRC and Wellcome Trust, but also the US National Institutes of Health (NIH) and the Economic and Social Research Council (with 19, 11, 8 and 5 mentions respectively). Next most prominent were RECs of various kinds. Thus, 15 respondents referred to 'MRECs',[11] 'COREC',[12] other individual RECs, and/or REC chairs or 'ethics chairmen' (R18: Epidemiologist, Uni) as sources of oversight, governance or guidance. Funders and RECs were seen by respondents as performing essential roles in relation to regulation and oversight, which we shall discuss in greater detail below (section entitled 'Actors: External and Internal "Oversight" Bodies').

Nine respondents referred to principles or frameworks relating to Caldicott Guardians. Several respondents referred to their own professional associations, Royal Colleges, the General Medical Council and/or

[10] Problems reported by many respondents in locating, accessing, understanding and being able to follow relevant guidance materials and governance systems are addressed below.

[11] At the time of our research, Multi-centre Research Ethics Committees (MRECs) were responsible for reviewing projects that involved multiple NHS institutions, collaborators or resources.

[12] The National Research Ethics Service, formerly the Central Office for Research Ethics Committees (COREC).

technical accreditation or standards bodies—the latter often in connection with verifying (or working towards achieving) compliance with very precisely identified 'ISO' standards.[13] Remarkably, however, in relation to the two principal, 'formal' regulatory bodies identified in Chapter Three, only five respondents mentioned the Human Tissue Authority (HTA); while only three identified the Information Commissioner or Information Commissioner's Office (ICO) as a source of guidance.

Sometimes, respondents named the actors directly. At other times, they referred to them indirectly, via documentary governance or advice produced by them that respondents had used (or attempted to use). Again, the MRC and, to a lesser extent, the Wellcome Trust featured most often. Examples here included MRC or Wellcome Trust 'guidelines' (six respondents), the 'MRC website' (R45: Clinical Scientist, Uni/Hos; R46: Epidemiologist, Uni/Hos), MRC or Wellcome Trust 'policy' (R8: Research Scientist, Repository), and the MRC 'clinical trial roadmap' (R30: Bioinformatician, Uni). However, the vast majority of governance or guidance sources identified—whether named directly as actors, or noted indirectly by reference to guidance materials produced—featured only once or twice.

## Analysis and Discussion

Three observations are worth highlighting from these findings. First, when we compare and integrate this sociological data with our legal analysis in Chapter Three, one striking discrepancy emerges. This is the significant gap between the plethora of 'official' (and unofficial) governance and guidance materials on offer, and reported patterns in their actual penetration and use. (We will return to this theme below.) Conspicuously sparser than one might expect are references to the two principal 'formal', day-to-day regulatory bodies, the HTA and ICO. This is despite both bodies possessing formal legal status, powers and responsibilities, and the extensive catalogue of 'official' governance and guidance materials that they have produced—not least, the HTA's statutory codes of practice, and the ICO's wealth of documentary and electronic resources. This raises questions about the accessibility and, perhaps the utility, of these resources.

Secondly, the large proportion of (usually) single references to a host of different actors or materials in a sense mirrors the densely populated nature

[13] ISO stands for the International Organisation for Standardization, which has produced and is responsible for over 17,500 international management standards. These include the 'ISO 9000' (R20) (quality management system requirements), 'ISO 27,001' (R19) (information security requirements), and 'ISO 17025' (R3) (general testing and calibration laboratory requirements).

of the regulatory space, as mapped out in Chapter Three.[14] Along with the 'laws' responses, in part, the wide variability and low frequency of responses may be a product of different interviewees having different practices, professional backgrounds, roles or responsibilities.[15] However, the sheer volume and diffusion of sources cited once more raises questions. For example, do these (and other similar) practitioners know where best to look for accurate, pertinent, authoritative governance and advice (insofar as these actually exist). Furthermore, it suggests that these practitioners do not voluntarily engage with 'official' bodies on a regular basis, or routinely seek out (or receive) clear, specific information or guidance about the legal requirements applicable to them, or where to go to find out more. (Again, we shall return to these matters below.)

Thirdly, additional findings support these interpretations about the restricted nature of practitioners' engagements with law and guidance, and suggest a high degree of uncertainty on their part. One such finding is the widespread blurring of categories—many respondents' appeared to be unaware of, or unable communicate (at least in the context of the interview), the legal differences between different actors in terms of their 'official' authority, status and powers. This underscores the general impression from the analysis that many respondents were uncertain over what 'laws' and 'official' requirements were applicable to them. Indeed, respondents often regretted the lack of any comprehensive, focused, authoritative, clear, consistent, readily accessible regulatory framework for biobanking in England and Wales. As we shall see in Chapter Eight, a number even expressed a positive desire for more regulation—such as through the introduction of a purpose-designed, 'official' biobanking governance, oversight and guidance framework.

## ACTORS: EXTERNAL AND INTERNAL 'OVERSIGHT' BODIES

### Key Findings

Respondents were also asked questions about 'oversight' of their 'genetic databases', to tease out what kinds of actors (or regulatory structures) they felt they had actually been overseen by, or were answerable or subject to, and about their awareness and experiences of oversight. Their responses have been analytically divided into two broad categories, which we have termed 'external' and 'internal' oversight mechanisms. For the most part, we did not ascertain consistent divergences within the dataset between different kinds of respondents. On this topic, however, our analysis of the data

[14] See ch 3 and Appendices 1–4.

[15] As one Clinical Scientist explained: 'we have a named person in the team that's responsible for clinical governance'. (R47: Uni/Hos)

did reveal a marked divergence between respondents from the public and private sectors.

*External oversight bodies*

Of the external 'oversight' bodies identified, RECs of various kinds (especially MRECs) clearly were dominant, with nearly half of all respondents (26 in total) mentioning them. Forms of oversight here included the need to get initial, prior ethical approvals for projects, attending meetings with RECs, addressing REC members' concerns or questions, and providing follow-up, progress or final reports. The second form of external oversight most commonly mentioned (10 respondents) emanated from research funding bodies. Perceived forms of oversight here included funders' project 'audits' (R3: Research Scientist, Uni; R47: Clinical Scientist, Uni; R40: IT Specialist, Repository), requirements to submit written reports, and requirements to attend various meetings with funding bodies (R14: Clinicial Scientist, Uni/Hos). But the main impact felt was through funders requiring compliance with their own preferred policies—especially the 'open-access' or 'public-access' agenda (a controversial innovation at the time of our research discussed in Chapter Nine). In the view of this Clinical Scientist who worked for a well-established longitudinal study, funders have power to set the 'ground rules' (R15: Uni/Hos). This included funders stipulating such matters as project management, rules of engagement, data handling, third-party access and the content of participant consent forms.

Overwhelmingly, it was public sector respondents who referred to RECs and funders as external oversight bodies. By contrast, private sector respondents, such as this Research Scientist, typically did not perceive themselves as being subject to any external oversight actors:

> I'm not aware that we have had external, certainly not external bodies [....] there is no external governance issues like over the processes. (R23: Pharma)

Another Research Scientist admitted, 'to be honest I've not heard of COREC' (R22: Pharma). Rather, as we shall see below, private sector respondents focused much more heavily on internal oversight structures.

Other external oversight methods identified by numerous respondents from all sectors related to fulfilling technical or professional accreditation and licensing-type requirements. Some were voluntary. Examples here included laboratory and other quality assurance audits by private accreditation organisations (such as monitoring compliance with ISO standards, as noted above). Others were non-voluntary audits, inspections, licensing and registration requirements, such as those imposed by the HTA, European Medicines Agency (EMEA), UK Good Laboratory Practice Monitoring Authority (GLPMA), and the US Federal Drug Administration (FDA). However, in terms of the leading 'formal' UK regulatory bodies and statutory governance frameworks, while 10 respondents referred to the HTA or HTAct in this context, only three mentioned the data protection regime.

### *Internal oversight bodies*

Turning to internal 'oversight' bodies or methods, two clearly predominated. One was institutional governance arrangements and supervisory actors located and operative within the institutions or organisations in which the respondents' biobanks or projects, or the respondents themselves, were based (directly referred to by 18 respondents). Respondents from across all sectors and institutional types made reference to these—including Primary Care Trusts, NHS hospitals, universities, charitable companies and private companies. Public sector examples included general internal audit, monitoring and control systems, the 'university lawyers' (R45: Clinical Scientist, Uni/Hos), Caldicott Guardians, and NHS Research and Development teams.

Within the private sector, oversight was seen almost exclusively as coming from, or being channelled through, three main internal sources. Each reflects a corporate infrastructure based on clear vertical and horizontal divisions of labour, decision-making, specialisation and responsibility. One oversight source was management hierarchies within the company. A second was various 'in-house' experts who were presumed to take care of compliance matters within their respective spheres, as this Research Scientist explained:

> In a company this size [...] what tends to happen is the lawyers translate the law into the internal policy and the policy gets transplanted into SOPs and then the individual groups follow the SOPs. (R22: Pharma)[16]

Similarly, bringing various in-house experts together was a key method for making informed decisions (and, by implication, ensuring their correctness and legitimacy):

> So decisions were made sort of with sort of science policy experts, we've got an ethicist in house, plus legal, plus the geneticists and the clinical trial folk. (R23: Research Scientist, Pharma)

The third internal source of oversight that private sector respondents spoke of using was standard form template documents, such as consent forms, which could be tailored for specific projects or studies. Again, as this next comment illustrates, these were presumed to embody what was legally and ethically required: 'I mean these templates go through hundreds of ethics committees so they get refined' (R23).

The second predominant internal source of oversight, as noted by 17 respondents, was from project-specific steering committees, boards or other such groups, set up especially by, or for, individual projects (as will be discussed

[16] SOPs stands for Standard Operating Procedures.

in Chapter Nine, such organisational structures were used in 'medium' to 'very-large' scale projects to govern 'access'). These were variously described as being 'steering', 'management', 'executive' and/or 'advisory' in nature. In terms of their remits, roles and functions, no uniform pattern emerges from the titles or vaguer descriptions given by respondents. Some bodies appeared to have fairly general roles while others were given more precise titles or descriptions—for instance, 'Science Committee' (R8: Research Scientist, Repository), 'Scientific Committee' (R15: Clinical Scientist, Uni/Hos, R36: Bioinformatician, Repository), 'Trial Management Committee' (R46: Epidemiologist, Uni/Hos), 'steering committee that would oversee requests for information' (R8: Research Scientist, Repository). Some bodies appeared to perform an admixture of roles. Thus, one respondent's project had a 'Data Monitoring and Ethics Committee' (R46: Epidemiologist, Uni/Hos); while, for another, the 'Law and Ethics Committee' played a fundamental oversight and advisory role: 'that's our first port of call' (R6: Epidemiologist, Uni).

Other forms of internal oversight mentioned by numerous respondents were regular project group, team, or partner meetings; projects' Standard Operating Procedures (SOPs); and at least seven different kinds of written contracts or agreements. The latter included material transfer agreements, memoranda of understanding or similar agreements between collaborators, intellectual property rights permissions (including patent licences), and third party access agreements.

## Analysis and Discussion

Reflecting on all of the findings described thus far, two overarching conclusions appear to be emerging. First, these data add to the growing body of evidence set out in this chapter that the governance framework described by our respondents depends very heavily indeed on 'informal' oversight methods. Especially noteworthy here are the 'gatekeeping' role played by RECs, controls and checks imposed by funders, and the prominence of internal institutional or project-specific oversight systems. Also significant is the difference between public and private sector practices. These findings confirm, and further refine, some of the key lessons drawn from our legal regulatory space analysis set out in Chapter Three.

As with the applicable 'laws', references by respondents to formal (or genuinely official) governance and guidance-providing bodies acting as 'oversight' bodies were relatively sparse. This contrasts with the dominant, central, and evidently crucial role played by RECs and funders, particularly for those working in the public sector. This is despite the fact that RECs have, at best, only 'quasi-formal' status,[17] and lack directly enforceable powers or

[17] See further ch 3 above.

sanctions. Funders' powers, too, derive essentially from informal means, most notably dictating the biomedical research agenda through issuing funding calls, vetting applications and controlling the purse-strings.[18] These patterns of engagement raise the question of why it is that RECs and funders figure so prominently in the accounts of some of our respondents. A primary explanation is rooted in instrumentalism. As we observed in Chapter Three, for many researchers, RECs and funders are 'gatekeepers'—only once their approval is gained can research begin. However, as we shall discuss more fully below, this pattern of engagement has important consequences for the ways in which governance is being enacted.

Secondly, our data analysis reveals that 'official' bodies (or, at least, bodies construed as being broadly 'official' in nature by respondents) do not figure prominently when practitioners are asked to discuss their practice. This may mean that such sources are not regarded as relevant or helpful. Alternatively, it may mean that they are not routinely accessed—for reasons that may be attributed to the working contexts and practices of practitioners, or to the nature, accessibility, or utility of the sources themselves. Whatever the reason(s) may be, in combination with other findings presented above and below in this chapter, the impression we gained from analysing the data is that most of the professionals who were interviewed in our study do not routinely or proactively turn to such bodies for guidance or advice. The general picture appears to be that practitioners are more likely to engage with, or to seek guidance from, regulatory bodies when they are compelled so to do—but even then, only to the extent required to negotiate their way through the immediate 'gatekeeping' step.

With respect to the difference between the public and private sectors captured in our analysis, in certain aspects, the latter could be characterised as 'self-reliant'. It is pertinent to ask what it is about the private sector that might induce or encourage the reliance on managerial hierarchies, in-house experts and standardised procedural templates that we have reported here. Answers would include the need to protect intellectual property, the organisational division-of-labour in many life science companies, and an ostensibly antagonistic process of regulation in which the State (in the form of regulatory agencies) acts as the final arbiter. If we accept that these are the leading 'drivers' of certain governance practices, then it is possible that some public sector projects increasingly will act in ways that are similar to their private sector counterparts. This could be the case particularly for public sector biobanks that are large-scale organisations, involve complex divisions of labour, have undertaken partnerships with private companies, or have been established with a strong incentive (or outright goal) of producing outputs that have a potential monetary value.

[18] See further ch 3 above.

SEEKING GUIDANCE: METHODS OF ENGAGEMENT

**Key findings**

The proposition the respondents interviewed in our study did not widely tend to engage with formal 'laws' or 'official' actors as sources of governance or guidance begs the question of where they do turn to for help (if anywhere) when they want or need regulatory guidance. Nearly half of our respondents (23 in total) indicated that they had proactively sought out relevant guidance, information or advice in some way. This section outlines the diverse forms and methods used.

Tellingly, only five respondents described having sought advice proactively, either formally or informally, from 'official' regulators or other external governance bodies. Mostly, this involved contacting REC staff or chairs. Just one, a Bioinformatician working at a newly established Repository, reported approaching the HTA and COREC directly for advice (R36).[19] More informally, one University-based Bioinformatician, who pioneered a web-based data-sharing platform, had arranged to meet with a Caldicott Guardian who worked within his organisation (R1); while a Clinical Scientist based in a University/Hospital setting reported that he would 'often pop out' to question NHS Research and Development staff or MREC members, since they were conveniently located in adjacent offices (R45). Three respondents reported seeking advice from their own internal, project-specific steering or other oversight committees.

Rather than looking to organisations, institutions or other such bodies, our respondents often seemingly preferred to contact other individuals, informally and ad hoc. For many, this was their primary (if not only) proactively sought source of information or advice. This included seeking legal advice, from lawyers and non-lawyers. Typically, respondents spoke to people whom they already knew, or had come across, through professional networks. People so contacted included immediate colleagues, project team members, external collaborators, and specific individuals known to be connected with research support organisations, funding bodies or professional associations. For Research Scientists working for pharmaceutical companies especially, 'in-house' specialists were key (as noted above in section 4.1.2); examples here included lawyers, bioethicists, scientists, 'the clinical study people' (R22), and the 'quality assurance department' (R23). Other individuals mentioned by respondents working in other settings included 'the [Wellcome] Trust lawyers' (R37: Research Scientist, Repository), various 'counsellors' (R31: Clinical Scientist, Uni/Hos; R43: Research

[19] This was in connection with the new 'research tissue bank' generic REC approval scheme, described in ch 3.

Scientist, Repository), and 'the IT people' (R32: Public Health, NHS). The above-mentioned Bioinformatician who developed a web-based data-sharing platform had sought and accepted legal advice from a university Computer Science academic who happened to have given 'a very good presentation' on data protection and anonymisation (R1: Uni).[20]

Several respondents looked to other researchers or research projects that they considered to be 'path-breakers' or 'trailblazers' for models to follow when under-taking new initiatives. For instance, one Bioinformatician at a newly established Repository had spoken to 'other existing tissue banks and found out what they normally write in their consent forms' (R20) in order to follow suit. Similarly, a Research Scientist in the private sector had searched another pharmaceutical company's website to ascertain its policy on banking samples, in order to use this as a template (R22: Pharma). Two respondents noted that their multinational project collaborators had proven to be useful guidance sources. On both occasions, the collaborators had prompted respondents to develop policies covering matters that they had not previously considered, by asking the respondents what their policies were on those things. This Repository-based Bioinformatician saw this as 'one of the advantages of the bigger places getting into the picture [...] in that they tend to be very hot on their policies' (R36).

A few respondents had attended biobanking conferences, presentations, briefings, or 'the right meetings' (R28: Clinical Scientist, Uni/Hos), and spoken to people there and asked questions. For one Bioinformatician at a newly established Repository, 'regularly going to meetings and talking to other people' was a fundamental way to 'become aware of what's out there' (R36). Four respondents identified attending professional training courses, or 'continuing professional development' (R31: Clinical Scientist, Uni/Hos), as a guidance source. Mostly, these were compulsory Good Clinical Practice and/or Research Governance courses, attendance at which was required by universities, the NHS, and by one biobank for all of its staff. One project had brought in external experts to give its staff specific, practical training (including teaching nurses how to deal with common questions from participants, and how to use the equipment). Another respondent referred more vaguely to attending 'some professional training type days run by pharma-type companies' (R30: Bioinformatician, Uni).

Aside from contacting other actors or social contacts, six respondents (working in public sector settings) had attempted to read laws or guidance documents directly themselves (none of these professed to having had any legal training). For one Clinical Scientist, reading materials was the primary method used: 'we've done a lot of, mainly a lot of reading ourselves' and I

[20] The advice was that you can 'boil down [the rules about confidentiality] to a fairly straightforward rule really'.

have 'tried to read as much as I can' (R45: Uni/Hos). Specific instruments tackled by various respondents included the DPA, HIPAA[21], relevant reports or discussion documents issued by funders or professional associations, and 'digests' of the Caldicott Report (R2: Clinical Scientist, Uni/Hos). Respondents reported mixed success in these endeavours, including some who had struggled. For example, one Repository-based Bioinformatician who was distinctly unimpressed with the EU Databases Directive[22] commented: 'People used to keep asking me to read the damn thing and I'd get to about page 4 and fall asleep.' (R29) Interestingly, several respondents had decided deliberately *not* to engage with relevant laws or guidance documents, despite knowing about their existence. For example, this Clinical Scientist was 'aware of other guidelines but I can't say that I've used them' (R2: Uni/Hos); while this Epidemiologist was 'not going to pretend that I'm an expert in this area, I'm not going to pretend that I spend hours and hours looking' (R24: Uni).

Visiting websites was a method used by five respondents. However, only five specific websites were mentioned in this context—and, with one exception, only by one respondent each. The websites so identified were those of the Department of Health, MRC, Nuffield Council on Bioethics, Human Genetics Commission, and a leading multinational pharmaceutical company. Once again, conspicuous omissions included such guidance-packed websites as those of the ICO, HTA, UK Clinical Research Collaboration (UKCRC) Regulatory and Governance (R&G) Advice Service,[23] British Medical Association, General Medical Council and Wellcome Trust. Also, as with reading the legislation, respondents did not necessarily find having to 'trawl' multiple websites a particularly efficient, productive or 'foolproof' method (R28: Clinical Scientist, Uni/Hos).

## Analysis and Discussion

An initial premise of our project was that the sheer breadth of potentially applicable laws and guidance could present a risk for those undertaking biobanking activities. With respect to this, at least two significant points stand out from the above findings. First, overall, no consistent pattern emerges as to respondents' appreciation and/or use of the enormous range of guidance sources (or actors) that exist.[24] Several demonstrated some knowledge of

[21] The US Health Insurance Portability and Accountability Act of 1996, 29 USC §1181.
[22] Databases Directive 96/9/EC [1996] OJ L/77/20.
[23] Although, note that this service was only launched UK-wide (following an initial pilot) from April 2007.
[24] See further ch 3 and Appendices 1–4.

the breadth of information, and information sources, that were available,[25] but, in the context of the interview, many respondents appeared to be largely unaware of the array of information that potentially had relevance to their practices.

Interestingly, however, at least some of those who had some awareness of the breadth of information described its very breadth as problematic or a cause of unease. For example, one University/Hospital-based Clinical Scientist spoke of needing to 'be aware of the things that are relevant to us' but then having 'to discard the mountains of information that are not relevant to us', while being aware that 'it's not foolproof and we are going to possibly miss things' and that he could not rely on his employers to tell him about relevant laws, or for such information to 'filter down' through the organisation (R28). Respondents working in University, University/Hospital and NHS settings particularly expressed concern over the sheer volume and confusing nature of the many guidelines and requirements applicable to them. By contrast, respondents involved with large-scale, multinational projects, and/or from multinational corporations, generally did not seem overly troubled. For them, dealing with multiple laws, frameworks, governance actors and guidance sources appeared to be an accepted part-and-parcel of the job.

Secondly, and taken together with our earlier findings, these data uncover a significant behind-the-scenes regulatory role played by individuals, through the operation of professional and social networks. Many respondents relied on informal, ad hoc interactions with other individuals or stakeholders as a primary source of guidance (especially their colleagues, or people seen as having relevant expertise, knowledge or prior experience), rather than turning to 'official' or other authoritative, external sources, or attempting to grapple directly themselves with the morass of laws and guidance on offer. This crucial—but ostensibly invisible—role played by individuals operating in social networks would very likely be missed, were one to rely purely on a legalistic mapping of the biomedical regulatory space (this is further discussed in Chapter Ten). This reinforces the impression noted above, that our respondents tend to limit their interactions with regulatory bodies to what is necessary, and do not widely regard (or call upon) such bodies as sources of helpful guidance, information or advice.

Several potential explanations may account for these findings. For instance, some practitioners may feel it is necessary—or, perhaps, just find it easier—to

[25] Thus, a University/Hospital-based Clinical Scientist (R28) reported having sought or received information from multiple sources, including 'trawling' websites, reading materials, receiving emails and correspondence from an NHS Trust and a professional association, and attending numerous meetings; a pharmaceutical company Research Scientist described having had 'a whole multitude of different input' (R34); while a Clinical Scientist working on a small-scale clinical study in a University/Hospital setting recalled there being 'lots of literature on it' (R14).

turn to informal actors or associates for help. Related to this, some may well see relying on other individuals, whom they perceive to have 'trodden the path' before, as a convenient shortcut technique. (As a method for enacting governance, the reliability of this process is at least open to question, as it seems to rest on the presumption that such individuals have done all the necessary groundwork, found out about the relevant law, and are able to encapsulate it succinctly and accurately.) Another potential explanation for practitioners' lack of engagement with regulatory bodies is that many may well be unaware of relevant guidance resources that are on offer.

Alternatively, those who are aware of guidance resources may feel (and understandably so) confused or overwhelmed by the sheer wealth of materials and actors that currently populate the regulatory space, and not know where best to look. This could easily induce a sense of 'paralysis'. Equally, practitioners may be unable to distinguish correctly between different sources, in terms of their relative authority, importance, status or relevance (as it cannot be assumed that they have the necessary training, knowledge or support that might enable them to do so). In this regard, it is noteworthy to recall that over half of our respondents did not report taking any proactive steps at all to seek relevant information or guidance about governance-related matters. Again, several reasons could account for this apparent lack of engagement. For example, it could be down to a lack of time. Where people are located within large projects or organisations with many staff and clear divisions of duties, it may not form part of an individual's role or responsibilities. Alternatively, some professionals simply may lack sufficient motivation or concern to know about the 'laws', or other external governance requirements applicable to them, to expend the necessary time and energy to look. (We shall return to examine the wide spectrum of respondents' attitudes towards governance in greater depth in Chapters Seven to Nine.)

## CONCLUSIONS

This chapter has presented findings about respondents' *awareness* of sources of governance and guidance about biobanking in England and Wales, and the nature and circumstances of their *engagement* with them. The data covered four broad topics: (1) respondents' awareness of formal 'laws' and other documentary governance sources; (2) their awareness of, and engagement with, 'official' governance bodies and guidance sources (which, in the event, embraced formal, informal and quasi-formal actors alike); (3) relevant external and internal 'oversight' bodies, as experienced by respondents; and (4) the methods of engagement used by respondents to seek guidance, information or advice.

Our analysis of the data has generated findings, which raise issues that are notable for those considering the governance of biobanking activities.

These include insights about respondents' general levels of awareness and understanding of the composition of the regulatory space and the current governance of biobanks, related activities and professionals. We have also begun to explore respondents' perspectives on key aspects of regulation. We shall return to this topic more fully in the following chapters. Meanwhile, our findings in this chapter largely confirm, refine and further enrich the key lessons learned from our legal mapping of the regulatory space presented in Chapter Three. However, as we have seen, comparing and integrating our legal and sociological work has also exposed several noteworthy divergences, gaps and mismatches, with potentially significant implications or ramifications for future biobanking governance. The principal findings, conclusions and themes to have emerged during this chapter—and some of their potential implications—are summarised below.

## Summary

When asked about the main 'laws', codes and guidelines governing their practice, respondents' answers differed significantly—both in terms of the broad range of instruments (or actors) identified, and the variable frequency with which each was mentioned. The HTAct and DPA featured most commonly. However, such references tended to lack detail, and were sometimes inaccurate in terms of what the statutes actually say. The vast majority of 'laws' or documentary governance sources identified (or alluded to) were mentioned only once. Very few 'formal' governance documents (including the leading Acts, legislative provisions, Regulations, Directives, statutory codes of practice, common law doctrines, and official guidelines) were mentioned at all. By contrast, nearly a quarter of respondents (often Research and Clinical Scientists) described 'good practice' or 'best practice' principles as key governance sources. Indeed, some even described such principles as carrying much greater significance than any 'laws' or 'rules'. While some answers about relevant instruments were specific, most tended to be fairly general. This included allusions to general legal forms and processes, and vague legal 'reference points'. Clarity of expression also differed markedly, ranging from precise to (more often) opaque and open to interpretation.

A conspicuously similar pattern emerged in relation to the 'official' governance or guidance actors that respondents reported having encountered, engaged with or attempted to use as sources of governance or guidance. Again, the respondents mentioned a broad spectrum of actors, representing nine distinct categories or institutional types. Again, too, the frequency of responses varied significantly. While RECs and funding bodies featured most commonly, the vast majority of sources were mentioned only once or twice. Very few respondents mentioned either the HTA or ICO, despite the fact that these bodies are two of the principal 'formal' regulatory authorities relevant to biobanking in England and Wales. Indeed, most of the actors identified were not 'official' bodies at all. Rather, the responses encompassed formal, informal and quasi-formal actors alike. Most respondents did not differentiate between actors, in terms of relevant legal differences in their formal authority, status, relevance, powers and so forth.

Both in respect of instruments and actors, most interviews were grounded firmly in the context of UK (and, to a lesser extent, European) laws. This was consistent with our sample of respondents recruited from a single jurisdiction, many of whom were not explicitly working across national borders. Typically, respondents who referred to foreign or international governance sources were engaged in multi-jurisdictional collaborations, received funding from overseas funding bodies, and/or worked for multinational organisations (such as pharmaceutical companies).

In combination, these findings on 'laws' and 'official' actors suggest a marked divergence between respondents in terms of their levels of understanding and/or knowledge of, and their engagement with, the key instruments and governance actors applicable to their practice. Overall, no consistent pattern emerges as to respondents' awareness of the enormous range of guidance sources and actors that actually exist. In part, this may be due to genuine differences—for example, how far certain laws or actors are relevant to differently situated professionals or biobanks; or how far certain practitioners actually need to know (or know in detail) about the regulatory requirements applicable to their circumstances. It may well flow also, in part, from the very complex, confusing, incomplete and unco-ordinated nature of the existing regulatory framework (as set out in Chapter Three). However, such explanations may not fully account for the very high degree of variation, inconsistency, unawareness and uncertainty among respondents that were interviewed in this project.

These findings also expose a fundamental underlying gap between 'apparent' governance and actual practice. The 'law-on-the-books' and composition of the regulatory space, when mapped and viewed through a legal prism, reveal that a plethora of governance and guidance materials and actors currently exist. However, at least in the context of the interviews that we undertook, reported patterns in their actual penetration, awareness, impact, and use in practice paint a dramatically different picture. In places, there was a lack of awareness of, confusion over, or uncertainty about what 'laws' or other 'official' requirements and governance actors pertain to practice, and precisely what is required of practitioners. Of our respondents who reported awareness of the breadth of potentially relevant laws and guidance, some viewed this as a source of problems and/or unease, whereas others considered this to be relatively unproblematic, or just a normal part of their working context. The latter typically were people who worked in large-scale, multinational projects, and/or for multinational corporations. These observations raise issues about how far day-to-day biobanking and related activities in England and Wales actually comport with all of the legal, ethical and professional requirements governing them. Judging from our data analysis, there are good reasons to suspect that such requirements may not fully be being met in practice. This could expose individual practitioners, their organisations, biobanks, and those responsible for running biobanks to potential criminal or civil legal liability, as well as other disciplinary sanctions and deleterious consequences. Accordingly, it is a matter that should be taken very seriously in the design of any future governance frameworks.

Returning to our empirical findings, when asked about decision-making and direct experiences of 'oversight', respondents identified two broad categories: 'external' and 'internal' oversight mechanisms. Mirroring our findings on instruments and actors, comparatively few adverted to formal bodies. Instead, for public sector respondents, two external, quasi-formal or informal actors—RECs and funding bodies respectively—overwhelmingly dominated. Respondents from all sectors also referred to external technical or professional accreditation requirements, and some to licensing obligations. The HTA or HTAct came up several times in this context. But the data protection regime barely featured at all. Internally, respondents from all sectors noted institutional governance arrangements and/or supervisory actors within their own organisations as being very significant. Several mentioned their own project-specific steering or other committees as well. But no uniform pattern emerged as to the roles, composition, remits, powers or functions of such boards, and only a few respondents reported having consulted with them directly. Unlike public sector respondents, private sector respondents focused almost exclusively on internal, rather than external, oversight. The three main forms identified (filtering down from company seniors; in-house experts; SOPs and other standard-form documentation) reflected corporate hierarchies. These involved, and depended upon, clear divisions of labour, expertise and responsibility.

In terms of proactively seeking guidance, information or advice about governance, just under half of all respondents reported having taken some steps. In so doing, they employed a diverse and idiosyncratic range of methods and forms. Most were informal, ad hoc, and (to the extent that we are able to judge this on the basis of the interviews) somewhat mixed in terms of their actual or perceived success. These included attempts at directly engaging with laws, reports and/or other guidance documents; visiting or 'trawling' websites; attending meetings, conferences or professional training courses; and discovering and emulating what previous 'trailblazers' had done. Barely any respondents mentioned seeking advice (whether formally or informally) from 'official' regulators or other external governance bodies. Instead, the overwhelming focus, or preference, centred on informal methods. The most favoured method was contacting other individuals—especially colleagues, and people seen as having relevant expertise, knowledge or prior experience. Contact usually was initiated informally and on an ad hoc basis.

For many respondents, consulting with or talking to other individuals was their primary, if not only, reported source of guidance, information or advice. Typically, respondents spoke to people whom they already knew personally, or had come across through professional or social networks.

Why respondents heavily favoured informal, interpersonal, and potentially unreliable methods for seeking guidance—especially in relation to complex, technical legal requirements—is an open question. It could be speculated that it was because either they believed it was necessary so to do, it was the only plausible or accessible avenue available, or it was simply easier and quicker than trying to engage with regulatory bodies. As there was also evidence to suggest that some were uncertain over where else (or where better) to turn, or were unable to differentiate accurately between the relative authority, status or importance of different sources, this finding carries important implications that any future governance framework ought to address. Over half of respondents did not report having made any attempt at all to engage with sources of guidance, or to seek out information or advice about the governance requirements affecting them or their practice. Given the evidence of uncertainty and misunderstandings over the legal requirements (summarised above), once again, understanding more fully and then addressing the reasons for this lack of engagement is important in terms of future governance. This is especially so, insofar as it stems from respondents not knowing where (or where best) to look for authoritative guidance, feelings of being overwhelmed, confused or paralysed by the sheer raft of different options available and the fragmentation of informational resources, unwarranted feelings of complacency about their legal or regulatory compliance, or basic unawareness that relevant rules or requirements actually exist and apply to them.

## Implications

Taking all of our data and findings thus far together, and extrapolating and generalising from them, a number of crucial, overarching, foundational conclusions emerge. Each carries important implications for biobanking and biomedical governance.

At least some professionals in our sample were uncertain, confused, and, at times, mistaken about the actual 'laws', other forms of governance, guidance materials and actors relevant to their practice. This raises a host of questions about the provision and communication of important information. For example, do other practitioners know where (or where best) to look for accurate, up-to-date, pertinent, authoritative governance, guidance or advice about the requirements applicable to them? Do practitioners generally have the skills and support networks that they would require to engage with relevant governance?

Compounding this, we have interpreted our findings to suggest that practitioners may not characteristically regard regulatory bodies as being particularly relevant or helpful sources of governance or guidance. Rather, we have suggested that they may limit their interactions with all laws, guidelines and regulatory bodies to the minimum necessary. Typically, this involves engaging with, and/or seeking guidance from, regulatory bodies only when compelled so to do—and then, only to the extent required to navigate successfully over the immediate, unavoidable hurdle or step.[26] If our interpretation is correct, this raises questions whether practitioners routinely, voluntarily or proactively look to such bodies (whether external or internal, formal or informal) for information, advice or help. If not, why is this the case and is it problematic?

Thus, we conclude that the effective level, quality and nature of oversight and control actually meted out to biobanking activities at the present time in England and Wales cannot be judged accurately simply by looking at the 'laws' on the books—either now or potentially in the future—or at the host of actors apparently at work. Rather, the governance system depends very heavily indeed on 'informal' governance, guidance and oversight structures. Principal among these must be included: (1) the crucial 'gatekeeping' role played by RECs; (2) entry controls, policy preferences, conditions and other requirements imposed by leading biomedical research funding bodies; (3) internal, institutional oversight systems; (4) biobanks' or projects' own steering, scientific, ethical or other such committees or boards; (5) the widespread role played by individuals, informally, privately, and ad hoc, through the operation of professional and social networks; and (6) the de facto power of 'trailblazers' to become, by default, norm-setters or standard-setters, especially in the absence of clear rules or regulations on point.[27]

If these findings are considered in terms of upholding good governance principles such as transparency and accountability, as well as ensuring legitimacy and quality control, while some of this informal governance or guidance activity is visible, or, at least, partially so (such as REC approvals, and funders' calls, decisions, policies and conditions), much of it is ostensibly invisible or 'hidden'. It takes place in private and behind-the-scenes—informally, ad hoc, uncharted, and effectively unpoliced.[28] The extent to which this is actually

[26] Examples here would include securing prior ethical approval from RECs; obtaining grants and satisfying funders' requirements; passing accreditation standards body inspections; obtaining HTA licences and such like.

[27] Note that this list should be regarded as provisional rather than complete. As we shall discover in ch 6, two additional, 'implicit' elements also require inclusion. These are: (7) the pervasive influence of forms of 'implicit' guidance that shape practitioners' attitudes, and positively motivate or negatively constrain their behaviour; and (8) the concomitant power of those actors (institutions and individuals) involved in educating or training practitioners, through their capacity to control, dictate, inculcate, perpetuate, reinforce and/or modify key aspects of 'professional culture'.

[28] This is especially the case with 'implicit' forms of guidance, discussed in ch 6.

'problematic' in the context under discussion is itself a matter of debate which will be considered further in the Conclusion. Nonetheless, as this chapter has confirmed—and as will become even more evident in the following chapters—only certain aspects of governance activity emerge from even the most careful, thoroughgoing, tolerably well-informed and practice-sensitive legal mapping of the regulatory space. This latter point underscores the importance of conducting tailored empirical research, so that governance recommendations and reforms may be properly informed, workable, and realistic.

# 6

# General Attitudes to Governance

*Susan MC Gibbons and Andrew Smart*

IN THE PREVIOUS chapter, we explored respondents' awareness of the laws, other governance-related instruments and regulatory actors relevant to biobanks and biomedical research in England and Wales. We also began to explore their engagement with these multifarious sources of governance. As we saw, respondents identified a wide range of what we might loosely term explicit, or extant, guidance sources. In this chapter, we continue to report the sociological research findings, by turning our attention to consider in depth the respondents' more implicit attitudes towards the regulation and governance of biobanking, associated activities and professions—and how such attitudes shape their behaviour and practice. The sociological research design has been described in detail elsewhere.[1]

During interview, the 49 respondents were asked (by CH) a series of questions about whether current governance reflects the requirements of practitioners, managers, users and other stakeholders, and what governance-related issues or experiences they had encountered (good or bad). In response, many made revealing comments—including reflections on what had influenced their decisions about how (or whether) to accept laws or guidance, and how they had chosen to interpret, construct, engage with or enact laws or guidance for their own purposes.[2]

This chapter is the first in a series of five chapters, each focussing on selected aspects of respondents' attitudes towards governance. Accordingly, the present chapter should be read in conjunction with the following two chapters (as well as with Chapters Five and Nine). Chapters Six, Seven and Eight each address significant aspects of practitioners' attitudes towards governance. But because each presents, at best, merely a partial and incomplete picture of the rich spectrum of intersecting—and, at first glance, seemingly divergent—implicit influences and attitudes at play, it is necessary to defer a

[1] See chs 1 and 4 above.

[2] The picture is further illuminated in ch 9, where Catherine Heeney and Andrew Smart examine access to data as a case study of enactment of governance.

detailed overview and summation of the principal ramifications and lessons to be learned from our data, findings and analyses in each chapter until the concluding section of Chapter Eight. There, as we shall see, taken together, our research findings paint a complex, multifaceted but highly illuminating and instructive picture of the current state of biobanking governance, insofar as it is understood, perceived, experienced and enacted in practice.

As increasingly will become apparent that picture carries many important implications for biobanking governance. Not least, analysing our data on attitudes helps to reveal where existing laws, guidance instruments, actors, and processes—or the lack thereof (actual or perceived)—are causing the greatest concern for, or impact on, working practices. Discussion of these issues also helps to flush out further potential 'mismatches' between the existing regulatory regime, and the understandings, perceptions, experiences, concerns, and behaviour of professionals engaged in biobanking activities. Furthermore, they provide an insight into the ways in which professionals are likely to interact with, and construct, governance in the future.

This chapter is divided into two substantive sections, each of which presents key findings and data on a specific topic, followed by analysis and discussion. These mirror the data analysis process, whereby data was thematically coded as relating to 'Attitudes to Governance' and subdivided into the various sub-codes reported here.[3] Building on the foundations laid in Chapter Five—where, as noted above, we examined explicit, extant sources of governance relevant to respondents—we return to the broad topic of governance sources. Here, however, our focus is on internal, 'implicit' influences. Thus, the first section below identifies three forms of 'implicit' guidance, principally deriving from respondents' practical experiences and professional cultures, that we found to have shaped materially respondents' attitudes or behaviour. The subsequent section then identifies five further key sets of underlying beliefs, assumptions and attitudes that had a material influence over their approaches to governance and decision-making. The chapter concludes with a short summary of selected observations on our findings.

## 'IMPLICIT' GUIDANCE

### Key findings

Outside the realm of explicit governance or guidance sources (as discussed in the previous chapter), the data analysis uncovered three additional factors that influenced respondents' reflections about the governance landscape. These were: (1) their professional background(s); (2) practical

[3] The analysis this chapter was developed, refined, and discussed by SG and AS, building on the initial coding completed by CH.

experience; and (3) thinking about, or interacting with, other stakeholders or actors. These three influences were coded together as 'implicit guidance', as they were (to differing degrees) embedded and tacit aspects of respondents' thoughts and actions. Although we were able to separate these three categories as being distinct from one another, and such separation is convenient for analytical purposes, they are, perhaps, better viewed as overlapping and interpenetrating. We examine each in turn.

*Professional Background*

In all, 14 respondents made reference to their professional background(s) during interviews. The first important feature of these references was that they often occurred when issues about disciplinary or professional boundaries were raised—something that was not unusual in some projects due to their interdisciplinary aspects. For example, individuals may come to see themselves as 'representing' their profession within the context of a project. One Clinical Scientist who worked in a project using a data-sharing platform saw people's professional backgrounds as being 'a very big component and particularly for me because I'm on the actual development team, I'm the only clinician' (R11: Uni/Hos). Equally, professional norms could be thrown into sharp relief in interdisciplinary meetings. A University-based Bioinformatician involved in a large collaboration reported having had 'definitely a sort of ethical moment' with statistician colleagues, when what was being described as technologically possible clashed with what the respondent saw as being ethically acceptable:

> And we had an incredible meeting where you know people were saying well you know it can be done, we've got the technology and we were saying yes we know it can be done just we're not going to do it which is quite ... [laughs] (R30)

Furthermore, working at, or across, traditional disciplinary boundaries could encourage self-reflection about working practices, as illustrated by this IT Specialist who worked on a program of sample and data collection:

> [O]ne of the things we're doing that I think is fairly radical for medical research, it's not radical in healthcare and I think to some extent our background has brought this with us, if we come out of medical research there are things that I don't think we'd be doing. (R40: Repository)

When viewed relationally, professional background thus can influence judgments not only about appropriate research questions or methodologies, but also about appropriate norms and values, and the responsibilities that individuals may feel towards promoting these.

Another notable aspect of this 'positioning' vis-a-vis other professions was that it informed some respondents' judgments about the competence of other people with whom they worked. Closely aligned professions (or colleagues) could be judged favourably. As one Clinical Scientist who worked on a small-scale clinical study observed:

> We of course understood the ethical issues of genetics, we were working closely with geneticists who understood the right way of approaching this sort of issue at the time. (R16: Uni)

However, more commonly, respondents doubted whether people from different professional backgrounds would be so attuned to legal or ethical issues. For example:

> Well yeah I guess that probably I have thought more about that [than] the average molecular biologist because I come from a psychology background and have been kind of, you know have had, have been through these kind of arguments you know but I think the biologists are reasonably well aware of it [...] (R29: Bioinformatician, Repository)

> We are following every piece of legislation and guidance that we can find. And actually in doing that we have had to take with us our colleagues in this organisation who may not actually have done that. (R40: IT Specialist, Repository)

> And the other thing is I think people don't realise that if you've got ethics [ie, research ethics committee (REC) approval] that's what you've got ethics for, you can't add bits and pieces to it and change the project. (R32: Public Health, NHS)

In slightly different ways, the three respondents cited above all cast negative aspersions. In the first, they are against the 'average molecular biologist' who has not been trained to be well-versed in ethics; in the second and third, they are against colleagues from other professions who are perceived to be less inclined to follow laws or guidance, or who appear not to have understood them.

Moving on, a second important aspect of professional background was embodied in the expression of values. Some respondents posited values as a collective disposition; this was commonly but not exclusively those who worked at in NHS and University/Hospital settings (unless labelled otherwise in the following extracts). Thus, the only Genetic Counsellor in our sample alluded to there being 'a general sense' of 'the importance of informed consent' in 'clinical genetics [...] in the UK' (R33). Other respondents expressed values as an individual imperative—for example, 'it would be morally indefensible for me as a doctor...' (R31: Clinical Scientist). The importance of these professional values was that they were seen as a way of enabling or constraining people's activities so that they 'do the right thing'. As a Genetic Counsellor expressed it:

> I suppose you know at the moment we're kind of relying on the fact that people have used professional judgment or common sense. (R33)

In terms of understanding the ways in which values might influence how people act, the notion of 'principles' (which was mentioned by four respondents) could be useful. When questioned about relevant 'laws', one University/Hospital-based Clinical Scientist who worked on a small-scale clinical study replied, 'I don't refer to anything; I'm guided by clinical ethical principals' (R48). Reference to principles arose especially in relation to the process of taking abstract legal or ethical ideas, and applying them in practice. This included both weighing up competing values, and applying them to a given context. As this respondent explained:

> Well yeah you have to just look at the principles and decide what to do [...] I've tried to apply the principles. As regard the law I think the law is very much in this country obviously by case law and then it's this thing of applying, as I said one principle against the other. (R19: Public Health)

Interestingly, two respondents—a Clinical Scientist (R11) and a Repository-based Research Scientist (R25)—used the term 'etiquette'. The latter, who worked on a programme of sample and data collection that would be made widely available, was particularly keen to assert the notion of 'scientific etiquette' as a distinct, high-level expression of group values and norms: 'I'm using "etiquette" carefully because I think that's what it is.' (R25) This careful word selection suggests something more than professional values, or any formal or codified legalistic relationships. It seems to underline a notion of socially shared conventions for good or acceptable conduct.

### *Practical Experience*

For many respondents, such as this Epidemiologist who worked on a well-established longitudinal study, practical experience was a key source of implicit guidance:

> [S]ome of it will just, in a sense we've always done it that way, from, sometimes you make it up yourself, you get it from other people, so I guess contact with external people and sort of ideas and so on and also ideas from journals and also in discussions with staff. (R6: Uni)

This brief extract encapsulates how practical experience has two, analytically discernable but interlinked, aspects—experiences of working practice, and experiences through learning. We will consider these separately in more detail below.

Turning first to *working practice*, in explaining their legal or ethical practices, respondents used four types of justification based on practical experience. One was simply claims of having 'experience' through past activities. As one Epidemiologist on another well-established longitudinal study put it, 'what's had an influence on our practices in part is our vast experience of releasing data to the world' (R24: Uni). A University/Hospital-based

Clinical Scientist similarly highlighted the potentially evolving nature of past experience as an implicit guidance source:

> Well I suppose you know because of the experience that we've built up over the years I think that's determined the way were doing it at the moment. It may be that we will have to change as things go on and may be lessons are learnt. (R9)

A second justification based on practical experience, expressed by those working predominantly in University settings, rested on claims about existing or embedded practice. Thus, two Epidemiologists used phrases like 'our approach has always been' (R18), or 'we've always done it that way' (R6). Such notions, however, also applied to practices that were perceived to have developed over time:

> I think it's due to the way that we've managed projects ourselves here in the laboratory and I think that we've developed practices from that and but yeah, that's probably how it's evolved really. (R3: Research Scientist)

> [B]ecause clinical genetics clearly got worked out, it clearly worked out at the time how you should approach an issue like this [...]. So you know that sort of stuff is pretty straightforward. (R16: Clinical Scientist)

As the latter extract indicates, where practice is embedded, it can come to be seen as self-evidently correct, uncomplicated and unproblematic.

A third type of justification based on practical experience was linked to specific people. This could be related to their abilities—for instance, 'if someone is good at doing this' (R6: Epidemiologist, Uni). Alternatively, it could relate to individuals' unique, prior work experience, or familiarity with particular work tasks or contexts:

> [X] has been involved in tissue banking before he came here so he has that experience. I have quality systems experience so I was aware of HTA, tissues and cells directive, that type of thing. (R36: Bioinformatician, Repository)

> A lot of the people here have drawn on previous expertise because we have a lot of people who have come from NHS laboratories and have been through various CPA accreditations. So I think people have drawn on their own experience as well as using guidelines that are available to laboratories. (R3: Research Scientist, Uni)

As the second extract once again illustrates, personal experience could work in tandem with other, explicit, guidance sources.

The fourth justification based on practical experience (offered by two Clinical Scientists in University/Hospital settings) invoked appeals to common practice. Thus, as one observed (in relation to dismissing a REC's concerns), 'if what we were doing was illegal every clinician in this country who does genetics must be breaking the law, so it was clearly a nonsense' (R47). Tellingly, this respondent justified his implicit belief that his conduct was lawful by aligning his practice to that of the profession generally, *even before* he had taken any steps to confirm the actual legality or otherwise of his conduct by consulting a legal adviser. That the intersection between practical experience and notions of law and ethics is blurry was further demonstrated by this other Clinical Scientist, for whom tacit knowledge had built up from cumulative experience:

> I've been doing this half my life at least, more than that now yeah I kind of grew up with it and you know what's ethical and what's not. [...] Now, that might be written down in the Data Protection Act somewhere but I didn't read the Data Protection Act to learn that, I just know what I will and won't show you. (R48)

In summary, when respondents explained or justified their working practices on the grounds of 'practical experience', this could mean past activities, existing practice, personal abilities and familiarities, or what was perceived to be common practice.

The other discernable source of implicit guidance relating to practical experience was from '*learning*'. This included both learning from one's own experience and learning from colleagues (as, for example, noted in the extract from R6 at the start of this sub-section). In respect of the former, this Public Health practitioner reported having 'learnt a lot from' struggling to explain things to colleagues from different disciplinary backgrounds about better ways to make people understand things (R32: NHS). Another respondent working in the NHS had learned to follow established protocols, because otherwise:

> [Y]ou'll come up against a brick wall further down in the process so you'll have to go back and do it again. So if you didn't do it you would pretty soon learn that actually it's in your interest to do it properly rather than go back. (R13: Clinician).

In respect of learning from others' experiences to inform emergent practices, one Bioinformatician at a repository lodging data recalled drawing on 'a few anecdotes' of other projects where a certain policy stance had 'almost always broken down' to avoid making similar mistakes (R29). This use of anecdotal evidence shows that learning can happen in a more general sense, in relation to shared knowledge within professional or working communities.

### *Thinking About, or Interacting With, Other Stakeholders or Actors*

Thirdly, respondents mentioned a number of actors who seemed to figure implicitly in their thinking about governance. As Chapter Five demonstrated, respondents discussed a wide range of interactions, with many different actors, in which their governance or guidance influence was explicit. But merely thinking about other stakeholders or actors, or interacting with them (including some already mentioned), also contributed in 'implicit' ways to shaping or constraining respondents' attitudes and behaviour.

Numerous actors or stakeholders relevant in this context did not otherwise perform any direct or explicit governance role vis-a-vis respondents. Most prominent within this category were patients, research participants, and potential future participants. As one respondent at a Repository observed, 'every single participant we've got and every potential participant that we might have we feel potentially are scrutinising what we do' (R40: IT Specialist). Other respondents similarly felt themselves to be constrained by potential 'scrutiny' from various stakeholders. These included 'our participants and the *Daily Mail*... [and] other biobanks' (R40), 'public scrutiny' (R36: Bioinformatician, Repository), commercial or NHS research partners, academic journals and judges.

Peers and professional communities also featured prominently in the responses of those with ties to University settings. For the following two University/Hospital-based Clinical Scientists, forms of implicit influence here included 'a very intense sort of scrutiny' from external examiners of doctoral students, because they 'will see precisely what's going on' (R14), and 'peer review' of published scientific papers (R14; R17). Awareness of professional community membership also carried an important reputational dimension. For this Bioinformatician, this factor was paramount. It influenced his ethical decision-making

> [M]ore than anything you know you don't like looking an idiot in front of your peers and we would generally consider unethical behaviour as fairly idiotic [...] (R1: Uni)

Thinking about patients and participants also influenced certain respondents—most notably, those with close ties to clinical settings and those involved in longitudinal research—due to deeply held feelings of 'relationship' or 'responsibility' towards them.[4] For one Clinical Scientist, this included a sense of 'duty to the funders and the people who took part to make sure that we don't squander the resource'; a two-part 'bond of trust' (to protect individuals' privacy, and to use samples only as permitted); and a concomitant 'bond of trust' and 'duty' to ensure that data are used correctly and never in ways that might harm individuals' interests (R31: Uni/Hos). Another respondent in a University/Hospital setting articulated a strong, implicit sense of ethical duty towards patients and others, which was closely intertwined with professional background and professional culture:

> It's a concern that the people that I have the primary responsibility to are the populations in e-labs, the views of those patients we were talking about. In my mind my ethical duty as a public health physician and an academic is to do the greatest good for the greatest number with the work we do and protect the interests of all of those, including the vulnerable and the concerned. (R41: Research Scientist)

[4] This issue is discussed further in relation to access in ch 9 below.

In relation to research participants more generally, an Epidemiologist involved in a well-established longitudinal study expressed concern that widespread, mandatory data-sharing—in particular, transferring biobanks into the public domain—could lead to such resources becoming 'unstewarded' (R6: Uni).

Other actors appeared to influence respondents both explicitly and implicitly. This category included funding bodies, RECs and universities. In addition to the extract already noted above (R31), this Epidemiologist who worked on a well-established longitudinal study felt a sense of responsibility to funders because 'they've paid for a lot of these collections and will pay for collections in the future. And I think their needs and concerns need to be considered' (R24: Uni). The words of this Clinical Scientist, who worked on a small-scale clinical study in a University/Hospital setting, clearly demonstrated how multiple actors can influence decision-making, whether explicitly, implicitly or in both ways:

> And you know I guess it's from experience that we've learnt that this is at the moment the best way to do it to be able to satisfy our ethics committee, to satisfy the university and the Trust for sponsorship as well as satisfy patients to be able to get the consent and so on. And so you know there is no one major influence, it is experience together with influences of several different stakeholders if you like. (R9)

Lastly, interactions, generalised 'discussions', and meetings with colleagues or research partners also functioned as a source of implicit influence in respect of governance. This process was reflected in phrases used such as 'discussions with staff' (R6: Epidemiologist, Uni); 'we'd all sit down and discuss what we want to do next' and 'we'd be happy to chat with them about stuff' (R27: Research Scientist, Uni/Hos); and, 'we of course discussed that' (R16: Clinical Scientist, Uni). Instances of 'talking' happened in a variety of circumstances: they could be especially intensive during the initial establishment of governance regimes (R36: Research Scientist, Repository), or at times of significant changes in practice. In respect of the latter, one Epidemiologist explained how their research team had 'agonised over' making changes to a participant consent form (R6: Uni).

## ANALYSIS AND DISCUSSION

Reflecting on these findings, at least four valuable insights emerge. The first is just how pervasive and powerful are the three distinct forms of 'implicit' guidance that we identified—namely, professional background, practical experience, and sensitivity towards other stakeholders and actors. Without having undertaken detailed, qualitative empirical research, it is unlikely that their significance for understanding the dynamics of governance in this context would have been exposed or become evident.

Secondly, and following on from this, the role played by the various forms of implicit guidance could well carry important implications for designing effective, realistic biobanking governance frameworks. An especially noteworthy influence was what we would term 'professional culture'—or, perhaps more accurately, 'professional cultures' as we noted some variability between respondents' who had different professional backgrounds or training, or who came from differently constituted working communities. Taking professional cultures as a whole, key aspects of that emerged from our data analysis included professional, disciplinary, or working community values and principles; shared practice; common or cumulative experience or knowledge; and various socially shared assumptions, conventions, understandings or 'etiquette' about what is legally, ethically or scientifically acceptable. Many of the examples set out above demonstrate that aspects of 'professional cultures' (broadly construed) often were fundamental. They shaped respondents' attitudes, decision-making, behaviour, presuppositions, interactions with others, and their opinions about others (including positive or negative disciplinary 'positioning'). Aspects of professional cultures tended to be tacit and deeply embedded. Consequently, they operated largely without many of our respondents apparently even being consciously aware of them or of their influence, let alone reflecting on, or questioning, them.

Thirdly, the importance of professional cultures highlights the powerful 'implicit' governance role played by those actors that have power to dictate, shape, inculcate, foster, maintain, alter and/or perpetuate professional cultural norms and practices. This was alluded to in Chapter Three. There, Scott's traditional list of 'regulatory space' resources was supplemented with the power to control professional education, the setting of good practice norms and the inculcation of professional cultures.[5] In Chapter Three, these powers were attributed to prominent educational bodies and professional associations—such as the General Medical Council, Royal Colleges, Academy of Medical Sciences and the British Medical Association. Through directing both the initial and ongoing professional training and indoctrination of biobanking professionals, the issuing of guidance materials to their own community members and the framing of 'good practice' or 'best practice' standards, such bodies wield a profound implicit as well as explicit influence.

However, accumulated evidence from the foregoing analysis, coupled with our findings in Chapter Five, suggests that other actors, too, need to be added to the list. On the educational side, this should include universities, medical schools, hospitals, NHS Trusts and any other institutions, employers or organisations (including larger-scale biobanks themselves) that engage in, or are involved with, relevant scientific, biomedical,

[5] See ch 3 above.

technological or other training. Crucially, certain key individuals, too, must not be overlooked. As we saw in Chapter Five,[6] informal interactions with individuals often formed respondents' primary or preferred (if not their only) source of guidance, information or advice. Thus, key individual people with whom practitioners interact—and also, perhaps, those whom they look up to, or who are regarded as role-models or exemplars of good practice—also should be added to the list. Such people may include, for example, eminent members of relevant professions, senior staff within institutions, expert advisers, supervisors and other mentors.

Finally, it is illuminating to revisit the subtle but significant role played by awareness of other actors or stakeholders who figured implicitly in respondents' thinking about governance. What is particularly fascinating to draw out here is the duality of influences involved—which we might loosely term 'negative' and 'positive'. On the negative side, many respondents appeared to feel constrained by an often vague or generalised, but very deeply held, sense of being subject to 'scrutiny'. Here, the characteristic effect seemed to be to dissuade them from even contemplating engaging in any questionable practices, lest some unspecified dire consequence ensue. Most prominent here were thoughts about participants, potential participants and peers. Although usually unarticulated, implicitly feared consequences here seemed to relate to disapprobation or shaming, a loss of public trust and support for their work, being taken to task, and reputational damage. Conversely, on the more positive side, feelings of interconnectedness or relationship with others—especially participants—appeared to generate within some respondents an equally deeply held sense of duty, or 'bond of trust'. Those with more 'personal' ties to study participants, such as clinicians or those involved in some longitudinal studies, often implicitly saw themselves as protectors or champions of participants' interests. Among other things, this positively motivated them to use research materials as well (and as productively) as possible, and, on behalf of participants, to resist efforts by others to engage in potentially harmful or ethically questionable activities.[7]

Both of these strands of implicit influence, taken in conjunction with our findings in the previous chapter, underscore the tremendous importance of social networks, as conduits of governance and guidance, and the power and impact of social control mechanisms. As we have seen—and shall see demonstrated even further below and in Chapter Nine—in many areas *social* relationships are powerful, effective and relevant forces of governance and guidance.

[6] See esp ch 5 above.
[7] This is discussed more fully in ch 9 below.

BELIEFS AND ASSUMPTIONS

**Key Findings**

During interview, many respondents made revealing comments about how they perceived law, guidance and governance in relation to their practice. Often short, and sometimes 'throwaway', such attitudinal remarks exposed various 'mental benchmarks', or further implicit influences, that respondents brought to bear on the process of engaging with, interpreting and deciding how to enact (or not) laws and guidance into their practice. Collectively, we have coded these influences as 'beliefs and assumptions'. We canvass below five different kinds of beliefs or assumptions that had a material impact on respondents' attitudes and conduct.

*Already Knowing*

Many respondents believed that they 'already knew', or had a good 'feel', for what they ought to do, or what the law required of them, without needing to look at the law itself to find this out. This sense of 'knowing already' clearly informed their decisions about governance. Commonly, it was put forward by respondents from a variety of professional background across the range of settings (other than pharmaceutical companies) to explain why they had not sought to engage with relevant laws or guidance directly. As this respondent at a Repository lodging genomic data put it:

> No they're, I, I guess that [the laws are] not obstructing us in the sense that we almost never come into contact with them which is a sort of sign that actually you know that you know. (R29: Bioinformatician)

As we saw in the previous section (in relation to sources of 'implicit' guidance), in many cases, this implicit sense of 'already knowing' stemmed from embedded professional cultures or practices, or from respondents' prior experience and expertise. Three examples capture these dual influences well:

> I don't refer to anything; I'm guided by clinical ethical principals. I've been doing this half my life at least, more than that now yeah I kind of grew up with it and you know what's ethical and what's not. (R48: Clinical Scientist, Uni)

> What else shapes how we do things? Law, ethics, I think that it's about in terms of, so some of it will just, in a sense we've always done it that way. (R6: Epidemiologist, Uni)

> So I understand quite a bit about the COREC rules and so on and also at an international level and so on. So not real knowledge but experience handling contracts and so on and just have a good feeling what to look for. (R37: Research Scientist, Repository)

Indeed, the belief was so strong for several respondents that so long as they felt compliant they would hope that all was well. For example, as this Clinical Scientist, who worked on a small-scale clinical study in a University/Hospital setting expressed it:

> Yes I don't know so much about the law. [...] I don't know, I mean I hope that the things that people follow in here are all well and truly within the law. It doesn't feel as though you're doing something, you're kind of skirting on the edge of the law that you have to consciously think would I be breaking the law if I did this. (R12)

As this extract demonstrates, interpolated with the belief about 'knowing already' could be an unquestioned assumption that the law must embody what respondents themselves believed to be appropriate, ethical or correct. The following examples are apposite:

> Now that might be written down in the Data Protection Act somewhere but I didn't read the Data Protection Act to learn that, I just know what I will and won't show you. [...] It probably is written down somewhere isn't it. [...] I'm sure it is written down somewhere so the answer to your question would be yes but do I have it written down on a piece of paper that I refer to daily? No. (R48: Clinical Scientist, Uni/Hos)

> Now sharing data with third parties is clearly an issue, nobody wants a research database to be, well we simply don't want identifiable data in the public domain in any way at all [...] but I think we know we don't. I think and I'm not sure that it's unclear, I think that it would be contrary to pretty well all professional and medical ethics and law to make data, named data or identifiable data of people's medical or personal characteristics, publicly accessible in any way at all without their explicit consent. (R25: Research Scientist, Repository)

Here, the first respondent 'just knows' what they would reveal, and presumes that this corresponds to the Data Protection Act 1998 (DPA), while the second similarly thinks that 'all professional medical ethics and law' must reflect what practitioners already 'know' should not be disclosed.

Interestingly, however, the Bioinformatician quoted at the start of this sub-section was prompted during the interview to have second thoughts about the safety of relying on this belief and its associated assumption:

> So I mean our interpretation of what we do is that there's very little law that impinges on us when it comes down to it [...] and you know that may be a naivety and maybe that we're breaking all sorts of laws all the time and we don't bloody know. (R29: Repository)

Another Bioinformatician, who had established a web-based data-sharing platform, had 'got the Caldicott Guardian here to come, and talk[ed] to him, to make sure that he agrees with what I thought was correct' (R1: Uni). While this respondent had taken steps to verify the accuracy of his assumptions about the legal position, the phraseology used is revealing. Once again, the implicit presumption is that the respondent's expectations were correct. All that was anticipated from the Caldicott Guardian was vindication and confirmation of this 'fact'.

*Doing Good*

A second belief that frequently coloured respondents' attitudes towards governance involved an image of themselves as 'doing good'. Several contrasted this positive self-image with a negative perception that biomedical regulations are premised on a mistrust of research and researchers. Respondents—typically Research and Clinical Scientists—who evinced this dual belief system tended to see both governance requirements, and the activities of key regulatory bodies (especially RECs), as erecting unnecessary 'obstacles' (R25: Repository) or 'barriers' (R15: Uni/Hos; R16: Uni) that 'hamstring' research (R16) (see also Chapter Four).

Paraphrasing from the data, the implicit starting point could be characterised thus: all that many respondents wanted to do was 'good'—most notably, benefiting patients, society and public health, through conducting research and making scientific progress. They 'knew' what was right and ethical, and what best would serve and protect their participants. Their motives and intentions were honourable. Therefore, a common outlook was that they should be trusted to do their job, and be empowered to do it without unnecessary hinderance. Thus, as two particular respondents put it:

> I trust that you know what I think as individuals we're very caring and aware of what we're doing and we want to do the best job we possibly can. (R40: IT Specialist, Repository)

> The researchers are doing, that they want to do good, they want to find things out. You know we're not out there to try and you know some of these other committee things that you read now it almost looks as if they're trying to screen out paedophiles you know. (R17: Clinical Scientist, Uni/Hos)

As the second extract indicates, against a self-image of 'doing good', forms of scrutiny, regulation or governance could be dismissed as ridiculous or unnecessary infringements. This perception is doubly underlined by reference to the rhetorical counterpoint of 'paedophiles'—a contemporary archetype for 'doing bad'.

By contrast, the biomedical research governance framework was commonly characterised as coming from the opposite starting point—namely, a presupposition that scientists, doctors, researchers and so forth are potentially harmful (to participants and the public), and so potential abuses must be prevented. Consequently, several respondents perceived regulatory systems and actors as being generally hostile towards research, overly sensitised by recent scandals, unfairly doubting of scientists and their integrity, and as posing unnecessary, over-cautious and excessively protective barriers to progress. We shall expand on such criticisms in Chapter Seven, when we report respondents' attitudes towards specific laws (especially the Human Tissue Act 2004 (HTAct)) and regulatory bodies (especially RECs). For now, the following extracts are indicative:

> What I think needs to happen is that we need to sweep away a lot of these people, they're in the way, they're not helping, they're not making, they're not making the world better for patients, they're not conserving their data in any particular way. What they're doing is forming really a barnacle on the backside of progress I'm afraid you know and basically what they're doing is stopping people getting access to things. (R31: Clinical Scientist, Uni/Hos)

> So there is a profound danger of people trying, sometime trying to be helpful and making maters worse. We need to diminish bureaucratic barriers and we need to work on the premise of (a) collective confidentiality and (b) the premise of trust that the patients actually trust the doctors, quite rightly and that you know that there will very rarely be abuses of that trust. (R44: Research Scientist, Repository)

The latter respondent thus views formal ethical governance as acting as an intermediary, inserting itself between professionals and research subjects or patients, effectively infringing on a relationship that otherwise (in this respondent's opinion) 'rightly' could operate on the basis of trust.

Tellingly, the same respondent was so aggrieved by 'the unethical nature of ethical committees'—which 'only go after people doing research'—that his Repository project team had decided to side-step the REC system altogether when an issue arose over whether or not they needed fresh consent for an additional use of participants' biosamples:

> And it's just ludicrous you know. And eventually we decided to, we discussed it among ourselves and we said well basically an exemption is to do diagnostic testing on them. So we decided not to bother bothering them because it was pretty obvious [....] I mean ethically it's a non-issue and yet we've got this massive sort of you know bureaucratic system. (R44)

This example of explicit avoidance of the regulatory system may, in practice, have been prompted by a number of interlinked factors. These may have included legitimate reasons for not approaching a REC. Nevertheless, whether or not this response contained elements of bravado, it amply illustrates this respondents' general disdain for the REC system as it related to the collection of data and samples his team had established.

### *Interpretation*

A fascinating spectrum of beliefs, attitudes and approaches emerged when respondents spoke about situations where applicable laws or guidelines were not clear-cut, and how they had dealt with this.[8] One response

[8] In ch 9, interpretation and enactment processes are further considered, where 'access' is examined as a case study.

to interpretative leeway was to implement guidance to the best of their ability: 'you try and do the best' (R6: Epidemiologist, Uni). More commonly, approaches to interpretation appear somewhat self-serving in orientation. Below are examples from respondents who believed, or assumed, that it was acceptable to interpret the law or guidance in such a way as would allow them to do as they wanted. For many, this, in turn, rested on implicit assumptions (outlined above) that what they wanted to do was self-evidently or undeniably 'good', and that their motives, activities and judgments all were underpinned by sound moral reasoning. Thus, this Bioinformatician at a newly established Repository commented:

> There are, well these are guidelines and polices which are more often interpretative nature, ie it's up to you how you want to put them into practice [...] and we have done that, that's what [...] most of these groups or organisations do is to interpret to suit what you want to do rather than following it to the letter. (R20)

Other respondents employed the notion of 'stretching the rules'. For example, in relation to doctors routinely sharing confidential data without patients' knowledge or consent, a Clinical Scientist involved in a small-scale clinical study explained:

> [O]f course one probably stretches the rules a bit in practice. Often shares, I mean [....] But doctors do all the time, you know we share clinical information because it's actually part of education. You know we have to share information because I say 'What have you done?' Now I don't feel that I'm breaking any ethical boundaries or confidentiality by doing that, but technically you could say I am. (R16: Uni)

Interestingly, this supposedly 'technical' violation of ethical (not to mention legal) requirements was felt to be justified, both by reference to normal, routine professional practice and the resulting educational benefits (one form of 'good') thereby achieved.

Interpretative judgments also were made about when governance structures were needed, and when not. This is discussed further in relation to the issue of access (in Chapter Nine), but for the time being it can be noted that these judgments tended to rest on respondents' own beliefs, preferences or self-selected criteria that were 'obvious' to them. Interpretative judgments were evident, too, in decisions about how to apply more formal guidance. As this Clinical Scientist, explained:

> I think that presumably has to be a level of interpretation and that's why the, the guidance has got to be clear enough that you can know where to draw the line between what's acceptable and not acceptable in your particular situation. [...] there still has to be rules and everybody's jobs are different but you still have to have a single document from which you work and decide which bits are relevant to you and which bits are not relevant to you. (R28: Uni/Hos)

In this respondents' opinion, guidance is effective when it empowers practitioners to decide—for themselves—'which bits are relevant'.

*Reassuring Others*

There was evidence that some respondents saw the 'true' function of governance processes as being simply to reassure or persuade others—especially RECs, participants and the public—of the rectitude and legitimacy of what they intended to do. A Bioinformatician at a newly established Repository commented thus:

> [S]o a lot of what we spent our time doing was reassuring people that you know we understand what these issues are and yes I have thought about them and here's what we've put in place. (R36)

We shall return to this theme of the 'messaging' function of governance in Chapter Seven, where it emerges as a major rationale for many respondents positively desiring governance.[9] Underpinning this perception may be an implicit belief among some that practitioners themselves did not in fact need to be governed per se. Rather, aspects of governance might be followed simply as an exercise in placating other stakeholders' concerns.

There were also respondents (all Research or Clinical Scientists in the following extracts, unless labelled otherwise) who saw jumping through the regulatory 'hoops' as being worthwhile. It could 'prove' to the wider world that their projects were sound, that they could be trusted, and/or that they 'were complying' with applicable guidelines:

> Okay we have no problems at all about basically jumping through as many hoops as we have to, to be clinically [compliant]. I've no issues with that whatsoever, I think it's very important. (R47: Uni/Hos)

For such respondents, explaining things to RECs (R23: Pharma), participants or the public (R15: Uni/Hos; R29: Bioinformatician, Repository; R42: Uni/Hos), was seen as important in order to garner and maintain support for their activities.

Others, however, were less sanguine—especially where they saw governance interventions as unnecessary, wasteful and over-protective. The views of respondents who worked at different repositories were particularly strident, in that they believed some concerns were, in their contexts, based on 'imaginary' harms (R25) or ethical 'non-issue[s]' (R44). In the words of this Clinical Scientist involved in a programme of data collection:

> [I]t's sort of convincing them that, or convincing them is the wrong word but you know explaining that actually this sort of process or project functions within [...] what I perceive as quite a well regulated framework and we have very few, or we've negligible you know concerns or issues [...] actually anything of that sort tends to be people think there are issues rather than there actually being issues. (R42: Uni/Hos)

[9] See ch 7 below.

*Selective Compliance*

Finally, respondents in a range of settings (not including pharmaceutical companies) had engaged in forms of 'selective compliance'. Decisions about how far—if at all—to accept and comply with governance or guidance rested on various underlying beliefs or attitudes. For example, three respondents spoke of governing themselves. As was noted in the previous chapter, a respondent from a repository lodging data phrased it thus, 'I think we make our own laws' (R38: Bioinformatician). Three others stated that they 'aim' to follow laws, governing policies or guidelines (R20: Bioinformatician, Repository: R28: Clinical Scientist, Uni/Hos); or 'attempt to be reasonably compliant' when 'the law says you have to do something' (R6: Epidemiologist, Uni). But the majority adopted what we might characterise as a 'pick-and-mix' attitude towards compliance, based on their own views, preferences, judgments, or perceptions. This had led some, for example this Bioinformatician, to follow only selected parts of relevant guidance:

> [T]here is a European charter that we're kind of in which is [...] a whole lot of ethical guidelines for science and [...] technically I don't think we are in it but we're sort of in the bits that we believe make sense. (R29: Repository)

Likewise, another respondents' team had 'adopted effectively what was not quite but very close to Good Clinical Practice standards to conduct the study' (R31: Clinical Scientist, Uni/Hos).

By contrast, the 'pick-and-mix' approach had led many more respondents to reject laws or governance outright, and instead proceed simply as they saw fit. Typically, this occurred where respondents judged a law or regulatory actor to be misguided, undesirable, flawed or contrary to what they believed that they ought to be free to do. Thus, one Clinical Scientist rejected guidelines covering genetics because of personal opposition to the notion of 'genetic exceptionalism':[10]

> [T]here are guidelines on consent and confidentially from the Joint Committee on Medical Genetics but I'm not a great believer in [...] I don't think genetics is a special case. (R2: Uni/Hos)

Furthermore, when discussing biosample storage time limits, this respondent was aware that 'there is a guideline from the Royal College of Pathologists, I think it says 15 or 25 years but there's no intention in genetics to start throwing away samples' (R2). This selectiveness was also apparent in other contexts. A Repository-based Research Scientist involved in a programme of data and sample collection, while discussing practices in

[10] For further discussion of 'genetic exceptionalism', see ch 7.

relation to intellectual property requirements where research materials are patented, admitted:

> Some people ignore it but lots of people don't. They're forced not to, often against their will by the rules that have been put in place in universities. (R25)

Also, this Bioinformatician based at a repository lodging data revealed that he had dismissed the EU Databases Directive 96/9/EC, because it 'was such a cack-handed and incompetent piece of work [....] I didn't lose much sleep over it but there were people who did' (R29).

An equivalent selective approach was taken towards regulatory actors. We have seen above various examples of respondents side-stepping certain actors or governance processes, and justifying this by reference to underlying beliefs. In making judgments about which actors' guidance to implement, for numerous additional respondents the esteem or regard in which they held regulatory bodies (or individual members) was crucial. One University/Hospital-based Clinical Scientist, for example, justified ignoring guidance from the General Medical Council because it is composed of, 'you know self righteous pompous old farts' (R48). Another encapsulated the general approach and underlying attitude well:

> I think you just know from the weight of the body [...] you look at the people that are on the working party so you know what weight to give to that document. (R28)

## Analysis and Discussion

Reflecting on these findings, three points particularly stand out as shaping reported attitudes and approaches to governance. First, from our analysis it is possible to characterise (or perhaps somewhat caricature an 'ideal type'[11]) of a certain professional 'outlook'. We stress that this is certainly not to claim that all respondents explicitly endorsed or expressed all elements of this outlook. However, its underlying themes were commonplace. As an ideal-type, we can encapsulate the core set of implicit attitudes underpinning this outlook, as follows: (1) practitioners already 'know' best what to do; (2) they do not need to look to the law (or guidance) to find this out; (3) the law (or guidance) is presumed to mirror what they already 'know'; (4) practitioners are trying to do 'good'; (5) their motives, intentions, interpretations judgments about what is 'good', and judgments about what and when governance is (or is not) necessary, are trustworthy, accurate, reliable and honourable; (6) practitioners therefore should be trusted, and empowered to arrange and conduct their activities as they see fit, free from unwarranted interference or hindrance. Where such beliefs

[11] M Weber, *The Methodology of the Social Sciences*, trans and ed EA Shils and HA Finch (Free Press 1997 (1903–1917)) 88.

and assumptions were revealed by our data analysis, it appeared that they operated without respondents questioning or testing them (although this might relate to their being expressed in the particular context of the interview). Moreover, they appeared to impact on how respondents went about enacting (or not) governance.

Secondly, respondents tended to be confident in their own abilities to judge correctly what is and is not necessary, appropriate, desirable, legitimate, 'good' and so forth. This applied both in respect of their biobanking-related activities, and their views about the kinds of governance, standard-setting or oversight required more generally. Equally strong in many responses was a positive self-image. Those we have labelled as Clinical and Research Scientists in particular characteristically saw themselves as agents for 'good'.

Thirdly, both in relation to inculcating core beliefs and assumptions, and cultivating high levels of self-confidence and positive self-images, we can discern once again the role of disciplinary or professional backgrounds, training and experience. These aspects of 'professional cultures' appear to profoundly shape our respondents' attitudes, judgments and decisions about governance. This helps to confirm our findings and arguments set out above, and in Chapter Three, about the pivotal role played by 'professional cultures' as a powerful form of 'implicit' influence.

We certainly do not claim that all respondents were breaching laws or guidance in a regular or systematic fashion. Nevertheless, at the extremes, there were admissions that pointed to practitioners deliberately avoiding, if not violating, relevant laws or governance, whilst feeling themselves to be justified in so doing. This appeared to be because—rightly or wrongly[12]—they perceive those laws or governance sources to be unwelcome, unworkable, wasteful, unnecessary or unwarranted 'obstacles' or 'barriers' to 'doing good'. Less extremely, the data analysis shows many practitioners using other methods, devices or justifications to manipulate, subvert or 'work around' governance systems, so as to ensure that such systems do not prevent them (again, rightly or wrongly) from doing as they wish. Examples here include rule-stretching, forms of selective compliance, creative interpretation, appeals to the notion of 'doing good' and subjective opinions formed about the merits of certain laws or regulatory actors. Finally, the data analysis shows a substantial proportion of our respondents simply operating, day-to-day, on the basis of an unquestioned presumption that the law must surely embody whatever they regard as ethical, acceptable or correct. Accordingly, they need not investigate the law further, or engage with it directly.

[12] Given the complexities of the existing regulatory framework (as elaborated in chs 3 and 5) it is at least understandable that respondents perceived that current measures undesirably hindered legitimate activities in certain cases, or failed to achieve the desired end results—and, thus, felt justified in avoiding or manipulating the system is necessary to make things 'work'.

Yet—and as the nationwide organ retention scandals that precipitated the HTAct so graphically illustrated—it by no means follows that practitioners' beliefs or assumptions about what is appropriate, 'normal', 'good', correct, desirable, ethical and so forth are mirrored by the law. Legally speaking, not even 'best' practice standards necessarily suffice. Legislation may well—and, indeed, in the biomedical context often does—impose more stringent requirements than professional or ethical guidelines might stipulate. Equally, as we saw in Chapter Three, in the absence of legislation, English judges have reserved to themselves the final say over what standards are legally acceptable.[13] For the courts, best practice is persuasive only; it is not legally definitive. Thus, in reality, it is quite unsafe for professionals engaged in biobanking activities simply to presume that 'the law', ethics, normal practice, best practice, and/or whatever they 'know', 'feel' or believe to be correct, are mutually synonymous or coterminous.

From a governance perspective, policymakers, lawmakers and regulatory bodies who wish to govern biobanking as effectively as possible cannot afford to ignore the insights and realities exposed by these findings. If our findings are more widely generalisable, then the impact of implicit beliefs, assumptions, and related aspects of professional culture over how biobanking governance is enacted (or not) in day-to-day practice may well be substantial, pervasive, and widespread. Doubtless, these tacit cultures of practice are also an unavoidable aspect of working contexts, but are themselves open to influence and guided change. To be genuinely effective, then, any future regulatory reforms may need to take these important undercurrents and influences into account, and factor them into the design, implementation and management of the governance framework.

## CONCLUSIONS

This chapter has presented our empirical research findings focussing on respondents' attitudes, beliefs and concerns regarding governance and guidance on biobanks and related activities in England and Wales. The data analysis here uncovered three powerful sources of 'implicit' guidance, and five core sets of beliefs and assumptions, all of which materially influenced our respondents' decisions about how, and how far—if at all—to accept laws or guidance, and how they had chosen to interpret, construct, engage with or enact laws or guidance in practice for their own purposes.

As noted in the introduction to this chapter, in order to capture a fully informed, accurate picture of practitioners' complex spectrum of intersecting attitudes, their interplay, the impact that they have in practice, and the resulting implications for biobanking governance, it is appropriate (and

[13] See ch 3 above.

essential) to defer a detailed overview and discussion of our principled findings and their ramifications until the end of Chapter Eight. For now, however, three incidental observations are worth noting.

First, the dual, pervasive, underlying role played by 'implicit' guidance, and by a particular outlook characterised by various implicit beliefs and assumptions—both of which may derive from aspects of 'professional cultures'—complements and reinforces our findings in Chapter Five. As we argued there, the current governance system depends heavily on (explicit or extant) 'informal' governance, guidance and oversight structures. However, as this chapter has shown, various forms of 'implicit' governance, coupled with the outlook mapped out above, also play a critical role. Consequently, we must expand our provisional listing of the principal structures involved in governance—namely, the list that appeared in the concluding section of Chapter Five[14]—to include two additional elements. In full, the list should read as follows: (1) the crucial 'gatekeeping' role played by RECs; (2) entry controls, policy preferences, conditions and other requirements imposed by leading biomedical research funding bodies; (3) internal, institutional oversight systems; (4) biobanks' or projects' own steering, scientific, ethical or other such committees or boards; (5) the widespread role played by individuals, informally, privately and ad hoc, through the operation of professional and social networks; (6) the de facto power of 'trailblazers' to become, by default, norm-setters or standard-setters, especially in the absence of clear rules or regulations on point; (7) the pervasive influence of forms of 'implicit' guidance in shaping practitioners' attitudes, and positively motivating or negatively constraining their behaviour; and (8) the concomitant power of those actors (both institutions and individuals) involved in educating or training practitioners, through their capacity to control, dictate, inculcate, perpetuate, reinforce and/or modify key aspects of 'professional cultures'.

Secondly, this reliance on 'informal' structures, and the powerful influence of 'implicit' governance and an underlying outlook materially shaped by professional cultures and interactions with others, together demonstrate the enormous importance of *social* networks in the biobanking sphere. Not least, such factors expose dependence on the power and operation of social control mechanisms to ensure legally, ethically and scientifically appropriate behaviour. As we have seen, practitioners may not generally be fully aware of the formal legal and other requirements touching their practice. Hence, they cannot be guided by them. Accordingly, whether or not their practice comports with such requirements effectively depends upon how far informal and implicit sources of governance or guidance operate, de facto, to redress the balance. This underscores the very significant governance role played by such mechanisms.

[14] See above ch 5, text to n 22.

Practitioners' sensitivity to perceived 'scrutiny' from participants and peers, and the impact of professional cultural norms, are two obvious illustrations. Here, then we can see that *social* forces are powerful, effective, and relevant factors in understanding governance as it is enacted in practice.[15]

Thirdly, given the governance system's reliance on informal and implicit governance structures, it follows that, to a very large extent, effective biobanking governance depends on practitioners respecting, upholding, and voluntarily adhering to, key systems, processes, norms and actors. But what do practitioners really think about the state of biobanking governance? How do they regard key laws and actors? What lessons have they learned from their experiences with them? What criticisms do they have? Do they have other general or specific concerns about the law or governance? If so, what? In the next chapter, we turn to examine each of these questions in detail.

[15] This intersection is more fully examined in ch 9 below, in the context of examining access as a case study.

# 7

# Attitudes to Particular Laws and Governing Bodies

*Susan MC Gibbons and Andrew Smart*

IN THE PREVIOUS two chapters, we explored respondents' awareness of the laws, other governance-related instruments and regulatory actors relevant to biobanks and biomedical research in England and Wales, and began to unpack the characteristic attitudes, beliefs and assumptions that materially influence how they perceive and engage with governance in practice. In this second chapter in our trilogy of chapters looking at attitudes, we continue to report our key findings regarding respondents' attitudes towards governance. Here, we focus on their general and specific attitudes, concerns and criticisms voiced, either relating to specific laws and regulatory actors, or to biobanking governance more generally. As we shall see below, and in the concluding section of Chapter Eight, analysing these data helps significantly in identifying where existing laws, guidance instruments, actors and processes—or the lack thereof (actual or perceived)—are causing the greatest concern for, or impact on, working practices. It also highlights more potential mismatches between the governance system as it stood at the time of interview, and the experiences, concerns and behaviour of practitioners.

The sociological research design has been described in detail elsewhere.[1] As noted in the previous chapter, during interview, the 49 respondents were asked a series of questions about whether governance reflects the requirements of practitioners, biobank managers, biobank users and other stakeholders, and what governance-related issues or experiences they had encountered (good or bad). In response, many respondents exhibited particular attitudes towards, or expressed concerns about, different aspects of governance. These canvassed both general and (sometimes very) specific issues and governance measures.

Much of the analysis in this chapter (and, again, in the next) draw on data that were coded to one of three 'Guidance' codes ('laws', 'advisory bodies',

[1] See chs 1 and 4 above.

'guidance'), and/or to an 'Attitudes to Governance' code.[2] Both kinds of data highlight specific attitudes about particular governance-related issues. Some data were cross-coded to both categories. Accordingly, many of the examples presented below capture issues that respondents felt particularly strongly about. Reflecting our data analysis process, whereby data was thematically coded and subdivided into the various sub-codes reported here, the chapter is divided into three substantive sections. Each presents key findings and data on a specific sub-code, followed by analysis and discussion. The first section reports respondents' attitudes and concerns about specific laws—focusing on the Human Tissue Act 2004 (HTAct) and intellectual property rights (IPR). The second section examines their attitudes and concerns about specific regulatory bodies and research governance systems—focusing on research ethics committees (RECs), the NHS, and funding bodies. The final section then canvasses five general, and three specific areas of concern raised by respondents, which relate to biobanking governance more broadly. The chapter concludes with some short observations. A full summary and detailed analysis of our findings and their implications appears in the final section of Chapter Eight, where we integrate our principal findings taken from all three chapters in the present trilogy.

One important topic of concern not addressed in detail in this chapter is access to data and/or biosamples (especially by third parties). This recent policy innovation—variously known as 'data sharing', 'open access', 'public access' and 'open source'—emerged very prominently during interviews as a major source of concern for respondents. Because of this, it warrants deeper examination than space here permits. Accordingly, in Chapter Nine Catherine Heeney and Andrew Smart explore access issues, using the topic as a case study, both to explore how 'access' is enacted in certain working practices, and to uncover the motivations that underpinned respondents' decision-making about, and management of, data and biosamples.

## ATTITUDES TO SPECIFIC LAWS

As we saw in Chapter Five, when asked about the main 'laws', governance systems, and regulatory actors relevant to practice, respondents' answers were highly variable. Most governance sources were mentioned only once or twice; and then, often in fairly vague terms. Here, we examine two specific areas of law that did attract a significant number of wide-ranging comments and criticisms—namely, the HTAct, and intellectual property rights (IPR).

By far, the HTAct was the law most frequently mentioned by respondents. It was also the one that they discussed most directly, in greatest detail, and, in

[2] The analysis this chapter was developed, refined, and discussed by SG and AS, building on the initial coding completed by CH.

many instances, with some feeling. The considerable range of views expressed reveals some illuminating insights into their attitudes and experiences—both with these laws specifically, and with forms of legal governance more generally. Some respondents also shared views about the Data Protection Act 1998 (DPA). As we saw in Chapter Five, the DPA was the second most frequently identified formal 'law' after the HTAct. However, the DPA did not feature prominently in respondents' thinking about governance. Few explicit comments were made about it; those that were tended to be unspecific. A characteristic example was a Bioinformatician who had developed a web-based data-sharing platform, who, after 'digging through' the data protection law, reported finding it 'fairly complex' (R1: Uni). Several respondents expressed concern that the DPA and HTAct clash in some of their definitions—for example, 'what is meant by anonymity' being 'different' under the two Acts (R21: Clinical Scientist, Pharma). Overall, however, as the DPA did not feature as a forefront governance consideration in respondents' minds, we do not examine it in detail here.[3]

## Human Tissue Act 2004 (HTAct)

### *General Attitudes and Concerns*

In all, 13 respondents, largely Clinical or Research Scientists plus an Epidemiologist with close ties to clinical work, commented on the HTAct.[4] Only two expressed praise. One, who worked in a repository lodging genomic data suggested that it was 'bringing more clarity', had 'made things simpler', and was 'quite straightforward'—at least, for those who hold only DNA (R37).[5] The other, who was based in a University/Hospital setting but also worked with a repository, commented that 'it seems very rational', and has been 'incorporated into the working lives of those that it affects, quite reasonably actually' (R42). Two other respondents were essentially neutral. The above-mentioned Epidemiologist perceived the 'biggest change' in practitioners' views and practices, after 'the whole Alder Hey thing came about', as having been that people 'began to think much more carefully about what to do' regarding samples, and he supported the HTAct's focus on making

[3] Note, however, that we shall return to both the DPA and HTAct in the section below entitled 'Areas of Concern', where we consider respondents' specific concerns about the currently bifurcated systems governing biological samples and data, and their attitudes towards regulating data.

[4] Note that, in many interview excerpts in this chapter, respondents used the abbreviation 'HTA' to denote the Human Tissue Act 2004. We have used 'HTAct' in this book, to avoid any possible confusion with the Human Tissue Authority, which is customarily known as the 'HTA' (and which we have abbreviated as such).

[5] Later in this section we discuss respondents' concerns over the scope of materials covered by the HTAct, and its exclusion of genetic material (eg, extracted DNA).

consent the 'gold standard' (R46: Uni/Hos).[6] The other respondent, who was employed in the private sector but had extensive prior experience of working in NHS settings, expressed the 'hope' that 'the virtual paralysis of human base tissue research' since 'the so called organ scandals' would 'be reversed' by the HTAct (R21: Clinical Scientist).

Overwhelmingly, however, when respondents reflected on the HTAct and its implications for practice, their views were critical. Indicative general comments included that it was an 'added layer of complexity' (R25: Repository), 'not a wholly helpful piece of legislation' (R27: Uni/Hos), and a 'potential nightmare' (R12: Uni/Hos). Respondents who were overtly negative blamed the HTAct for having triggered an unfortunate culture change, leading to harmful shifts in people's practices, based on excessive caution and fear. One, for example, claimed that 'all of the anxiety leading up to it, post Alder Hey' had made tissue samples much harder to obtain (R12: Uni/Hos). Another saw the HTAct as akin to punishment, arguing that current structures of collaboration and exchange now had become unworkable, all because 'there were a couple of criminals who illegally obtained and kept tissue and now the rest of us have to pay the cost of that to the detriment of the system' (R27: Uni/Hos).

Like these two respondents, several others whose research was predominantly small-scale clinical studies and conducted in University/Hospital settings exhibited strongly negative feelings, or had a range of concerns, about the HTAct. The cost of obtaining licences was a marked concern (as discussed in the section below entitled 'Areas of Concern', in conjunction with other resource-related concerns). The majority, however, raised more specific, isolated problems or difficulties. Issues mentioned by lone interviewees included: a lack of initial consultation, disproportionality, and confusing licensing criteria (R27); a lack of clarity over audits, and quandaries over when best to obtain an HTAct licence (R45); and difficulty choosing a 'designated individual' (R9). In relation to available guidance on the HTAct, this University/Hospital-based Epidemiologist, complained that '[t]he information is quite fragmented and it's presented in huge reports that we have to really go through to pick out, try and find your specific scenario' (R46). Two more widespread areas of critique concerned the materials covered (or not) by the HTAct, and the HTAct's implications for post-mortem samples. It is illuminating to look at each in turn.

### *Scope of Relevant Materials*

Without doubt, the single piece of legal knowledge most firmly understood and conveyed by respondents during interview was the fact that DNA is not

[6] On the backdrop to the HTAct, particularly the nationwide hospital organ retention scandals, see, eg: D Price, 'The Human Tissue Act 2004' (2005) 68 *Modern Law Review* 798; K Liddell and A Hall, 'Beyond Bristol and Alder Hey: The Future Regulation of Human Tissue' (2005) 13 *Medical Law Review* 170.

covered by the HTAct's licensing requirements. For example, as this Research Scientist at a repository holding genomic data put it: 'as long as you're handling DNA samples [...] you don't need licence and it's quite straightforward' (R37). Yet, the scope of biological materials covered—or excluded—by the HTAct was a key matter of concern (and sometimes confusion) for many. Two principal, interrelated issues emerged: (1) inconsistency, irrationality, and gaps, particularly given the exclusion of DNA; and (2) the lack of differentiation between 'different' materials covered by the HTAct.

Before presenting our findings on each issue, it is worth pausing to outline the legal position. With one limited exception,[7] the HTAct's core consent and licensing provisions cover only human tissue that falls within the statutory definition of 'relevant material', as that term is elucidated, interpreted and applied by the Human Tissue Authority.[8] The crux of the definition is that 'relevant material' encompasses only material 'which consists of or includes human cells'.[9] Some materials are explicitly excluded—namely, gametes (sperm and ova) and embryos outside the body; [10] hair and nail from living persons;[11] and cell-lines and any other human materials created outside the body.[12] Genetic materials, such as extracted DNA and plasma-extracted DNA, are indirectly excluded, by virtue of the fact that they are sub-cellular or acellular in nature.[13] Thus, they do not consist of, or include, any human cells. When discussing this topic during interview, no respondent used the statutory term, 'relevant material'. Instead, comments were (perhaps unsurprisingly) couched in non-legalistic or scientific terms. Commonly, respondents referred to very precise tissue or sample types.

Turning to the first of the two principal issues to emerge from the data, numerous respondents criticised perceived inconsistencies, irrationality or gaps in the scope of 'relevant material' covered by the HTAct. As noted above, these were largely Clinical or Research Scientists and often working in University/Hospital settings (extracts in the remainder of this sub-section and the next are only labelled where they were otherwise). Typical comments included describing it as 'bizarre' (R15), 'silly' (R27) and, 'really silly' and not 'logical' (R12). Central to such criticisms was concern over the HTAct's blanket exclusion of genetic material—especially DNA. Thus, one respondent thought it 'seemed slightly bizarre' to include 'specimens from which you might obtain DNA', yet exclude the DNA (R15). Another suggested that

7 The HTAct, s 45 offence relating to non-consensual analysis of DNA rests on a definition of 'bodily material' (s 45(5)) that is broader than the definition of 'relevant material'.

8 HTAct, s 15.

9 HTAct, s 53.

10 HTAct, s 53(2)(a).

11 HTAct, s 53(2)(b).

12 HTAct, 54(7).

13 In animals and plants, the DNA polymer is located inside the cell nucleus, which is a sub-component of the cell.

'DNA based information' (or 'the data') should be handled 'the same way' as 'biological material' because the two are not 'qualitatively that different from each other' (R12). To another, it seemed 'sort of [...] not counterintuitive really but you would, you might expect that DNA would fall under' the HTAct (R42). A fourth suggested that epigenetics, genetics and 'free DNA serum' all should have been included (R45). A fifth criticised the HTAct for not making clear whether or not RNA samples are included, given that DNA is excluded (R37: Repository). Meanwhile, a sixth thought it 'silly' that 'some samples are differently treated than others', especially when '[i]t's all potentially information isn't it?' (R27).

For the latter Research Scientist, who worked on a small-scale clinical study, one major concern was the detrimental impact on patients' relatives of the HTAct having made it *more* difficult for them to get permission to access tissues, as opposed to DNA (in terms of obtaining consent):

> [S]o tissue and DNA is treated different, dissimilarly, I think that's the problem. We can do tests on tissue which give more genetic information than a test on DNA and yet the two are regarded dissimilarly. And there's much greater controls over tissue than there is over DNA okay. (R27)

Whereas most respondents seemed to think that DNA should be regulated at least as stringently as cellular tissues, if not more so,[14] on practical, clinical grounds this respondent implicitly espoused the opposite view (by stating that 'they can do tests on tissue which give more genetic information' (R27)).

Two further respondents also raised practical concerns. These related to the HTAct's 'ambiguous' (R42) position on DNA and the storage licence exception. In terms of the formal legal position, under Regulations made pursuant to the HTAct, relevant material taken from a living person may be held for up to 48 hours without a licence, where the person storing it is intending to use it for transplantation, and the storage is for less than 48 hours.[15] Both respondents were worried about inadvertently holding tissue samples without a licence if, for logistical reasons, they could not extract DNA within the 48 hour statutory time limit allowed before a storage licence must be obtained—for example, where a sample 'gets to us on a Friday, it's a bank holiday on Monday, we're not able to extract the DNA until Wednesday' (R27). Significantly, however, their comments appear to evidence a more widespread confusion among practitioners, regarding the actual scope and applicability of the 48-hour legal window for holding relevant material without a licence. According to the Regulations, that window appears to apply only to relevant material held pending transplantation—not

[14] Respondents' conflicting views on 'genetic exceptionalism' are discussed below.

[15] Human Tissue Act 2004 (Ethical Approval, Exceptions from Licensing and Supply of Information about Transplants) Regulations 2006 (SI 2006/1260), reg 3(3)(b).

to material taken and held for any other purpose.[16] By contrast, anecdotal evidence (including from the two respondents cited here) suggests a belief in practice that the exception applies to all relevant material, across the board. In this respect, the correct legal position is, indeed, 'ambiguous'.

Having revealed a significant number of concerns about the nature and application of HTAct, it is interesting to note that three respondents speculated about the possible rationale(s)—or the lack thereof—for aspects of the HTAct's ambit. One was utterly at a loss to comprehend why, 'quite bizarrely the Human Tissue Act doesn't seem to extend to cell cultures for some reason I don't understand' (R15). The 'impression or [...] interpretation' of another was that it was 'just a pragmatic decision' on the part of 'whoever was dealing with the logistics of administering' the legislation to exclude DNA (R42). This had two presumed aspects: first, that there was 'probably far too much out there to consider including DNA at the same time'; and secondly, that the HTAct's principal focus was on 'pathological and diagnostic samples rather than research samples per se' (R42). Meanwhile, in relation to the 48-hour storage licence exemption window (which, initially, was only going to be 24 hours), a third respondent attributed its doubling in length during the legislative process to 'some politician deciding they didn't want bad press because it was impinging on transplantation so they gave them a 48-hour window' (R27). Notably, this suggests a perception that the law-making process involves intersections between various actors—here, a discernable biomedical group (a transplantation 'lobby'), politicians and the media.

The second principal set of critiques relating to the scope of 'relevant material' was rooted in doubts over how the HTAct treats 'different' kinds, and/or amounts, of relevant material as equivalent. For some respondents, implicated here was a notion that different emotions may attach to different tissues or samples. For example, a 'problem' identified by one Clinical Scientist who worked on a small-scale clinical study was the fact that:

> [L]ittle microscopic slides are treated the same as a baby's brain or something they're not the same emotively but they're going down the line of saying no you might have to have the same licence and ethics approval for you know samples of urine or whatever for goodness sake. (R12)

This extract—typical of many—is indicative of a strongly held belief that there are significant qualitative and quantitative differences between 'different' kinds of tissue, that should be (but currently are not) reflected in the law. The extract set out immediately above alludes to three possible parameters of 'difference': (1) sample type (whole organ; urine; tissue slide); (2) ontological status (a deceased child's brain vs bodily waste); and (3) quantity (little microscopic slide vs whole organ). Elsewhere during interview, the same respondent likewise differentiated between

[16] See text to n 15 above.

'whole organs or anything very emotive' on the one hand, and 'blood samples, DNA samples, microscope slides of people's heart, bits of heart' on the other (R12). Two further respondents also raised concerns that implicated size or quantity—although, in slightly different ways. One, who was also involved in small-scale clinical research, was fearful of violating the HTAct, despite 'the fact that it's a tiny little lump of tissue' (R27). The second, who worked on larger research studies, criticised the HTAct's exclusion of 'big databases, big serum banks' that hold only genetic material (R45).

### *Post-mortem Tissues*

In respect of post-mortem tissues (that is, tissues taken from the bodies of people after they have died), two principal concerns emerged. First, three respondents working in the context of small-scale clinical studies blamed the HTAct for making post-mortem samples far more difficult for researchers to obtain. Different reasons were put forward to account for this. One suggested that it was 'getting harder all the time' and 'really quite a problem', because '[p]eople are not archiving samples, they're not keeping tissue blocks' any more (R12). This respondent attributed the problem to a culture of excessive caution in the wake of the HTAct, 'and all of the anxiety leading up to it, post Alder Hey' (R12). Another suggested that 'we simply won't be taking brain out of a hospital whereas we might have done before', because the respondent's university department had decided not to get an HTAct licence, due to the 'huge amount of bureaucracy' involved (R14). Meantime, it was also argued that, 'it's tissue from the dead that gets tricky', due to complexities over rights, responsibilities and boundaries of different professionals, the need for a licence, and fear of breaching the HTAct by retaining tissue for too long without a storage licence (R27).

The second HTAct concern specific to post-mortem tissue concerned obtaining consent—particularly the practicalities and timing of obtaining consent from family members in the immediate aftermath of death. For example, one of the above-mentioned respondents involved in small-scale clinical studies felt that the HTAct had created a 'tricky' ethical dilemma, because 'the only practical way to do it would be to take the sample and then go and get consent from the family' (R12). Otherwise, by the time consent had been obtained, scientifically speaking 'it's too late' to obtain a useful sample (R12). Yet, under the HTAct, consent must be obtained first.

These specific concerns pertaining to post-mortem tissues are illuminating because they highlight the ways in which new legal frameworks present everyday practical and ethical quandaries for some practitioners—in these instances largely Clinical or Research Scientists, with particular concerns expressed by those involved in small-scale clinical studies. It is a matter of debate whether the HTAct is actually the root cause of these problems, and whether the regulatory burden is—in actuality or normatively speaking—excessive.

Nevertheless, these expressions of concern do highlight that such perceptions exist among our sample of respondents.

**Intellectual Property Rights (IPR)**

As we saw in Chapter Five, several respondents each variously referred to a range of different aspects of IPR laws and practices. The patents system and material transfer agreements (MTAs) featured most prominently. Only two, Bioinformaticians both employed at a repository holding genomic data, gave positive accounts of patenting. One advocated patents as a 'well developed mechanism for giving protection and there's a safe patent in the public domain which is a good thing' (R29). The other recognised the importance to scientists of having the means to protect 'rights in terms of credit' (R39). However—and like the majority of respondents who spoke about IPR mechanisms—the latter respondent went on roundly to condemn the patents approach. Four specific problems were noted: the legal and other 'transaction costs'; the need 'to negotiate everything'; the fact that, 'if you've got to negotiate, you've no idea how long it's going to take, you've no idea how much it's going to cost you and that produces uncertainty'; and the fact that patents sometimes can be used as a 'defensive mechanism' simply to exclude others, thereby hampering research and delaying scientific progress (R39).

For another respondent at a different repository, a Research Scientist involved in a programme of sample and data collection, the 'massive lack of clarity still [...] because of the intellectual property arguments' was seen as engendering 'inertia' and fear for researchers, who

> find that there's some company that claims to have patented a whole chunk of stuff which is probably garbage but the reality is that unless you know it's garbage you feel that you can't go and work there in case you get taken to the cleaners. (R25)

The same respondent also painted a picture of researchers who

> want to work in an area [...] and they're suddenly presented with a forest of legal agreements and obstacles and I think people are genuinely discouraged sometimes. (R25)

Both for this respondent and several others, their strong perception was that the current IPR framework detrimentally impacts upon practice. As just illustrated, one concern was that patents dissuade people from embarking on research. Another Research Scientist based at a repository holding genomic data described IPR as a 'major problem' and something that 'could have blocked things and we try to keep it away' (R37)—notably, by avoiding altogether working with potential collaborators who hold patents. In this respondent's experience, 'once [intellectual property] gets in then things always get more complicated' (R37). Accordingly, negotiating the wording in the agreements 'could take a very long time... [and] delay things substantially' (R37). It was thus better for this respondent simply to avoid it if possible.

The repository-based Bioinformatician who identified MTAs as a relevant form of governance also noted that they were 'still a lot of hassle'—especially in contrast to data-sharing agreements (R39). This was due to 'the hassle that gets created from different MTAs for materials that could also be similarly standardised, it would put some lawyers out of work but [...] I don't really care about [that]' (R39). Interestingly, the need for 'legal' advice, and need to involve 'the lawyers', were dual aspects of dealing with IPR that several respondents commented negatively upon. The general attitude that emerged from our data was that, once the law or 'the lawyers' had to be involved, or the wording of agreements or 'paragraphs' had to be negotiated legally, research processes slowed down 'substantially', and became far more protracted, problematic and complicated (R37).

In relation to IPR, then, it is noteworthy that the concerns relate specifically to the formal, legalistic boundaries on sharing materials—and that legal interventions in ownership are interlaced with issues about routine practice, scientific progress, ethics and professional careers (which we explore further in Chapter Nine below).

## Analysis and Discussion

From our data analysis, we can draw out five key insights about respondents' attitudes towards specific laws, the legal framework and 'legal' governance measures more generally. First, few specific laws were discussed in any detail. This accords with our findings about awareness in Chapter Five, as does the clear predominance of the HTAct. By comparison, the absence of the DPA as a forefront consideration—despite respondents generally being well aware of its existence—is a notable omission.

A second striking feature is the fact that the vast majority of comments made about specific laws were negative—sometimes, quite vehemently so. This was in spite of respondents being prompted in the interview to reflect on both what was good as well as bad. Interestingly, too, as several of the HTAct and IPR comments illustrate, our analysis shows that concerns were aired not only by respondents who were overtly critical of particular laws, but also by those who broadly supported them or their underlying goals.

Thirdly, respondents' attitudes about specific laws were the product of a range of factors or influences. Primarily, they seemed to be rooted in, or shaped by, respondents' own direct, practical experiences of trying (successfully or otherwise) to implement those laws, or anecdotal knowledge (or impressions) of colleagues' experiences. This helps us, perhaps, to explain why only two respondents praised the HTAct, whereas others who commented were critical of it (and, likewise, the IPR regime). In relation to the HTAct, key criticisms rooted in negative practical experiences (usually of Clinical Scientists involved in small-scale clinical studies) included tissue samples (especially post-mortem tissues) being much harder to obtain; costs, bureaucracy, and uncertainty relating to licences; confusion over whether certain tissues constitute 'relevant material' or not; and confusion and anxiety over the 48-hour storage licence exception window. In relation to IPR, key criticisms deriving from practical experiences (usually of Bioinformaticians based in Repositories) with the patents system included major bureaucratic 'hassles' and 'transaction costs' (the need for negotiation; involvement of lawyers; resulting costs, delays and complexity), and a significant perceived lack of clarity over the legal requirements. This led to uncertainty, unpredictability and a fear of getting 'taken to the cleaners'. In relation to both the HTAct and IPR, then, where respondents aired specific views, the law appeared as a 'barrier' to getting on with doing the science, or emerged as a source of anxiety. Only rarely did it surface as something that facilitated practice, or made respondents feel secure.

Normative or principled reflection also played a role in shaping some respondents' attitudes towards specific laws. In respect of the HTAct, such attitudes included widespread perplexity, incredulousness and/or condemnation over the scope of materials covered and excluded; the overly cautious, fearful climate engendered by the HTAct (commonly attributed to its origins); and certain presumed rationale(s) thought to underpin aspects of the legislation. In relation to IPR, very noticeable was a strong undercurrent of hostility towards patenting in principle. Thus, some respondents saw patents as a purely 'defensive' mechanism; a spurious, undesirable device used to 'block' their access to materials; as stymieing potentially beneficial research; and as probably 'garbage' in any event. Clearly, the IPR system did not command an appreciable degree of trust, faith or support among our respondents; although, it may be pertinent that the bulk of our respondents were working for the public sector. Indeed, overall, our data drawing on both sorts of factors show that a majority of respondents did not believe that either the HTAct or the IPR regimes had been (or can be) integrated into their own practice—or biobanking practice more generally—with a tremendous degree of success.

These observations interlink with a fourth key insight—namely, the very great extent to which many respondents either had found, or firmly believed, that specific laws had had a concrete, detrimental impact on practice. In addition to problems reported with the HTAct, it is noteworthy that two respondents reported actively having rejected or avoided potential research projects or collaborators, simply to avoid IPR. That whole legal regime, and the consequences of involving 'the lawyers', both were viewed in a particularly unfavourable light. Overall, both the HTAct and IPR system were seen by some as 'obstacles' or 'hassles' that hamper, rather than facilitate, biobanking-related activities. (We shall return to this theme again below, where we note further expressions of concern about governance measures 'stifling' research in various ways.)

Fifthly, it is illuminating and important to take proper stock of respondents' substantive criticisms of the HTAct. The exclusion of genetic material from the HTAct's core licensing (and consent) provisions was well known by respondents—particularly as it had the effect of relieving many of them from the need to comply with those parts of the HTAct. Nevertheless, the overwhelming majority were still mystified by, and highly critical of, the HTAct's ambit. One major target for criticism was the blanket exclusion of DNA from the scope of 'relevant material'. It is notable that respondents should so strongly have criticised the exclusion of DNA. Including DNA within the HTAct would have meant that many of them would have been subject to the legislation, with its additional workload (to the extent that they would have had to actively address the practical issues of compliance). These comments underline that their grounds for making criticisms of governance extend beyond pragmatic concerns about workload. Another major substantive criticism worthy of note is the HTAct's uniform treatment of what many respondents perceived to be qualitatively and/or quantitatively 'different' kinds of tissue, all of which currently fall within the scope of 'relevant material' and are treated alike.

In taking stock of these critiques, helpfully, our data analysis suggest at least some underlying reasons for them. In relation to DNA, for example, two respondents pointed to the equivalent 'informational' potential of both DNA and biological materials as a reason for handling both in the same way. In relation to the HTAct failing to differentiate between 'different' kinds of tissue, the data analysis has uncovered at least three potentially significant grounds which many respondents felt the law should embrace and reflect—namely, (1) tissue type; (2) ontological status or 'emotiveness'; and (3) amount or size. How far such factors actually ought to be accorded legal significance is, of course, a matter for policymakers and lawmakers to decide. However, based on our findings, it is a question which at least some practitioners in certain contexts obviously feel strongly about. Thus, it is an issue which should be given serious thought in future considerations of governance.

Overall, our analyses of respondents' attitudes towards specific laws offer some invaluable insights—and carry potentially important implications—for the future governance of biobanking and related activities. Not least, they highlight a number of key areas where practitioners claim to be struggling with existing laws, and where better, clearer, or more effective, governance or guidance may well be beneficial. The data analysis also exposed several areas in which two major, existing legal frameworks fail to align with practitioners' perceptions, experiences and understandings, and thus reportedly conflict with day-to-day logistical realities of biobanking practice. In each case, policymakers, lawmakers, and relevant regulatory actors may well wish to pay careful attention to the findings, investigate further, and, if warranted, address the various problems, concerns and mismatches identified.

## ATTITUDES TO REGULATORY BODIES

This section examines respondents' attitudes towards specific regulatory bodies and associated biomedical research governance systems. Consistent with our findings in Chapter Five, the bodies most frequently discussed by respondents were research ethics committees (RECs) of various kinds (including COREC[17]), NHS actors, systems or staff, and leading research funding bodies. Before reporting on each, one methodological point should be noted. As the interviews progressed, the very significant role of both RECs and research funding bodies became more apparent to us. For this reason, in later interviews, respondents were prompted on RECs and/or funders, if they had not mentioned them already themselves. As a consequence, the data that follows blend opinions that were offered spontaneously and views that were solicited more overtly.

### Research Ethics Committees (RECs)

Well over one-third of our respondents expressed views about RECs; these were largely Clinical and Research Scientists working in various settings (unless the extracts in this section are labelled otherwise, they are from Clinical or Research Scientists). Their attitudes ranged from broadly positive to extremely negative. However, virtually no positive comments stood alone. Almost all either were accompanied by, or couched within the context of, much more extensive catalogues of complaints. Thus, some respondents were torn between the two positions.

Turning first to broadly positive comments, numerous respondents gave muted praise to RECs. Two felt that RECs were 'usually reasonably good' (R23: Pharma); while a third recalled that, on one occasion, a REC had made a constructive suggestion that had led to 'quite a good change' to a participant consent form (R32: Public Health, NHS). Two Clinical Scientists working in University/Hospital settings acknowledged that systemic levels of bureaucracy had 'got better' (R31); that processes were now 'quite streamlined and standardised' (R12); or described RECs as helpful. But such instances of praise were tempered markedly by associated critiques. As two typical responses were framed:

> [T]hey tell me that LREC forms are much easier than they used to be and that is actually true but they are still insane really compared to the level of damage that they're supposed to be trying to avoid. (R25: Repository)

[17] The Central Office for Research Ethics Committees (COREC), now the National Research Ethics Service (NRES).

> [Y]ou know ethical oversight is absolute necessity but the way that LREC perform and continue to perform is just dreadful. (R6: Epidemiologist, Uni)

Several respondents, again predominantly Clinical Scientists who worked in University/Hospital settings, commented favourably on situations where they had had the opportunity to discuss things directly, or engage more fully, with RECs. One, for instance, felt 'very fortunate' in having spent many years interacting with a single REC, which had been 'involved at the outset in the planning and the consultation and so they had a very good understanding of what the project was about in the local area, the setting of the decisions we were trying to make and the aims and objectives' (R42). Regretfully, however, that 'close link' had been lost when the REC system was reformed,[18] meaning that the project now had to deal with several different RECs, without any ongoing working relationship or established knowledge base. Whole other respondents echoed this view, one contrasted the greater levels of engagement and mutual understanding possible with internal institutional review bodies, against 'the anonymity of it and the huge numbers that are involved with the RECs', where there is also no 'feeling of ownership' or motivation on the part of RECs to find out more about projects (R17).

Turning to the negative attitudes expressed, here the sheer range of topics covered, and the depth of feeling often conveyed, were very marked. In summary, despite recent reforms, the REC system still was widely thought to be overly bureaucratised, time-consuming, costly, duplicative, wasteful, increasingly stringent, overly protective, distant, out-of-touch, remote from communities and researchers, lacking in adequate knowledge and expertise, inconsistent and often ineffective. This raft of criticisms can be divided into three broad categories.

First, REC procedures were seen as being too bureaucratic and wasteful. Thus, one respondent at a Repository undertaking a programme of sample and data collection felt that both RECs and NHS clinical governance arrangements had 'become openly bureaucratised, rather tedious and serious disincentives to the point where they are almost certainly counterproductive' (R25). Other respondents based in a range of settings variously described aspects of the REC process as 'pretty traumatic' for certain colleagues (R24: Epidemiologist, Uni); 'more and more stringent'

[18] For brief histories of the REC system in England and Wales, and major reforms over the past decade, see eg: S Kerrison and AM Pollock, 'The Reform of UK Research Ethics Committees: Throwing the Baby Out with the Bath Water?' (2005) 31 *Journal of Medical Ethics* 487; E Cave and S Holm, 'New Governance Arrangements for Research Ethics Committees: Is Facilitating Research Achieved at the Cost of Participants' Interest' (2002) 28 *Journal of Medical Ethics* 318; SMC Gibbons, 'Are UK Genetic Databases Governed Adequately? A Comparative Legal Analysis' (2007) 27(2) *Legal Studies* 312, 333–38 and references cited therein.

(R42: Uni/Hos); 'laborious' and 'the major stumbling point' (R34: Pharma); and 'a real barrier' and 'just dreadful' (R6: Epidemiologist, Uni). Excessive cost, time and resource wastage were particular problems. Two extracts are indicative:

> [I]t took us a year and a half to get permission and we had to send out 2 cubic metres of paper to ethics committee [...] Now we've actually overcome that now to some extent with COREC, though I still give up the will to live. [...] The application form for COREC is 75 page application form, it takes about 4 hours [...] And it's just ludicrous you know. (R44: Repository)

> [I]t's a lot of paperwork, it's a lot of time that goes into all of these things [...] not doing the research, not using your resources to do what you're funded to do. (R17: Uni/Hos)

REC application forms—originally designed for traditional, hypothesis-driven research projects and clinical trials—also were criticised heavily for being not suited to new biobanking activities (especially longer term, prospective, multi-purpose, multi-user, genetic and/or research-based biobanks), being unwieldy and being too generic to assess the specifics of a study. Finally, respondents complained of duplication. This arose both where RECs' functions overlapped with other governance regimes (such as HTAct licences), and where approvals from multiple different RECs—in the UK and/or abroad—had to be obtained. Several respondents reported encountering complications from having had to incorporate numerous local variations in ethical requirements. (We shall return to concerns over duplicative governance systems and multi-jurisdictional problems in the next section.)

The second broad head of complaint—mentioned by virtually every respondent who commented on RECs—was that RECs lack adequate knowledge, specialisation and/or expertise. Scientific, local and legal understanding all were highlighted. Many respondents felt that REC members' lack of knowledge or understanding commonly led to two major problems: (1) RECs raising unwarranted concerns, 'odd' or 'bizarre' questions, ethical 'non-issue[s]', or 'barriers'; and (2) RECs missing their ethical targets through being 'out of touch'. Looking more fully at each point, several respondents accused RECs of lacking a proper understanding of the science—especially genetics and epidemiology. Specific complaints included RECs having no scientific basis for their judgments (R6: Epidemiologist, Uni); failing to 'get it' (R30: Bioinformatician, Uni); being 'not naive but...' (R42: Uni/Hos); and lacking the specialisation needed to understand the scientific reasons for the study design (R23: Pharma). Interestingly, the latter private sector respondent wished that the problem of 'odd queries' coming back from RECs—that 'you kind of think, I'm not really sure where that's coming from'—and the 'difficult relationship' between applicants and RECs, could be improved through applicants being able to 'talk directly to the ethics committees' or 'to whoever asks the question':

> And then it's probably something that we can solve in ten minutes but it may take three iterations of ethics committee to review, to actually understand what the question is about. (R23: Pharma)

As a direct consequence of RECs allegedly lacking proper understanding, three Research Scientists, two of whom worked at Repositories and a third who worked closely with a Repository, complained that RECs guard against 'imaginary' harms (R25); 'inflate [things] to be a major problem but in fact it's fine' (R42: Uni/Hos); raise ethical 'non-issue[s]' (R44). Other respondents, including two from the private sector, credited RECs with being 'often quite well meaning' (R12: Uni/Hos) and 'genuine' (R23: Pharma) in their concerns to protect participants, but that they nevertheless risked 'overkill [...] in terms of what jeopardy you're putting the patient with regard to the research' (R34: Pharma). REC over-protectiveness was widely seen as highly wasteful—both through respondents needing to spend time and resources assuaging unwarranted concerns, and through their having to explain to RECs either what they 'should' know already (including, in one case, a legal interpretation), or what is 'really' important (R23: Pharma) in ethical terms.

Several respondents alluded to concomitant failures by RECs to identify the correct ethical issues. Often, this perception was interwoven with the attitude that RECs are 'out of touch' (R49: Uni), remote from researchers and local communities (the general public), and, hence, unable to grasp the true expectations, needs, and wishes of participants. Here, respondents believed that they, rather than RECs, correctly understood participants' needs and expectations (which, invariably, aligned with their own views);[19] and, moreover, felt confident to speak for participants. Thus, as one Clinical Scientist who worked on a small-scale clinical study put it:

> [P]atients generally worry a lot less than we might I think, or ethics panels might about all the niceties of consent and privacy and data coordination, they just want progress, they are really very strongly motivated. (R12: Uni/Hos)

As this Epidemiologist who worked on a well-established longitudinal study argued, 'we have some understanding and trust' with participants, who (according to the respondent) say:

> [D]o the genotyping, find things out, make the world a better place, that's what we signed up to do. Whereas [RECs are] sort of protecting people in a way that nobody actually wants or if they do they'll have withdrawn from the study. (R6: Uni)

[19] Note that we shall return to this perceived alignment in ch 9 below, in relation to access to data and samples.

Rather than REC approval—or, indeed, any other regulatory imprimatur—most respondents who held these views saw participant support and satisfaction as the key measures of good practice, such as the following Clinical Scientist:

> [I]f people don't want to participate in our project they're free not to, if they do they're enthusiastic about it, they're really excited about participating in the project so for me that is the bottom line of it all. (R17: Uni/Hos)

Thirdly, numerous respondents complained of having experienced inconsistent decision-making, interpretation of laws, questions or ethical concerns from RECs, or of having received conflicting opinions. In the phrase of one respondent, they were 'quirky' (R23: Pharma). Differences had emanated not only from different RECs, but also, on occasion, from the same REC. For instance, one Bioinformatician observed that 'they go through waves of wanting stupid amounts of information on a consent sheet and then saying well no that's now too long' (R30: Uni). A Clinical Scientist bluntly described RECs as 'bizarre', 'counterproductive' and 'a nightmare, just because they're so idiosyncratic and difficult to work with' (R17: Uni/Hos). Individual REC members, too, were regarded as problematic by some respondents:

> [O]ne of our great bugbears is RECs, I mean these are just full of these sort of ill-informed people [...] sort of well meaning but [...] (R6: Epidemiologist, Uni)

> [Y]ou get one guy on the committee and that person can just make it, make life miserable for you and not the help the ethics of the situation, that's what bothers me. (R17: Uni/Hos)

Two respondents spoke of individual REC members getting 'a bee in their bonnet' (R17: Uni/Hos; R23: Pharma).

As is apparent from our findings, complaints about RECs were commonplace. While not wishing to suggest that any concerns raised might lack substance, they should, however, be read in the light of the findings recounted in the previous chapter—namely, about some respondents' underlying beliefs, assumptions and attitudes towards governance, and the perception that governance can sometimes unreasonably infringe on their practice.

## NHS Actors and Systems

NHS actors—notably, Primary Care Trusts (PCTs), NHS Research and Development (R&D) offices, and other NHS research governance systems and staff—ranked second only to RECs in terms of the number and range of attitudes expressed. Around a third of respondents commented (again the extracts below are largely from Clinical and Research Scientists, unless labelled otherwise). Their comments here were among the most vitriolic. Only two respondents expressed praise—in both instances, for the Caldicott

Guardian system. One thought that 'the Caldicott framework probably works quite well actually' (R2: Uni/Hos); while the other, a Bioinformatician who had developed a web-based data sharing platform, described a particular Caldicott Guardian as having made things 'all fairly straightforward' (R1: Uni). Overwhelmingly, though, attitudes were negative, and frequently quite hostile.

One Clinical Scientist castigated NHS-based data protection officers for giving 'totally different' interpretations of the law, and for interpreting the law in ways that were capricious and likely to harm patients (R31: Uni/Hos). This respondent felt that 'people should not interpret laws themselves [...] There needs to be a standardisation of that, okay' (R31). Speaking of NHS research governance systems and staff more generally, the same respondent went on:

> What I think needs to happen is that we need to sweep away a lot of these people, they're in the way, they're not helping, they're not making, they're not making the world better for patients [...] What they're doing is forming really a barnacle on the backside of progress I'm afraid you know and basically what they're doing is stopping people getting access to things. (R31)

In this respondent's view, these obstacles also were the direct cause of 'sponsors, which are largely industrial, of those trials [...] voting with their feet', and 'taking their trials to other countries' (R31).

Several respondents criticised NHS R&D offices—both for lacking competence, and for hampering, rather than facilitating, research. One Clinical Scientist, for example, described them as being 'sort of overwhelmed and understaffed, and if you approach them with something that doesn't really fit one of their pigeonholes [...] then they don't really know what to do with you' (R42: Uni/Hos). This respondent added, moreover, that he thought staff could favour their own self-interests, describing them as 'tentative' and reluctant to take on anything novel, in case it might prove 'contentious' or 'throw up problems' (R42). Likewise, a Repository-based Research Scientist claimed to have 'constantly run into [NHS] people who are badly trained or just lazy or misinformed, getting in the way of access to clinical data in case it's bad for their position' (R44).

Overall, NHS and PCT research governance requirements were seen as being 'expensive' (R44: Repository); 'quite a slow process' (R8: Repository); taking 'an enormous amount of time' and 'much more bureaucratic' (R9: Uni/Hos); and a great deal of 'paperwork' (R30: Bioinformatician, Uni; R47: Uni/Hos). Additional problems arose where respondents needed to obtain multiple NHS R&D approvals from offices located within several PCTs (due to the lack of any centralised, national approval system). One Research Scientist based at a Repository involved in a nationwide programme collecting data and samples had found dealing with the different offices 'very much a mixed picture' (R8). Meantime, a University/Hospital-based respondent, who undertook small-scale clinical research, explained that 'trying to get approval from every hospital that you're going to recruit patients from' took 'literally you know two years', and joked 'I used to have black hair before I started this actually' (R9). In the event, over 350 individual signatures had to be obtained:

> So that means you know that we have to get a hundred different approvals and we have to fill out the form a hundred times. And you can image the administration that involves, the bureaucracy that involves and the time it involves and so you know so the line time between sometimes actually getting the funding and actually doing the study is getting longer and longer. So that's one major issue which is a huge, huge problem. (R9)

In this respondent's view, 'by doing it that way [without a central R&D office] they're not actually improving patient safety or improving governance processes they are just creating a whole layer of bureaucracy' (R9: Uni/Hos).

**Funding Bodies**

Although less frequently discussed in detail than RECs and NHS systems, funding bodies still attracted a wide and revealing range of comments from a variety of respondents. As with RECs (although to a much less severe degree), attitudes were both positive and negative. A few respondents expressed praise regarding advice provided. One Clinical Scientist said that funders were flexible, and that 'there's always a degree of latitude' (R14: Uni/Hos). Several funding bodies were said to be acting with 'good future vision' by a Research Scientist involved in programme of data collection (R41: Uni/Hos). Their supportive attitude was contrasted with that of the Department of Health, which was, in the view of this respondent,

> a fly in the ointment [...] it seems, it amazes me, it seems so out of touch with what the people on that [named] council estate want and what we know we need to develop for a big and better health service with closer relationships with science. (R41: Uni/Hos)

But the majority of respondents who commented criticised funders for providing insufficiently explicit or detailed guidance about their requirements—for example, explaining precisely how DNA banks should go about setting up arrangements to handle third-party access requests (R18: Epidemiologist, Uni; R19: Public Health, NHS). As one respondent put it, the Medical Research Council 'is rather fond of diktats just telling you, "you must document your study you know just do it", which is not very helpful really' (R30: Bioinformatician, Uni).

A significant area of contention (which we shall discuss in Chapter Nine) related to the issue of 'access'. One lone respondent—a Repository-based

Bioinformatician (R10)—enthusiastically welcomed the recent adoption of 'open-access' data-sharing policies by leading UK funders. Other respondents who commented, however, either criticised or vigorously opposed this step. One Clinical Scientist who worked on a well-established longitudinal study, for example, felt that the Wellcome Trust had built up 'very clearly a head of steam [...] at the highest level', and had pushed the open-access agenda onto researchers, being too domineering, but without figuring out the correct 'political balance' between different funders, or how 'some compromise' might be made (R15: Uni/Hos). General complaints included that the introduction of open-access requirements was premature and insufficiently thought through, lacked consultation and lacked sensitivity to the possible need for flexibility or compromise. The retroactive effect of the new ground rules, priorities or policies was a particularly contentious aspect. Respondents were especially aggrieved where funders had not provided any additional funding, other practical support (such as archiving services), or recompense. As this University-based Bioinformatician put it:

> Now I'm being asked to deposit data [...] and there's no suggestion that I'll get funding to help me, there's no suggestion I'll get time to do it, there's no suggestion I'll get any payments for doing it either. [...] just said [...] this is a requirement for us to fund your next project that you have done this. So which is a big stick rather than a carrot and a carrot would be preferable. (R30)

Some complaints related to opinions that funders were seeking to be overly controlling (R15: Uni/Hos). One Clinical Scientist who worked on a small-scale clinical study spoke negatively about the influence that funders have over research, with their 'set of priorities' and requirements for 'translatable research' (R14: Uni/Hos). This respondent felt that funders wielded too much power in setting the research agenda, through refusing to fund other things. At the micro-level, however, excessive control over individual projects did not emerge as a major issue.

Finally, the above-mentioned Clinical Scientist who worked on a well-established longitudinal study also identified a 'multi-funder problem' (R15), which can make managing projects with several different funders very difficult. In their case, the problem manifested in two ways: where several co-funders all 'demand a say in the governance arrangements' (R15) but their policies differ or conflict; and where co-funders each wish to place their own nominee on the project's steering committee or other internal oversight body.

## Analysis and Discussion

Two sets of invaluable insights and lessons emerge from the analysis of data on respondents' attitudes towards regulatory bodies. One reflects respondents' attitudes towards particular actors. Through these, we can identify a range of key perceptions, and specific, concrete, practical issues warranting attention. The other encompasses various common or cross-cutting themes. These emerge when we step back and reflect on the data as a whole. Overall, our summaries of respondents' attitudes towards, and experiences of, RECs, NHS actors and systems and funding bodies make for salutary and somewhat sobering reading.

In relation to the first set of insights and lessons, while several positive comments were made about RECs, these were outweighed comprehensively by negative opinions and complaints. Generally speaking, respondents endorsed in principle the need for biobanking activities to have ethical approval. However, three major concerns (usually expressed by Clinical and Research Scientists) were bureaucratic wastefulness; RECs lacking knowledge, specialisation, and/or expertise (not least, leading to them focusing on 'non-issues', over-protectiveness and missing their ethical targets through being 'out of touch'); and inconsistency. One potential improvement suggested was to increase the opportunities for practitioners to interact directly with RECs, and (where appropriate) develop enduring working relationships. Interestingly, one respondent who advocated such closer links also acknowledged the importance of RECs maintaining their independence and impartiality. Increased intimacy inevitably raises the spectre of 'regulatory capture'. Nevertheless, the data analysis reveals a widely, deeply and often strongly held belief that the approach (as at the time of our research) simply is not serving the needs of those undertaking biobanking activities.

In terms of the implications for governing biobanks, then, this strongly suggests that the existing biomedical governance framework—especially with its de facto, heavy reliance on RECs (as described in Chapter Three), rather than any tailor-designed system for governing biobanking activities—is neither well suited to, nor adequately equipped for, reviewing, approving or overseeing biobanks or other longitudinal, larger-scale, complex, multinational or highly technical projects. Having said this, our discussion previously in Chapter Six suggests that (for some, at least) *any* regulation is likely to become a source of complaints and dissatisfaction. Thus, one challenge in investigating further the need for reform, and in formulating any fresh governance measures, is to identify from the common complaints those that really are signs that the current systems of practice are inappropriate.

The relatively high volume of references to NHS actors is unsurprising, given that the majority of our respondents came from the public sector. What is noteworthy, however, is just how overwhelmingly critical—cynical, even—so many respondents were about NHS governance. Especially significant findings here include the perceived lack of consistency and standardisation in decision-making; and administrative difficulties, inefficiency, duplication, cost, paperwork, and other bureaucratic burdens, particularly where multiple NHS R&D approvals had to be obtained.

Respondents usually blamed a lack of centralisation, or of regional or national coordination.[20] This has important implications for governing certain biobanks (notably larger-scale projects and geographically dispersed networks), as they commonly involve NHS patients or participants from multiple hospitals or geographical areas. Another key finding is the accusation of incompetence, and a cultural fear of novelty, on the part of NHS R&D staff. The former claim—in effect, that 'other people lack the right knowledge'—is what we might expect to find (whether or not true) where actors have misaligned interests or diverging interpretations of what is relevant or important. The latter accusation taps into deeper organisational and social issues. According to some respondents, all these factors materially hamper research.

Finally, in respect of funding bodies and biobanking, two findings are particularly worth drawing out. One is the perception that funders do not always provide sufficient guidance about how grant-holders might actually implement their requirements in practice. A second is the very marked opposition to how 'open-access' data-sharing policies were introduced. If other practitioners in England and Wales broadly share our respondents' views, then our findings suggest that at least some may well feel aggrieved that funders effectively 'foisted' or 'forced' their agenda onto the research community—not least, by exploiting their power (also explicitly criticised by respondents) to control the research agenda, and set priorities that reflect funders' own interests and wishes. This tension is further explored in Chapter Nine.

Stepping back to reflect on the data overall relevant to this point, a set of six significant common or cross-cutting themes can be extrapolated. Like our characterisation (or perhaps somewhat caricatured 'ideal type'[21]) of a certain professional 'outlook' in our analysis in Chapter Six, we do not claim that all respondents explicitly endorsed or expressed all elements of these themes, but the underlying ideas were commonplace. First, existing governance actors and systems actually block, rather than facilitate, research. RECs and NHS systems particularly are tarred with this brush. Secondly, key public sector regulatory systems can be disorganised, cumbersome, inconsistent, highly bureaucratic, sometimes incompetent, wasteful and to be avoided wherever possible. Thirdly, and consequently, those governance structures are often viewed as counterproductive disincentives to research, that drive research projects (and associated investment and expertise) abroad. Fourthly, certain governance actors—notably, RECs—are thought to be 'out of touch' with what participants, communities and the public want (and being perceived to be 'out of touch' materially reduces regulatory

[20] Note that, since we conducted our interviews (May 2006 to November 2007), both the REC and NHS R&D systems have been modified, including through greater centralisation.

[21] M Weber, *The Methodology of the Social Sciences*, ed and trans EA Shils and HA Finch (Glencoe, IL, Free Press, 1997 (1903–1917)) 88.

actors' authority, relevance, status and influence). Fifthly, in place of 'out of touch' regulatory actors or structures, for some respondents it was the 'real' (or, at least, the perceived 'real') interests, needs and expectations of those people—namely, research subjects and patients—who actually take part in research studies that were the 'bottom line' or benchmark for deciding how to behave. Sixthly, here, some respondents tended to align their own interests (especially in being free to pursue their work unhindered) with those of participants, communities and the public. Once again, as we saw in Chapter Six, some respondents exhibited a high level of self-confidence in their ability to say what other stakeholders needed or wanted, and to speak for them.[22]

We cannot safely speculate over how far our respondents' beliefs (at least as there were expressed in the context of the interview) about other stakeholders' 'real' interests, needs or expectations were accurate (or indeed, how widely such attitudes are reflected in the research community at large). The obvious danger is that, at least on occasion, they may have imputed or projected erroneously their own attitudes, perceptions, interests, expectations and so forth onto others. Wherever the true position lies, taken together, these six cross-cutting themes have considerable implications for governing biobanks in the future—especially if such governance is to be achieved in ways that command support, respect and, ultimately, adherence from practitioners. They also helpfully illuminate some key potential flashpoints, or grounds upon which practitioners may choose to avoid, subvert, undervalue or simply ignore governance measures, based on their own attitudes, beliefs and interests.

## AREAS OF CONCERN

Having addressed specific laws and regulatory actors, we turn now to other substantive areas of concern for our respondents to do with the biobanking governance framework. We begin by looking at five general, overarching concerns. We then examine three specific topics raised by respondents—namely, jurisdictional difference; distinguishing between biological material and data; and special concerns relating to regulating data.

### General Concerns

Five significant overarching or general governance-related concerns emerge from the data: (1) stifling of research; (2) laws lacking clarity; (3) problems associated with multiple governance systems; (4) 'genetic exceptionalism'; and (5) the resource implications of laws or regulations. We consider each in turn.

[22] See further ch 9 below.

*Stifling Research*

Seven Clinical or Research Scientists (who worked in University/Hospital settings unless specified otherwise) argued that current laws, guidelines or governance frameworks stifle or 'hamstring' (R16: Uni) research, patient care or scientific progress. In one instance, this outcome was viewed as an unintended consequence of good intentions:

> [T]hings like the Human Tissue Act [...] they're all very, very anti-research, they may be meant well they have a very bad knock on. So again I just sense that they're often quite well meaning but [...] you can over protect and end up just not making any, missing opportunities to make scientific headway. (R12)

Others, however, showed less goodwill—as demonstrated by phrases noted above, such as 'barnacle on the backside of progress' (R31); 'not actually improving patient safety or improving governance processes they are just creating a whole layer of bureaucracy' (R9); and 'wall of legislation' (R44: Repository). Further concerns over bureaucracy or conflicting laws impeding practice were embedded in opinions of two respondents who worked in the context of small-scale clinical studies about how laws curtail the movement of tissues (R14; R27). In all such cases, the concerns centred not simply on a lack of 'help' offered by laws or actors, but on a perception that laws or actors are actual impediments or obstructions.

An interesting related concern, aired by two Bioinformaticians based at a Repository lodging data, was that current governance frameworks preclude 'serendipitous discovery' (R29)—a form of progress often praised by scientists for having led to many important, but unexpected, findings (including in genetics). Restrictions over accessing materials, and limiting their use on a study-by-study basis, were particularly blamed for shutting down potentially fruitful avenues of inquiry: 'it really closes down serendipity' (R10). One even advocated a researchers' 'right to roam', framing it as a matter of 'high moral argument':

> [W]hat you want in science is to enable serendipitous discovery and you know the argument [...] I would make is kind of an analogist to the right to roam [...] if you discover something that you think should lead you to another database [...] then you don't want to be hampered by the whole, the barbed wire fences that [IPR] licensing would put in the in the way of that right to roam. (R29)

*Lack of Clarity*

A second general governance-related concern was that current laws and legal frameworks lack clarity. Some respondents made direct claims that certain laws—for example, the DPA—are unclear. Thus, one Clinical Scientist claimed that 'nobody was absolutely certain of the law [...] and it is, it is unclear isn't it?' (R28: Uni/Hos); while this Public Health practitioner thought that 'sometimes it's not very clear, when you're reading through the Data Protection Act and you're reading the documentation [it] isn't clear or straightforward as to what you can do or what you can't do' (R32: NHS).

On other occasions, concerns about clarity were couched in terms of uncertainty over interpretation. One indicative comment (by a Repository-based IT Specialist referring specifically to the Data Protection Directive 95/46/EC[23]) was that '[i]t's very, at the moment very open to interpretation' (R40). Meanwhile, a Research Scientist based at a different Repository reported that 'the lack of clear cut answers in the area of practical application' was 'frustrating' (R37). Both for this respondent, and for the next, concerns over clarity and interpretation were compounded where multiple (potentially inconsistent, as well as individually unclear) sources of law, legal actors or jurisdictional requirements applied:

> Much more challenging has been actually the regulations surrounding it and in various parts of the world we've had either individual ethics committees or states or whole countries where it's unclear [...] (R49: Clinical Scientist, Uni)

Numerous concerns related to the lack of any single, clear, legal definitions of much-used concepts or terms. Prominent examples cited were 'high quality sample', 'consent', 'informed consent', 'confidential', 'confidentiality', 'anonymous' and 'anonymity'.

Interestingly, some respondents recounted examples of where they had taken governance-related steps that, 'officially' speaking, probably were unnecessary. Despite the potential wastefulness involved (both for respondents and systemically), this was done for various reasons. One reason, expressed by two respondents who worked at different Repositories, was a desire to appear to be as legally 'compliant' as possible (R8: Research Scientist; R36: Bioinformatician). However, a few respondents also spoke of having taken steps just to be on the 'safe side', when it was unclear what was legally required. This also was commonly seen as the easiest practical option. Thus, one Clinical Scientist spoke of obtaining fresh REC approval, after new Medical Research Council guidelines had come into effect, because 'we thought it was just wise', and 'we just found it a bit easier' (R47: Uni/Hos). The same respondent further observed that, while 'in law apparently you don't actually need' written consent from relatives to take tissues post-mortem, because 'there's least clarity [...] a lot of pathology departments ask for it [...] And in fact we just go and get it now because it's actually quicker and easier' (R47).

[23] [1995] OJ L/281/31.

*Multiple Governance Systems*

Around a quarter of respondents voiced concern over multiple governance systems. Two distinct aspects emerged: (1) the perceived wastefulness and bureaucracy associated with having to complete multiple, overlapping procedures; and (2) situations where conflicts or inconsistency arose between different laws, actors, or systems.

In section 3 above, we identified four areas where multiple governance systems posed problems for respondents—namely, multiple REC approvals; multiple NHS R&D approvals; the 'multi-funder problem'; and inconsistent legal interpretations given by NHS data protection officers. Of these, the one most often singled out for criticism (by nine respondents) was the need for multiple REC approvals. This included not only from RECs within the UK, but also, in three instances, from RECs abroad as well. Negative consequences reported by Research and Clinical Scientists working in different settings included delays, costs, bureaucracy, inconsistent decision-making in three cases (R23: Pharma; R44: Repository; R47: Uni/Hos), and a lack of standardisation (including procedural). As one Clinical Scientist put it, it can be 'quite tricky' where different legal and/or ethical standards apply in different countries (R11: Uni/Hos). Concrete examples included differing rules over consent, and differing sample retention time limits As this respondent observed, incorporating all of the 'variations country by country' into protocols and consent forms for multinational projects 'does actually get quite complicated' (R22: Pharma).

*'Genetic Exceptionalism'*

Whether or not laws and governance should treat 'genetics' as being special or exceptional was a troubling issue for numerous respondents. This hotly debated question divides opinion—as, indeed, our data reflect. Because DNA is essentially immutable, intrinsically personally identifiable, potentially predictive, and may carry implications for individuals, family members and whole populations, proponents of the 'genetic exceptionalism' viewpoint believe that genetic samples and genotype data should receive higher levels of legal protection, oversight and regulation than other biosamples or sensitive personal information.

Three respondents made comments that implied that genetic samples and data are worthy of special attention in respect of governance (it is perhaps notable that two of these were Clinical Scientists who worked in University/Hospital settings). One singled out DNA from other biological samples as imposing particular 'responsibilities', because, from it, 'you could generate huge amounts of genotype data very readily' (R15). Another claimed that 'genotype data [...] characterises the person' and 'can have important personal implications for the person' (R31).

Accordingly, guarding what can be done with it is 'very important' (R31). A third respondent, a Bioinformatician at a Repository lodging genomic data, cursorily raised the issue in terms of more generic concerns about genetic discrimination (R39).

Another Bioinformatician also noted the potential risk of 'discrimination' or 'misuse' of genetic information held in large databases (R30: Uni). By contrast, however, this respondent believed 'all round that genetics is over-hyped' (R30: Uni). Similarly, three other respondents—including only one Clinical Scientist—also clearly were in the 'anti-essentialist' camp. The Clinical Scientist rejected special guidelines for consent and confidentiality relating to genetics, because 'I don't think genetics is a special case. [...] So I don't think there may be particularities about genetics that we need to take account of' in terms of governance (R2:Uni/Hos). Meantime one Research Scientist argued that 'people need to see' the 'genetic data [...] as just clinical data, it's just information' (R23: Pharma), and another thought that:

> [I]t would be a mistake to get into genetic exceptionalism. I think genetic data can be sensitive and they can be non-sensitive and the same is true of psychometric data and I can't quite see why one should get oneself in a tizz about the one hand and ignore the other. (R25: Repository)

A more ambivalent University/Hospital-based Research Scientist, who worked on a small-scale clinical study, observed that many believe that 'genetic data [...] is somehow richer', and seems to get 'more to the heart of what someone is which I'm not sure is true but that's the perception' (R27:). On balance, this respondent concluded that 'all medical information is confidential' and 'should be subject to the same level of control'. Yet, while in general believing that 'there's nothing special about genetics [...] it does have little peculiarities that have particular strengths, not quite the right word, particular stresses'. Moreover, in practice, the respondent was 'particularly careful with genetic data, more because of the sensitivity of the work I suspect' (R27).

### *Resource Implications*

Several respondents raised concerns over the resource implications of finding out about and implementing various laws. The cost of HTAct licences was a prominent target. Here, the University/Hospital-based Research Scientist quoted immediately above (R27) highlighted three specific areas of concern. One was the cost for organisations that have multiple 'premises', and so have 'to pay goodness how many licences to the HTA for every premises at £6,000 a go'. A second was how such licence costs would be funded: 'and nobody has given us any extra funding to do that'. A third concern was over how the HTAct deems different buildings on single 'sites' to be different 'premises', requiring separate licences: '[e]ven though it's just across the road and it's on the same site it's six premises [...] So that's six plus six plus six plus one times 6,000 for the licences to cover all those premises'.[24] As another responded working in a University/Hospital context indicates, such financial concerns could have very practical impacts on scientific practice, including delaying the inception of work:

> [W]e are already ready for HTA, it's only a question of money, it's £6,000 per year licensing and so the later we start paying the 6,000 the better so it is not so much the licensing whether we start it now or later but it's more like the money. (R45: Clinical Scientist)

Other laws also triggered resource-related concerns. This Research Scientist, whose Repository project worked across several different countries, observed that, 'to understand the whole thing we could have [...] said let's analyse the various laws and then kind of put together what is there [...] And we probably could have spent the three years of the funding just with doing that' (R43). Because of the prohibitive time and cost entailed, they had simply decided *not* to engage with the laws. As the respondent explained, investing so much time and money into trying to ascertain and integrate all of the different laws 'was not the way to go', because 'the patient organisation [...] wanted us to do things right away, make biomaterials available.' (R43).

Another respondent highlighted the financial implications of implementing the EU Clinical Trials Directive 2001/20/EC[25]—both for future studies, and for projects already underway when it came into effect. Again a respondent who worked in the University/Hospital sector undertaking small-scale clinical research described such funding worries as having 'had a major impact on us', been 'a major, major factor which has affected us over the last five years enormously', and as 'creating a major issue for us' and 'a major issue for a lot of other researchers that I talk to' (R9: Clinical Scientist):

> [Y]ou have to build in some of the funding and costs associated with it into the grant applications now you know and we're struggling with the previous grants we've had, we didn't actually foresee any of this. Well you know we foresaw that there would be some problems, we didn't foresee the scale of the problems. (R9)

## Jurisdictional Differences

As previous sections have shown, various areas of concern for respondents implicate issues of jurisdictional difference. Several respondents whose

[24] Note that the HTAct licensing scheme, as administered by the Human Tissue Authority, in fact differentiates between 'main site' licences and 'satellite' licences. Licences for the latter are substantially cheaper. Thus, in 2009–2010, main site licence fees for research cost £6,000, compared with £900 for satellite sites.

[25] [2001] OJ L/121/34.

work involved multiple jurisdictions (often, but not always, those at Repositories or Pharmaceutical Companies) clearly understood that they needed to be 'aware of all these laws and [...] apply all of them, comply with all of them' (R21: Clinical Scientist, Pharma), and be 'very [...] adherent towards [...] local and applicable laws' (R35: IT Specialist, Repository). For four respondents, applicable national laws were not seen as being 'so [...] different' (R45: Clinical Scientist, Uni/Hos) from one another that they posed particular problems. For example, in relation to clinical trials, one respondent who worked for a pharmaceutical company felt that while 'all these laws [...] can cause problems', the organisation had 'plenty of expertise of applying multiple national laws and ethical standards and in fact there is [...] a lot of harmonisation in that area' (R21: Clinical Scientist). Another Clinical Scientist was of the view that 'most of the laws and regulations' across nearly 20 countries were 'remarkably similar'—not least, because 'one country copies another country's regulations' (R49: Uni). A third respondent, whose Repository project worked across, and drew data from, several different countries, spoke of 'harmonisation on a European level that's pretty much similar standards in all the European countries', with only 'small differences' (R43: Research Scientist).

However, 10 respondents (including people working at three different Repositories) drew attention to specific governance-related challenges that had arisen where 'there's different jurisdictions that can complicate things' (R35: IT Specialist, Repository). Those challenges sometimes arose within, but mainly across, national borders. There were three broad types. First, the basic requirements of knowledge and understanding often were problematic. Understanding could be hampered at the level of knowing the different status of legal instruments in different places—for example, 'teasing out what's actually legislation and what's actually just research guidance' (R23: Research Scientist, Pharma). Identifying what laws might apply, and how they relate to one another, was another difficult, resource-intensive obstacle (R43: Research Scientist, Repository). So, too, was confusion over how certain laws apply differently in different jurisdictions, such as the HTAct in England and Wales vs Scotland (R35: IT Specialist, Repository). Interacting with colleagues abroad, where each has a 'slightly different regulatory framework', could cause difficulties, through people having different understandings and concerns about the law, and the consequent 'need to manage [multiple] sets of relationships' (R40: IT Specialist, Repository).

A second category of challenges stemming from jurisdictional difference involved potential conflicts—either between legal instruments in difference places, or where national laws conflicted with the desired practices of the study. A private sector Research Scientist complained that 'Sweden, Denmark and France have had sets [of] research guidance mainly that have conflicted' (R23: Pharma). This respondent (and one other) had experienced problems due to differing legal restrictions over the collection of biological materials, or over their removal from nation-states, such that they had 'ended up in a difficult situation' due to 'the biobank law' in Sweden, which meant that 'they didn't like samples leaving Sweden' (R23).

A third kind of challenge—some illustrations of which also have been noted already—involved trying to integrate or incorporate local or national legal (and ethical) variations into study protocols and other key documentation (for example, participant consent forms). This respondent, who worked at a Repository reflected on practice since the EU Clinical Trials Directive 2001/20/EC,[26] and wryly commented:

> [I]f you want to create a level field and you [...] don't want to get criticised what you do is you level up. So if the French required that everyone wears a beret when they're doing research then you say European level everybody should wear a beret. And if the Germans say everyone should walk backwards on Christmas Day then you add that too. And all they did was add every national quirk until they ended up with an almost impossible bureaucratic barrier. (R44: Research Scientist)

Respondents from a range of settings who were dealing with jurisdictional difference revealed a range of strategies or techniques. The more successful included drawing on accumulated 'expertise' (R21: Clinical Scientist, Pharma), and working co-operatively with others—for example, ongoing engagement with an American lawyer to sort out 'issues of how you send' NHS data to the USA, where 'they don't have a Data Protection Act like we do' (R32: Public Health, NHS). Less successfully, many respondents had opted—or felt compelled—to adopt exclusionary strategies. The effect was to exclude studies, research activities, collaborators, or settings that did not 'fit'. Thus, one University-based Clinical Scientist ended up using no 'genetic material from [a particular country], simply because that's [that county's] government rule or it's a regulation' (R49). Similarly in the private sector, this Research Scientist reported having taken 'a fairly hard view', and decided that, to avoid the potential for 'error', 'if it doesn't fit our process [...] if we can't use this system for whatever reason we don't collect' (R23, Pharma). Where exclusionary techniques were used, these respondents appeared to blame the underlying jurisdictional differences for being detrimental, both 'to the science we can do' (R23), and to the populations excluded.

### Biological Material and Data

As outlined in Chapter Three, an essentially bifurcated statutory system currently provides much of the regulation for human biological material and personal data in England and Wales. Put very simply, one set of

[26] [2001] OJ L/121/34.

provisions (notably the DPA) covers data; while a separate, largely parallel, but non-identical system (notably the HTAct) covers biological material. We were interested to find out how far—or, indeed, whether—this dual, conceptually differentiated, legal approach reflects the understandings, perceptions, expectations and practices of professionals. Accordingly, the interview schedule included prompts for respondents to comment on the distinction(s) that they perceived (if any) between data and biological material, and how they thought the two should be regulated.

Their responses reveal a wide spectrum of often fundamentally divergent attitudes. Individual opinions about the 'divide' (if any) were very mixed. Broadly speaking, they fell into three camps—in favour of recognising a 'divide'; against it; and ambivalent. However, several different rationales underpinned respondents' judgments in each camp. Genetics also emerged as a blurry or 'grey' area.

Three respondents, who all worked in University/Hospital settings, argued that data and biological material could be considered (in some ways) equivalent—and, accordingly, should be handled legally in 'the same way' (R12: Clinical Scientist). As noted above (in relation to the HTAct's exclusion of DNA), this respondent described 'data' and 'biological material' as not 'qualitatively that different from each other' (R12). For another, equivalence flowed logically from the fact that '[i]t's all potentially information' (R27: Research Scientist). For a third, who was involved in a programme of data collection, the answer to our question about whether the law should recognise any distinction was simple and clear cut: 'No.' (R41: Research Scientist). When invited to elaborate, the rationale offered was equally succinct: '[s]amples are data it's just that they're, some of the data hasn't been unlocked' (R41).

Other respondents in the University/Hospital sector tended more towards favouring a division. Some drew very sharp distinctions. One Clinical Scientist involved in a well-established longitudinal study, for example, argued that samples and data 'clearly' differ, on at least three grounds—custodial arrangements; accessibility; and special public sensitivity over tissues (including the pre-HTAct nationwide organ retention scandals, and organ trafficking):

> I mean the data requires a different sort of custodial arrangement than biological samples, it's a lot easier to access data than it is biological samples and if we can go back to the reason for the HTA's possible alarm about body parts being traded and so on. So you know they're very clearly in the public mind, there's a difference between biological samples and data. (R15)

Another Clinical Scientist argued that 'it's right to have two different types of legislation because they are different questions that you're trying to solve' (R31). In this person's view, the DPA and HTAct 'have two different purposes'. For the DPA, it is 'protecting the privacy of the individual and the use of the data'. For the HTAct, it is ensuring ethical conduct such that biomedical professionals 'don't do something with that sample that you haven't got permission for' (R31).

Other respondents, too, pointed to practical reasons for drawing a distinction, such as the fact that data can be held for much longer than samples, differing consent requirements, and the need to have 'procedures in place' for dealing with 'biohazards' (R21: Clinical Scientist, Pharma). This Epidemiologist, involved in a different well-established longitudinal study suggested that 'the difference' lay, 'in a sense', in the fact that 'biological samples, particularly DNA, is [...] information which can be reinterrogated' (R6: Uni). However, this respondent was particularly ambivalent, both over whether data and tissues should be treated differently by law—'I'm not sure they should, they are currently but I'm not sure they should be'—and whether the two really are intrinsically distinct:

> [S]o I guess there is a difference. [...] I guess I don't have strong views. [...] I mean I guess there are issues of things like it is biological, it is a piece of somebody that somehow at a very visceral level. [...] so I guess that may be at some level even though rationally I can't [explain it]. (R6)

Others also intimated that a distinction may lie in the special emotional nature, or ontological quality, of biological material. As the only Genetic Counsellor in our sample put it:

> There is something about, I don't know there's something and I don't know what it is, there's something that feels different about [...] sample information also the tying up of the two together feels different for lots of people. (R33: NHS)

The implication underlying such perceptions was that biological materials should be subject to different rules than data—and also, potentially, to more stringent or rigorous legal protections. Yet, a few respondents advocated precisely the opposite view. For example, one Clinical Scientist emphasised that 'the data has got to be protected as much as the sample', if not more so:

> I think data is almost more worrying than samples, the protection of that is that it's definitely as worrying if not more worrying because that data actually tells you something specific whereas a sample as yet may not tell you anything specific, it could be just misused. (R28: Uni/Hos)

Similarly, even those respondents who—as noted above—delineated strictly between data and samples, raised potential 'grey areas' in relation to safeguarding genetic data. For this Clinical Scientis*t*, the data contained in genotypes 'characterises the person' and 'can have personal implications for the person, so what you do with that data is very important' (R31:Uni/Hos). Another, took a less personal (more research focused) view, and argued that they would not wish to see the law 'draw too great a distinction between responsibilities of somebody having a DNA sample and the responsibility of someone using the data generated from a DNA sample' (R15: Uni/Hos). This respondent, who as previously mentioned was involved in a well-established longitudinal study, argued that, with the right resources DNA can be used to generate 'huge amounts of genotype data very readily' (R15).

Interestingly, one openly ambivalent respondent suggested that 'it doesn't really make [...] sense' to think of treating biological material and data differently. This was because (in the biomedical context, at least), 'biological materials have no intrinsic value unless they have some data attached to them. A chunk of tissue is just a chunk of tissue' (R21: Clinical Scientist, Pharma). This notion—that tissue lacks value without data—was echoed elsewhere. Notably, it was one of three reasons given by respondents for a desire to see (greater) harmonisation between the DPA and HTAct. Thus, one Clinical Scientist observed that 'a sample is no good without the data. So there are common fields there' (R2: Uni/Hos). Accordingly, it was 'in the interests of everybody there should be some harmonisation between' the DPA and HTAct (R2). A second reason for wanting the laws 'to be convergent' (albeit not identical), and for why 'the continuum needs to be there', was because of the 'risks to individuals' and 'legitimate concerns' over 'abuses' of data, whether data 'derived from those samples linked to an identifier, or the data alone' (R41: Research Scientist, Uni/Hos). This respondent, who was involved in a data collection programme, was if the opinion that the answer lay in treating samples, under law, as 'potential data'. The third reason for advocating harmonisation, identified by a Research Scientist at a Repository involved in a programme of sample and data collection, was to ensure that 'samples and the data therefrom' are 'managed in a coherent way' in practice (R25). This respondent's fear was that a knock-on effect of the 'whole bundle of new bureaucracy around samples' following the HTAct would be that people would start treating data differently in practice—and 'that needs sorting out' (R25).

## Regulating Data

Finally, numerous respondents raised special concerns over regulating data. The three dominant themes were: (1) balancing the need to maintain confidentiality (especially participant anonymity) against conflicting interests; (2) potential mismatches between what is expected, desired, or assumed to be possible, and actual techno-scientific capabilities; and (3) complaints that data are over-protected, in law and in practice.

Looking first at conflicting interests, the predominant concern here related to tensions between obeying anonymisation requirements on the one hand, and the desire to identify and recontact participants, for various reasons, on the other. Four respondents were especially troubled. Despite recognising that 'the patient's identity has to be protected, absolutely'

(R20: Bioinformatician, Repository), they believed that de-anonymising data so that participants can be recontacted sometimes may be 'something positive' (R15: Clinical Scientist, Uni/Hos). One such 'positive' situation was in this latter respondent's well-established longitudinal study where he felt it would serve participants' own wishes: 'the cohort have been asking for a long time' for 'feeding back', because 'they want to see what's happening with the information they've provided' and to 'be assured that something really has happened' (R15). Alternatively, identifying participants may serve the interests of researchers, by permitting follow-up—for example, in the former respondent's newly-established Repository project where 'you want to follow up on the patient to press the outcome whether the patient was recovered, whether the cancer relapsed, now these type of things' (R20). Two respondents noted the sometimes difficult distinction between research and the clinic, indicating a particularly strong ethical duty to provide feedback where significant information about research participants' health comes to light. As one, a University/Hospital-based Clinical Scientist, put it:

> I feel very strongly these people have donated something and they have been so trusting, as a doctor of course I feel uncomfortable if we did not feed back something that we knew there was some potential for making things better. (R45)

Similarly, the other observed that, when 'working with clinical samples [...] there are occasionally such startling results that you have to [...] find this person again', in order 'to report something back to someone, even though you said you wouldn't' (R30: Bioinformatician, Uni). In practice, then, given this prospect that researchers might discover things that participants 'should really know about [...] you're not going to break the tie completely' (R30).

Techno-scientific capabilities—or, perhaps more accurately, constraints—posed a second set of problems and concerns in terms of data protection. With respect to the withdrawal of data from studies, for example, this Repository-based IT Specialist who worked on a programme of data and sample collection admitted that promises made to participants in a consent form were, technologically speaking, impossible to meet once the data had been anonymised:

> The original current consent statement is that when, if you want to withdraw fully then your samples and data will be destroyed. And this is where the IT side of things I think is starting to show that people haven't, the science community hasn't got the basic understanding of IT and the implications of it. (R40)

The same respondent also repeated the view that a key difficulty was 'an implicit assumption' on the part of certain non-IT professionals, 'that once you get that data, storing it in a database is a piece of cake really. [...] And I wouldn't want to say that' (R40).

Two Clinical Scientists challenged the very feasibility of anonymising genetic data at all (R31: Uni/Hos), one of whom highlighted, in the context of his well-established longitudinal study, 'a big problem in assuring [...]

cohort members that there isn't going to be harm done by some inadvertent linkage of genotype information to them, or by implication to their families, now or in the future' (R15: Uni/Hos). As well as harming individual participants and their families, this latter respondent feared that any such damage to trust might impact negatively on the whole study (an issue that is discussed further in Chapter Nine).

Finally, three respondents specifically argued that data is over-protected—either in terms of current governance requirements, or as a matter of practice. In the context of small-scale clinical research, this Clinical Scientist perceived a misalignment between existing laws and the interests of research subjects (and researchers):

> I mean generally speaking I just worry that we're over-protecting, over-cautious in the way that we look after families, simply because they don't usually see it as in their interest to not be able to move freely and just get the information that we need and make it available. (R12: Uni/Hos)

A second respondent, this time a Research Scientist at a Repository, blamed over-protectiveness on confusion and uncertainty stemming from the DPA, with practitioners therefore being overly anxious not to get 'caught out' by violating it inadvertently:

> [P]eople who do research and people who govern research are anxious about the legalities of data protection and so they end up over-stating data protection in order to make sure they don't get caught out with having said something was okay when it wasn't. (R44)

Another Research Scientist (R27: Uni/Hos) also expressed concern that data protection laws are confusing and widely misunderstood. Notably, both of these latter respondents (R44, R27) cited the 2002 child murders in Soham, Cambridgeshire, as a point of reference.

Overall, one comment from a Clinical Scientist captured well a view, implicitly shared by many, that there is a need for scope within the law to negotiate the boundaries of confidentiality and data protection—not least, so that research is not unduly hampered. Such scope to balance competing interests is needed to manage the 'difference between being difficult and obstructive and the protection of the patient' (R28: Uni/Hos).

## Analysis and Discussion

Several illuminating insights emerge when we reflect on our respondents' five overarching, general governance-related concerns—namely: (1) the stifling effect of many governance-related rules or processes; (2) problems over the lack of legal clarity, and its consequences; (3) multiple governance system issues; (4) whether or not the law should reflect 'genetic exceptionalism'; and (5) the resource implications of various laws. Notably, these concerns relate both to the substantive content and structure of governance requirements, and to their implementation or enactment. In addition to posing practical problems, they provoked expressions of confusion and uncertainty—even anxiety or fear—among some respondents. They also draw attention to a widespread perception that governance measures themselves frequently impede, obstruct or even preclude—rather than support, facilitate or foster—legitimate biobanking-related activities.

This broad theme of governance being obstructive recurs in the three key, specific areas of concern that emerged from our data analysis—namely: (1) the ramifications of jurisdictional difference; (2) whether (and, if so, how) the law should reflect any distinction(s) between biological material and data; (3) and specific issues over data protection, data regulation, and balancing competing interests. Our findings here reveal instances of respondents, consciously and deliberately, either taking unnecessary regulatory steps, or adopting (or finding themselves subject to) excessively stringent standards. This commonly occurred because laws or applicable governance requirements were unclear, open to interpretation, unknown or misunderstood, and the respondents were anxious, therefore, not to fall foul of them accidentally. Moreover, sometimes respondents opted for exclusionary techniques to side-step governance-related difficulties altogether. These findings parallel similar exclusionary strategies used to avoid IPR (as noted previously in this chapter). In all such cases, the impact of complexities or flaws in the governance regime was to foreclose biobanking-related activities—activities that, otherwise, these respondents would have wished to undertake.

Multiple governance systems—both domestic and multi-jurisdictional—and jurisdictional difference were two, often intertwined, sources of difficulty. Our analysis uncovered three major challenges posed by 'jurisdictional difference': gaining understanding and knowledge of the various laws (and managing relationships); conflicts between different jurisdictions; and conflicts between different national laws and the aims of a given study. Associated problems included wastefulness (due to duplication, bureaucracy, and a lack of standardisation); conflicting or inconsistent decision-making; and managing incompatible, or cumulative, legal or ethical requirements. Problems over the resource implications of finding out about and implementing various laws or governance requirements—especially HTAct licences, different laws in different countries and retroactive changes—also affected a significant proportion of our respondents. As well as the associated stress or worry expressed over how they would fund their legal compliance, some respondents reported various negative impacts on practice. Significantly, all three of these specific areas of concern emerge as cross-cutting themes—in particular, mirroring similar concerns voiced in the previous section in this chapter.

Responses to the questions about the intersection between biological samples and data, the nature of each, and how each should be dealt with under the law, uncovered a strikingly rich—and diverse—spectrum of views.

Overall, individual opinions about the 'divide' (if any) between biological samples and data were mixed (some for; some against; others ambivalent) and these did not readily map onto professional or differences in working contexts. Our analysis of this theme demonstrates that these varying viewpoints were underpinned by a range of (potentially conflicting) forms of practical and ethical reasoning. These included logical or rational arguments, practical or pragmatic grounds, and more ontological or emotional reflections. Leading examples included actual physical characteristics (data being more easily 'accessible', samples being hazardous); potential characteristics (both samples and data being 'information'; samples being 'potential data'); claims about the relationship between the two (samples lack value without data; both must be managed coherently); the perceived purpose of legislation (protecting privacy vs ensuring ethical professional practice); public perceptions or concerns; comparative risks (abuse of data); and the 'special' emotional or ontological quality of biological material, especially some types such as whole organs. Interestingly, while several respondents sensed that some important difference may well lie in the latter, some found this extremely hard to articulate. Also interestingly, there was evidence that DNA samples, genetic material, and genotype data all raised potential 'grey areas' or blurriness for some respondents—including those who distinguished sharply between biological material and data. The frequently ambivalent comments expressed here correspond with respondents' more general views over whether or not the governance framework should adopt a stance of 'genetic exceptionalism'. Again, individual differences in opinion about this did not easily map onto differences in profession or workplace setting.

No clear consensus emerges from our sample of interviewees over whether biological materials and data generally should be subject to the same or different rules—or which (if either) might warrant higher, or more stringent, standards of protection. However, one point that many respondents did agree upon was the desirability of (greater) harmonisation between tissues and data regulation (often framed by explicit reference to the HTAct and DPA). Supporting reasons included to avoid conflicts, manage risks and ensure coherent practice. Overall, our findings strongly suggest that the currently bifurcated, conceptually sharply differentiated, legal approach to regulating biological materials and data does not fully align with the understandings, perceptions, expectations, or needs of practitioners engaged in biobanking activities (or, indeed, other stakeholders). Nor does it accurately reflect, or adequately manage, the realities and complexities of practice in many respects.

We have already identified (earlier in this chapter) numerous examples of such mismatches, problems, or gaps in relation to the HTAct. In relation to the DPA, in this section several specific concerns similarly emerged. Our analysis uncovered three dominant themes—namely, tensions over conflicting interests; mismatches between what is sought and what is possible; and complaints of data over-protection hampering research. Analysing these concerns reveals an equation, in which the interests or values underpinning the data protection regime (notably preserving confidentiality) need (or need better) to be negotiated or balanced against: (1) the needs of researchers and (perceived) wishes of participants; (2) the affordances offered by techno-scientific capabilities; and (3) the danger of 'over-protecting' data (especially where this is driven by misunderstanding, hypersensitivity or anxiety).

## CONCLUSIONS

This chapter has presented the second tranche of empirical research findings covering respondents' attitudes, beliefs, and concerns regarding governance and guidance on biobanks and related activities in England and Wales. Building on the foundations laid in Chapters Five and Six, the data here covered three substantive topics: (1) respondents' attitudes towards specific laws, focussing on the HTAct and IPR; (2) their attitudes towards specific regulatory bodies and research governance systems, focusing on RECs, the NHS, and funding bodies; and (3) five general, and three specific, areas of governance-related concern.

While we must defer a detailed overview and analysis of our principal findings and their implications to Chapter Eight, three observations are worth recording here. The first is that certain dominant, fundamental overarching or cross-cutting themes can be distilled from the data analysed in this chapter. When those analyses are consolidated, at least eight predominant, significant topics of concern for our respondents—and, potentially, for professionals engaged in biobanking activities more generally—come to the fore. These may be encapsulated as follows: (1) the law's treatment of biological material; (2) the law's treatment of data; (3) confusion, anxiety, and detrimental knock-on effects in practice caused by certain laws lacking clarity (notably, the HTAct and DPA); (4) resource implications (including implementing governance requirements in practice, the 'bureaucracy' of approvals processes, ongoing oversight systems, and funding retrospective, retroactive rule changes); (5) multiplicity problems (especially multiple governance systems; multiple sources of law, legal actors, or jurisdictional requirements; multi-funder issues; and aspects of jurisdictional difference); (6) problems or tensions due to mismatches between the governance framework, competing conceptualisations of the nature of 'biological materials' and 'data', and the realities (and limitations) of what is possible, or makes sense, in practice; (7) mismatches between the governance framework and key stakeholders' needs, interests, expectations, understandings or desires; and (8) unwarranted obstacles, stifling or hampering of biobanking-related activities, due to governance requirements (including laws, regulatory actors, governance systems and processes) being unnecessary, disproportionate, inappropriate, wasteful, poorly organised, poorly drafted, unclear, confusing, overly complex, overlapping, duplicative, inconsistent, incoherent, overly stringent and/or unduly inflexible.

Secondly, taken in the round, our respondents' comments during interview reveal a wide and illuminating array of attitudes and concerns. Each carries important implications and potential lessons for how the governance regime for biobanking may be improved. Many of the views expressed by respondents raise matters that specific policymakers, lawmakers or other regulatory actors may wish to consider and act upon (as set out above, and reiterated in Chapter Eight). Certainly, the complaints of our respondents should be taken seriously and investigated, even if it must be acknowledged that they may not be representative of all biomedical scientists working in the area; or, indeed, be reflective of the kinds of concerns that other stakeholders or actors may have—including those such as patients, research subjects, ethicists, lawyers, policymakers, lawmakers and so forth.

Thirdly, however, while the data consolidated and encapsulated immediately above expose widespread dissatisfaction, concern and hostility towards governance, they do not, in fact, paint a complete or accurate picture of practitioners' overall attitudes or characteristic outlook. Viewed in isolation, the data in this chapter may well convey a distinct impression that professionals engaged in biobanking activities are predominantly, overwhelmingly, and perhaps even implacably hostile to governance, whatever its shape or form. But this is by no means the case—as we shall discover in the following chapter.

# 8

# Preferences for Governance

*Susan MC Gibbons and Andrew Smart*

IN THE PREVIOUS chapter, we explored a wide range of criticisms and concerns expressed by respondents about governance laws, actors and systems relating to biobanking activities. These covered both general topics and more specific issues. If taken in isolation, the data contained in Chapter Seven (and in previous chapters, too) might well convey an impression that professionals engaged in biobanking activities are predominantly, and overwhelmingly, hostile to governance—in any shape or form. However, this is by no means a complete picture. The reality is much more subtle, nuanced, multifaceted and complex. Despite many practitioners holding serious (and it might be argued legitimate) concerns, it is not the case that they are opposed to governance per se. On the contrary, as this chapter—the final instalment in our trilogy of chapters on attitudes—will show, many respondents expressed a positive desire for governance. This included a desire, in many instances, for their own activities and projects to be subject to governance or oversight. Likewise, during interview, many respondents put forward concrete suggestions for possible improvements that might be made to the current biobanking governance framework. Once again, such improvements (or others) would be welcome.

The sociological research methodology has been described in detail elsewhere,[1] but notably for this chapter during interview, the 49 respondents were asked questions about suggested improvement to governance. Much of the analysis in this (and the previous) chapter is drawn from data that were coded to one of three 'Guidance' codes ('laws', 'advisory bodies', 'guidance'), and/or to an 'Attitudes to Governance' code.[2] Reflecting our data analysis process, whereby data was thematically coded and subdivided into the sub-codes reported here, this chapter is divided into two substantive sections, each presenting our major findings and data on a specific topic, followed by analysis and discussion. The first section reports on expressions

[1] For a full account, see chs 1 and 4.

[2] The analysis in this chapter was developed, refined and discussed by SG and AS, building on the initial coding completed by CH.

of a positive desire for regulation, and the reasons given for this. The subsequent section then reports and reflects upon the concrete suggestions put forward by respondents, of possible ways to improve the existing governance framework, and the reasons or explanations given. Finally, bringing together, integrating, summarising and systematising all of the principal data, findings and analyses from Chapter Six, Seven and Eight, together with relevant insights drawn from previous chapters, a final section concludes by presenting a detailed account of our major findings relating to respondents' attitudes towards governance, reflections on the lessons to be learned, and important implications for any future governance reforms.

## POSITIVE DESIRE FOR GOVERNANCE

### Key Findings

During interview, many respondents complained that biobanking activities are over-governed, or of feeling over-burdened or constrained by regulation in their practice. Memorable examples include claims that laws or regulatory actors 'hamstring' research, preclude 'serendipitous discovery', over-protect data, hinder progress or are 'very very anti-research'. Yet, at the same time, just over one-third of respondents expressed—unprompted—a positive desire for governance. This was not simply an abstract desire for governance, or for 'others' to be regulated. In a majority of cases, respondents expressed a positive, personal desire that their own projects, and they themselves, be subject directly to governance of some kind. Often, the desire was very strong. Respondents wanted not only governance for the biobanking sector per se, and/or changes to improve the current framework, but also, in some cases, an even greater level of governance than currently exists.

In this section, we examine respondents' expressions of a positive desire for governance, together with the reasons volunteered. We group these together under five broad categories: (1) wanting professional 'comfort' or support; (2) ensuring 'appropriateness'; (3) reassuring others and demonstrating legitimacy; (4) fostering participation and 'trust'; and (5) the practical benefits of governance. In the subsequent section, we continue this theme, by outlining respondents' concrete proposals or suggestions for improving the subsisting governance regime, together with the explanations or rationales offered for those.

#### *Professional Comfort and Support*

Several respondents (particularly those working with NHS patients, or with long-standing relationships with research participants) spoke of governance as providing a welcome source of comfort, protection, 'professional support', or 'confidence' that their arrangements and activities were 'appropriate', 'valid', 'well-founded' and so forth (R11: Clinical Scientist, Uni/Hos). Frequently, this was tied in with providing assurance to research participants—another prominent reason given for wanting governance, discussed below. Thus one University/Hospital-based Clinical Scientist, who had become a 'custodian of samples', felt 'extremely comfortable that there is this legislation [the Human Tissue Act 2004 (HTAct)] in place, I suppose to reassure participants that we're doing the right thing' (R42). This Epidemiologist, who (in his words) 'collaborated' with research participants on a well-established longitudinal study, had drawn confidence from completing the research ethics committee (REC) process:

> Cleary everything went through an MREC and they were overseeing what actually happened and making sure that the procedures [we] were adopting, were appropriate [...] So when all of that is in place and you find that your respondents are actually willing to collaborate as well then you feel fairly confident about the process. (R24: Uni)

A third respondent, who worked on small-scale clinical research in a University/Hospital setting, was happy being subject to governance, because it afforded a source of protection:

> [T]he legalities and clinical governance of freedom of access to medical records which we're, which we are of course obliged to work within. I've got no problem with that, if a patient wants to know what we've done, protects all of us in the end. (R27: Research Scientist)

As we have seen at various junctures during the previous chapters, many respondents expressed unease in relation to the current regulatory framework. Some fears stemmed from a lack of clarity in specific areas of the law—notably the HTAct, intellectual property rights (IPR), data protection and confidentiality. Some fears stemmed from feeling overwhelmed by the sheer morass of laws, guidance and governance sources on offer. But other respondents were apprehensive because they had discovered—in this case when initiating a novel data-sharing project—the lack of any clear or 'obvious' (R11: Clinical Scientist, Uni/Hos) regulatory framework setting out exactly what they needed to do, how, and when.

One notable area of concern was the management of participant-identifiable data. This NHS-based Clinical Scientist felt that '[i]t is a big responsibility dealing with a lot of patient-identifiable information' (R32). Another respondent working in the NHS put it thus: 'we're making these decisions, we're actually taking massive risks [...] having to determine for ourselves [...] the best way of handling [patient-identifiable information]' (R19: Public Health) in the absence of any clear, applicable regulatory framework. This respondent went on:

> That's what [named colleague] means by personal risk, there is nothing, no one we can go to and say well actually Security for Biobanking for Dummies says based on legislation this, that and the other we must do that. (R19)

It could be surmised that this reference to 'Security for Biobanking for Dummies' alludes to the *For Dummies* series of instructional books (published by John Wiley and Sons), and implies that the risks are compounded by the lack of basic, understandable and easily accessible information about governance which could be used to justify decision-making. Underlying this view was a concern that novel or prototype initiatives should set a rigorous governance template that others could follow, because 'what we do and the way we do it will form the way it's done around the world, purely and simply because we're kind of at the sharp edge' (R19).

To overcome such regulatory gaps and assuage their fears, some respondents looked to alternative sources for assurance or comfort, including self-created governance arrangements. For example, this Clinical Scientist who worked on a web-based data-sharing platform relied heavily on the project's own internal advisory board:

> Well really to give you the confidence that yes what you are doing is appropriate and you have taken the appropriate soundings and this is a valid way to go forward [...] and so I found it very valuable in terms of giving me the professional confidence to sort of keep moving the project forward, that we were doing, staying reasonably in the middle of the road. (R11: Uni/Hos)

The same respondent also had followed the REC route—despite believing it was a poor fit—because 'there wasn't an obvious alternative', and it was 'the only way we could get sort of government approval professionally':

> Well right at the start of the project the main difficulty was actually knowing what regulatory framework we should be operating under because [...] there didn't seem to be any regulatory framework that would give us any professional support [...] So the only way we could get sort of government approval professionally was to go via MREC route. [...] So it was because I couldn't find any clear framework for how you set up the database [...] and no one led me to an obvious framework of how to do it and how you'd configure it and how you'd regulate it. (R11)

### *Ensuring Appropriateness*

Three respondents (all of whom were less directly involved with NHS patients) spoke of governance procedures as a valuable mechanism for ensuring that their activities were more broadly 'appropriate' (R24: Epidemiologist, Uni)—in the sense of according with societal views or public sentiment. In the words of one Research Scientist at a Repository lodging genomic data:

> We are happy to get guidelines from ethics committees, from regulatory agencies because we want to be in compliance with what society, how society thinks we should conduct our business. (R37)

Similarly, in the absence of any specifically tailored governance framework, a private sector Research Scientist spoke of having 'all these inputs' as being beneficial, 'to make sure that we're aligned with public sentiment, public concerns, with best practices as it were and you know that's the best we can do' (R34: Pharma).

*Providing Assurance, Demonstrating Legitimacy*

The word 'appropriate' came up repeatedly. Numerous respondents, across different working contexts, drew on the concept of 'appropriateness' in the sense of seeing governance—and especially regulatory body approvals or accreditations—as a crucial means for providing assurance of 'appropriateness' to other stakeholders, or for demonstrating their project's legitimacy or credibility. Key stakeholders mentioned (or implied) were participants, potential participants, biobank users (such as external researchers), funding bodies, peers, and 'society' or the 'public' generally. Here, respondents desired robust governance at least in part to enable them to certify—and, thus, to present to others—their ethico-legal competencies. One University/Hospital-based Clinical Scientist, for example, was particularly pleased that a study was 'NCRN badged' (R47).[3]

Two respondents spoke explicitly in terms of using governance to 'prove' themselves in some way. In addition to being 'NCRN badged', the previous respondent's project also had obtained fresh REC approval—probably unnecessarily—in part, because 'we wanted to prove we were [...] complying with MRC guidelines' (R47). Likewise, a pharmaceutical company employee, describing how 'thorough' the company had been in following de-identification and data-coding procedures, explained that 'it is a very sensitive area and I think they want to be sure, I think they want to prove to the world that they're being as thorough as they can be' (R22: Research Scientist). A third respondent, who worked at a newly established Repository, also had taken the REC route, 'although officially we didn't need REC approval [...] because we're setting ourselves out as open to public scrutiny and honest broker we wanted that approval' (R36: Bioinformatician).

Three respondents described having technical standards to follow, regular audits, or an 'accredited quality control system' (R1: Bioinformatician, Uni) as essential means for biobanks to demonstrate legitimacy and quality. This included demonstrating the technical quality of their collected data and biosamples. For this Clinical Scientist, who had acquired an otherwise inappropriate REC approval for a novel data-sharing endeavour (as discussed in Chapter Seven), a supplementary reason given for so doing, 'was trying to find some way of quality assuring the project [...] and at the end of the day the only real benchmark we could get for that was an MREC'

[3] The National Cancer Research Network (NCRN), established by the Department of Health, provides an infrastructure for the NHS to support clinical trials, other studies and research.

(R11: Uni/Hos). Meanwhile, this Research Scientist saw regulatory approvals both as a mark of credibility for projects, and as an important source of professional 'recognition' for the individual practitioners involved:

> [I]f you pride yourself on your work and that you want to be seen as a centre of excellence in the work you're doing you want some form of recognition so that people believe that that's what you're offering. (R3: Uni)

As this last extract indicates, while the desire for approval may not be motivated by a wish simply to *appear* trustworthy, nevertheless one function of an approval or certification process is that it enables people to make judgments about trustworthiness.

### *Fostering Participation and Trust*

Many respondents viewed governance as a vital element for securing the public support and voluntary participation upon which biobanking and biomedical research fundamentally depend. Two Research Scientists working in contexts in which public trust remains a sensitive issue described various governance measures as strategically helpful, to foster participant recruitment or general support. Despite seeing RECs as 'the major stumbling point', and 'the clinical investigators we work with' as 'the next stumbling point', a respondent working in the private sector nevertheless considered having to obtain agreement from both actors to be strategically useful: 'if we have those two on board then the patient is never a problem' (R34: Pharma). The other respondent, who worked on a national programme of sample and data collection, thought that given the prevailing climate of sensitivity surrounding biobanking and genetics, and in the absence of any dedicated, unitary biobanking approval process, it 'helps' to tick as many regulatory boxes as possible:

> And it actually helps us I think if we can say look you know we have MREC approval, we have a Human Tissue Authority Licence, we've been inspected by a government agency which says that we are doing this you know to high ethical standards, we have the [named protocol], and [the named policy document] which is very, very clear about how this information can be used. I think that actually helps us. (R8: Repository)

Several respondents who desired governance to attract research participants (or prospective participants) spoke in terms of engendering 'trust'. The general tone of these comments was that having laws or governance per se, and being able to show compliance with them, helps to engender 'trust' in professionals, projects or processes. Such 'trust' was seen as being important for managing their relationships with patients, participants and members of the public. A Genetic Counsellor, for example, spoke of 'wanting people [who] go in research to know that they can trust us rather than a kind of reactive in terms of thinking about breaches that have happened and having your wrists slapped and therefore try and do it retrospectively' (R33: NHS). This Clinical Scientist, based in a University/Hospital setting but who also worked with a Repository, saw merit in having legislation in place (specifically, the HTAct), 'to reassure participants that we're doing the right thing and we're not exploiting them in any way or anything that they've donated to us' (R42: Uni/Hos). Meanwhile, a Repository-based *Bioinformatician* wished that biobanks had an oversight body that operated like the Human Fertilisation and Embryology Authority, because 'it would provide a level of reassurance to a whole bunch of people that you know scientists aren't allowed to be completely loopy which is the worry' (R10).

### *Practical Benefits*

Finally, respondents in various settings believed that various practical benefits flow from having recognisable processes of governance. Two described external governance input—specifically, questions or suggestions from RECs (R23: Research Scientist, Pharma; R32: Clinical Scientist, NHS)—as a source of improvements to template documents and a consent form respectively. Others pointed to internal governance and management procedures within their workplaces as sources of important benefits. Despite complaining over the time, paperwork, and record-keeping involved, three respondents noted positive gains from having detailed, explicit procedures laid down in Standard Operating Procedures (SOPs) or technical standards. One, for example, thought it provided 'a very good paper trail' and 'very good method' to 'go back and [...] just double check that you haven't got something wrong' (R3: Research Scientist, Uni). Other perceived benefits included that 'everything is much more streamlined and everybody gets the same information, everybody gets the same training' which can help in preventing people from 'deviating' from expectations (R3); facilitating development over time by starting an 'improvement cycle' (R40: IT Specialist, Repository); and, ensuring continuity when personnel change and reducing errors (R19: Public Health, NHS).

One (usually implicit) reason why respondents desired changes to the current governance framework—including for some additional laws or regulatory provisions—was to remedy or redress the specific guidance issues, complaints, criticisms, gaps, or other concerns identified throughout this chapter. Touching on this, two University/Hospital-based Clinical Scientists alluded to problems associated with individual projects or practitioners having had to 'reinvent the wheel' (R28; R31), by formulating their own, ad hoc, individual governance solutions, methods or procedures. In the words of this repository-based Research Scientist, it is 'quite quite tiring. Well tiring is not the right word but we don't have yet a working solution if you like which is uniformly accepted' (R37). The implication was that having more formal governance would help to alleviate such difficulties.

**Analysis and Discussion**

Reflecting on these findings, five important insights emerge. First, it is noteworthy that so many respondents positively desired to have clear governance for their biobanking activities—not only implicitly, but frequently explicitly articulated, often with some force, and unprompted. Moreover, these respondents wanted it not simply in abstract terms, or as governance for 'others', but directly applicable to themselves, and their own practices and projects. When we unpack their five principal reasons given for this, a fascinating and complex picture emerges of their various underlying contexts, motivations, beliefs, attitudes, expectations and perceptions. These relate not only to respondents' views about governance per se—what it should be, what it should do, its core purposes, functions, value and utility—but also about themselves, other stakeholders, their relationships with them and the role of certain regulatory actors.

Secondly, the positive expressions of a desire for governance appear to be underpinned by a very particular vision of what biobanking and biomedical governance should seek to do. Our data analysis demonstrates that, for many respondents, 'governance' has—or, ideally, should perform—two fundamental purposes or functions. Put very succinctly: (1) the primary, twofold purpose of governance is to confirm legitimacy, and to convey a 'message' of legitimacy to other stakeholders; and (2) a subsidiary (but, nevertheless, still important, expected, and valued) purpose of governance is simply to spell out for practitioners, in practical, accessible and legally compliant terms, precisely what they need to do.

From a practitioners' perspective, being provided with clear routes to compliance can assist them towards being able to gain the legitimacy embodied by the law or guidance to which they are subject (and then to be able to proclaim this). This demand for clarity accords with a refrain that we have found to be commonplace when biomedical scientists interact with lawyers and ethicists. To paraphrase, the scientists sometimes say: 'Just tell us what you want us to do, so that we can do it and then get on with the science'. There is, however, a tension between such demands for clarity from non-science professionals or experts, and our finding in Chapter Six that some scientists claim to 'know best'. There is, thus, a potential for discordance between the desire (and/or requirement) for legitimation, and the willingness and/or ability of some scientists to comply with the governance regimes to which they are subject.

Expanding on these observations, it is illuminating to reflect on how respondents tended to interpolate the first four principal explanations put forward for wanting clear governance—namely, wanting professional comfort; ensuring 'appropriateness' (both in a normative sense, in terms of aligning with 'public sentiment', and as an instrumental tool to support recruitment or funding); reassuring others (by demonstrating 'appropriateness', validity, legitimacy, credibility and so forth); and fostering participation and 'trust'. What emerges from each of these, and their interplay, is a strong sense that, for the respondents that commented, at root 'governance' principally exists—or, where currently absent or deficient, should exist—to provide a crucial tool. That tool is crucial to *confirm* practitioners' (perhaps pre-existing) knowledge or belief that their projects, purposes and practices are 'appropriate' (including normatively), and to *demonstrate* that appropriateness, legitimacy and so forth to those stakeholders that really matter. The most important stakeholder audience for this governance message is participants, prospective participants and the 'public'. But biobank users, funding bodies, and peers also are important 'target audiences', as practitioners need to interrelate with them, and depend upon them in various ways, too. Thus, a wide group of stakeholders forms the target audience for biobanking governance—governance as submitted to by professionals and projects, and as carried out, and verified, by relevant regulatory authorities.

This commonplace perception that the main purpose of governance is (or should be) to provide a confirmatory device or symbolic badge of 'appropriateness', accords with earlier findings in this series of chapters.[4] In particular, it recalls our conclusions about respondents' beliefs, attitudes and professional self-image in Chapter Six. There, we characterised an outlook among some respondents whereby they appeared to feel that they 'already know' what to do; are trustworthy, moral agents of 'doing good'; and so governance should serve to facilitate biobanking activities whilst 'reassuring others' that all is well. It also accords with our findings in Chapter Six (and also in Chapter Nine) that some respondents believe that their own interests align or are concurrent with the interests of participants.

In addition to serving its primary, twofold purpose, our findings also demonstrate that governance may also be expected to perform a second fundamental purpose or function—namely, to spell out, in practical, accessible and legally compliant terms, what practitioners need to do. Thus, alongside normative grounds, our analysis identified a wide range of instrumental benefits from having governance. These, too, lay behind a strong, positive desire to be governed. Being able to demonstrate legitimacy and quality was considered very important for some respondents. Having clear governance systems—and (again) being able to prove one's compliance with them—were viewed as being crucial in order to be seen as a 'centre of excellence', to prove a biobank's credentials, to secure participant and public 'trust', to ensure that 'people believe' that claims about what is being offering are true, and to assist with recruitment. Other practical benefits included continuity, error-avoidance and standardisation.

[4] This was also a finding from a study of the UK Stem Cell Bank. See N Stephens, P Atkinson and P Glasner, 'The UK Stem Cell Bank: Securing the Past, Validating the Present, Protecting the Future' (2008) 17(1) *Science as Culture* 43.

Within this wide range of different benefits, in addition to more strategic goals (such as fostering participation), it is noteworthy that some respondents placed considerable value on having clear procedures or standards to follow. Specifically, they wanted governance to provide them with clear frameworks—especially frameworks containing the 'nuts and bolts' for how to set up and run databases; manage participant-identifiable data; and design and operate database IT systems in a unified, standardised way—that embody all of the relevant legal requirements. Tellingly, as we shall see again in the next section, the positive desire for this type of governance or guidance was very marked indeed.

Thirdly, an influential cross-cutting theme within our data analysis is how some respondents widely perceived biobanking to be a particularly 'sensitive' area. Often, this belief underlay a positive desire for more stringent or enhanced governance. 'Sensitivity' had two aspects. One was linked to recent scandals (such as those that prompted the HTAct), and the perception that there is a heightened public sensitivity over genetics. Thus, we found examples of respondents welcoming new legislation to reassure people that scientists are not abusing people's trust or donated material, or being 'completely loopy'. The other aspect of 'sensitivity' involved the newness of biobanking. Thus, we found examples of respondents speaking of the need for robust governance to 'prove' to the world that biobanks and associated practices are legitimate, and for 'path-breaking' initiatives, in particular, to be governed closely from the outset, so that correct benchmarks are set for the future.

Fourthly, it is noteworthy that several respondents looked to governance systems and regulatory actors—particularly RECs—to ensure 'appropriateness' at a normative level. This was framed in terms of biobanking activities aligning with 'what society wants', 'public sentiment', or 'public concerns'. It is argued that underlying this rationale is an implicit perception, or expectation, that RECs and 'regulatory agencies' do (or could, and should) accurately represent, embody and enact accepted public or societal norms. Underlying that perception, in turn, may be an even deeper presumption that such settled norms actually exist in the first place. Despite many respondents being highly critical of RECs, and dubious about other regulatory bodies, for at least some, their implicit understanding of the nature of such bodies is that they represent, and can mediate or 'speak for', 'society' or the 'public' view.

Fifthly, it is striking to compare the regulatory space analysis in Chapter Three with the list of stakeholders that were important to respondents here, because they comprise the core target audience for the governance 'message'—namely, participants, potential participants, 'society' or the 'public', as well as biobank users (such as external researchers), funding bodies and professional peers. In Chapter Three, aside from funding bodies, none of these stakeholders featured directly as 'regulatory actors'. None have any obvious regulatory power, or ability to control the six key regulatory space resources that were identified.[5] Yet, our findings here suggest that such stakeholders have considerable implicit significance to professionals, due to their interrelationships and interdependencies. It would be going too far to suggest that this renders such stakeholders 'regulatory actors', or gives them an effective 'voice' or power of their own in the biobanking governance regulatory space. The extent of their implicit influence here fundamentally depends upon (and is determined and defined by) what practitioners believe the interests, concerns, perceptions, preferences and so forth of such 'silent' stakeholders to be. As we have seen, this, in turn, is shaped (and limited) by how far practitioners perceive their own interests to align, elide, or coincide with the (supposed) interests of others. Nevertheless, at least drawing the comparison serves to highlight the fact that, for at least some of our respondents—and, perhaps, practitioners more widely—the interests and concerns of key 'silent' stakeholders are important matters that the governance framework properly should take into account and address.

## SUGGESTIONS FOR IMPROVEMENT

### Key Findings

Towards the end of their interview, after having discussed biobanking governance needs generally and any problems or concerns, respondents were asked what changes they thought should be made to improve the current governance system or guidance. They were also asked what an appropriate set of legal instruments or guidelines for biobanks might look like. These questions prompted just over one-third of respondents to make explicit, concrete suggestions for possible ways to improve the governance framework. Many respondents offered up a range of ideas, often covering multiple different topics. Some identified several options for resolving particular problems. We encapsulate below the principal, concrete suggestions made for improvement—and the reasons given for them—organised under six broad headings: (1) political and policy steps; (2) governance frameworks; (3) specific law changes; (4) oversight and governance bodies; (5) guidance sources and dissemination; and (6) alternatives to governance.

[5] The six key resources were: (1) formal legal authority; (2) possession and control of information; (3) possession of wealth; (4) organisational capacities; (5) the ability to publish or disseminate one's views and preferences effectively, and to persuade others to accept them as being authoritative or persuasive; and (6) control over professional education, the setting of good practice norms, and the inculcation of professional cultures.

*Political and Policy Steps*

At the political and policy level, respondents made specific suggestions. One Public Health practitioner argued 'that with the ethics there's only very much a biomedical ethical framework, there isn't a public health ethical framework' (R19: NHS). We interpret 'the public-health ethical framework' to include ideas of the common good, public benefits and cost-effective use of resources. Accordingly, this person suggested establishing 'a public health ethical framework to balance [...] the biomedical one', since the latter is 'all individualistic' (R19). Underpinning this proposal was the more specific assertion that 'we haven't got the wider policymakers understanding the difference between screening and testing very well', coupled with a desire to see a more communitarian set of 'principles'—rather than traditional, 'individualist ones'—become accepted as the relevant factors at play within the genetics and biobanking fields (R19).

Reflecting a wish that was implicit in many of the responses we have reported and analysed above, a Clinical Scientist involved in a well-established longitudinal study called on the major UK funding bodies to agree a limited set of common policies, procedures and requirements (R15: Uni/Hos). In the interview, this request was particularly grounded in relation to issues of open access. As noted already, issues associated with open access were discussed at length by a number of respondents, and a more detailed account of their experiences, concerns and proposals is presented in Chapter Nine. The same respondent (who identified the 'multi-funder problem', discussed above) also called on those bodies to clarify between themselves the 'political balance' between different co-funders; for example, in terms of which of them should be involved directly in governing individual biobanks or large-scale projects (R15).

Several respondents called for greater consultation. They proposed that policymakers, lawmakers and regulatory bodies should consult with—and act on the input from—key stakeholders, when formulating policies, governance and guidelines. Key stakeholders identified were researchers and scientists (in particular, to ensure that rules do not preclude opportunities to make scientific progress); affected patient support groups; participants and patients; the wider community or public; and (in relation to formulating open-access policies or rules) data-holders, the original collectors of datasets, and secondary researchers or data-users. One Bioinformatician at a newestablished repository suggested that a big representative organisation—'such as the Wellcome Trust [...] or it could happen at the Department of Health level'—should bring together 'people from various other organisations' and 'researchers' to 'forge [...] dialogue' to develop uniform IT security standards (R20). This should include 'clear directive[s] as to how you should isolate [...] patient-identifiable data from other patient data and how you should store it, how you manage at the hardware and the database level';

what needs to be done when 'exporting data'; uniform standards for coding data; and minimum standards for traceability (R20). Finally, this University-based Epidemiologist emphasised the need for proper empirical research:

> So the other thing I would say about guidelines is they should be driven by empirical evidence that that's what people really feel, either qualitative or quantitative surveys because at the moment [...] people just with strong opinions on ethics committees say oh people don't like that [...] it should be based on empirical demonstration that people don't really like that. (R6)

As this latter respondent specifically notes (and this is implicit in many of the other examples in this sub-section), a question was raised about the extent to which policy decisions (or the initiatives that create them) are sufficiently inclusive of all relevant 'voices'. This is not to accept at face value the views of these respondents that the organisations named above, or others, have not engaged in appropriate consultations—or, indeed, have not considered questions about credit and so forth. Rather, it is simply to note that these biomedical scientists appear unsatisfied with key policy processes or outcomes, and their complaints are being couched in terms of fairness, communication and inclusion.

### *Governance Frameworks*

Another suggestion was the introduction of various forms of governance frameworks. As just noted, one Repository-based *Bioinformatician* (R20) called for a clear, detailed framework for designing, setting up and running biobank IT systems—especially setting out uniform, standardised data security standards and procedures, not least for handling participant-identifiable data. In terms of the actual content of such a framework, as well as covering the matters noted immediately above, another *Bioinformatician* at a different Repository urged that the standards set for 'computer security' should be proportionate, taking:

> [T]he same approach to security as one would take to [the] physical building in which we store the information and you know this is not MI5 [...] so one needs to get some balance about that and [... not] go off the deep end. (R10)

Another popular suggestion was to create a clear, common regulatory framework for governing sample and data collections—both on the national and, some suggested, international scale. This was offered by seven respondents working on medium, large and very large-scale projects, including two working at a particular repository, others involved in data-sharing projects and a well-established longitudinal research project in University or University/Hospital settings and a Public Health practitioner. It was argued that such a framework should be coupled with detailed guidance at the 'operational level' (R19: Public Health, NHS) about what is required, and how to implement the 'higher level' framework requirements (R36: Bioinformatician, Repository) in practice. For example, respondents variously spoke of having a 'clear framework for how you set up the database [...] and how you'd configure it and how you'd regulate it' (R11: Clinical Scientist, Uni/Hos); having a regulatory framework to follow, together with guidance on the 'fine detail' (R19), that covers 'the day-to-day technical, how you do your job' (R24: Epidemiologist, Uni); and having consistent, 'generic rules' or a 'generic solution' (R24), with the possibility of tailoring this to the specifics of individual contexts (R15: Clinical Scientist, Uni/Hos; R24: Epidemiologist, Uni; R28: Clinical Scientist, Uni/Hos; R44: Research Scientist, Repository). One respondent—mirroring the findings reported by Catherine Heeney in Chapter Four—noted that such a common framework would have 'to be flexible because many genetic collections [...] vary tremendously' (R24). Similarly, this repository-based Bioinformatician—who argued that developing 'a framework for how [...] databases are eventually used, get used and eventually destroyed [...] is something that's very, very important'—also acknowledged that '[c]reating a common framework for genetic databases in commercial organisations, in charity organisations, in common [public] organisations is going to be very, very challenging' because of their enormous variability in practice (R20).

Aside from providing certainty, clarity, and convenience, avoiding 'reinventing the wheel' (R28: Clinical Scientist, Uni/Hos), and filling gaps in the existing law and regulatory matrix, among the reasons given for wanting a common governance framework of 'generic rules' was so that 'future' projects 'can be designed around it' and 'embedded' within it—that is, designed from the outset to be in compliance (R20: Bioinformatician, Repository). Among the calls for both a framework and associated practical implementation guidance, this Public Health practitioner stressed that 'you need to sort out the services as well' as having 'just ethical statements' or 'saying this is the policy':

> [Y]ou know people like sexy to do policies, sexy issue guidance but actually [...] this is about implementation on the ground, that doesn't attract status or anything. But ultimately [...] you know in practice, put your money where your mouth is and make it happen rather than just pronounce something. (R19: NHS)

In terms of the precise form of the framework-plus-guidance that might be created, one respondent suggested designing a flexible 'road map', 'checklist', 'step-by-step guide' or 'wizard'—perhaps akin to the Medical Research Council's clinical trials tool kit—that people could work through, with prompts, showing what is expected and what must be done in order to comply (R30: Bioinformatician, Uni). Another referred to having a 'crib sheet' covering any given 'common situation' (R28: Clinical Scientist, Uni/Hos).

*Specific Law Changes*

In Chapter Seven, we highlighted a wealth of problems relating to specific laws—notably, the HTAct, IPR and the Data Protection Act 1998 (DPA). A number of respondents made concrete suggestions for specific law changes, addressing seven different areas. Two sought new laws setting out property rights and ownership in respect of human biological materials. One, a repository-based Bioinformatician, suggested this must include law on how to 'value' samples and collections financially for the purposes of buying, selling or transferring biobank resources (R20). The other, a Clinical Scientist involved in a well-established longitudinal research project, wanted the new law to clarify who 'owns' samples or collections, especially where multiple institutions are involved (R15: Uni/Hos). As the latter put it, 'it's time to sort out who owns what' (R15).

Seven respondents who worked in a variety of settings called for clear, authoritative, legal definitions of commonly used terms. Some wanted clear UK law on such points. Others sought wider, international standardisation of terminology and definitions (perhaps via an agreed lexicon). Laws were specifically sought defining the concept of 'open access', and also the following terms: 'high quality sample', 'consent', 'informed consent', 'confidential', 'confidentiality', 'anonymous', and 'anonymity'. In a related vein, one Bioinformatician suggested legislation or 'some sort of legal framework' to define, and stipulate the parameters of, the concept of 'broad consent':

> I think legislation on how broad consent can be would be useful because we have a problem you know in that we will be doing things [with samples] that [...] were inconceivable when we recruited. (R30: Uni)

One underlying reason given for wanting formal, legal definitions was the fact that having differing, and uncertain, legal interpretations—both domestically, and as between the UK and other countries—makes collaborations difficult (R30):

> Because there is no formal interpretation it impedes us when we have to work with people who take a very different stance. (R30)

Accordingly, 'it would be useful to have a real law that said define what counts as confidential or not in this country' (R30). Similarly, a Research Scientist who worked in a well-established longitudinal research project had found it problematic, in dealing with a REC, 'debating their interpretation of the MRC guidelines as opposed to our interpretation of the MRC guidelines' on the meaning of 'consent' (R5: Uni).

Continuing the theme of standardised meanings, where terms must be subject to interpretation, a University/Hospital-based Clinical Scientist called for standardised instructions or guidelines for particular regulatory actors, to ensure that they interpret laws and governance requirements in a consistent manner, and to overcome problems of inconsistency. Specifically targeted here were RECs and NHS data protection officers:

> [RECs] should not interpret laws themselves, there should be a simple set of frameworks for how those laws are interpreted. Data protection we've now got data protection officers throughout PCTs and various things, totally different interpretation from different people. There needs to be a standardisation of that, okay. (R31)

To overcome problems with material transfer agreements (MTAs)—such as delays, transaction costs and the involvement of lawyers, as identified in Chapter Seven—one Bioinformatician at a repository lodging genomic data suggested extending the 'science commons' notion, and using 'creative commons licences' in place of MTAs (R39). Three other suggestions for new legislation were made by one respondent each. Another Bioinformatician at the same repository strongly advocated introducing 'appropriate legislation' to prohibit genetic discrimination, 'that said employers, insurance companies or whatever just is not allowed to use your genetic information in this manner'—again, to provide an 'extra level of reassurance' for the public (R10). A Bioinformatician who worked in the University sector sought a law precluding police access to, or searching of, biomedical and research databases for forensic purposes—especially to reassure participants and make the consent process less complicated (R30). A third respondent, an Epidemiologist, called for a legal requirement that all publicly funded data collectors must lodge their data somewhere (rather than destroying it) at the end of their projects (R18: Uni).

### *Oversight and Governance Bodies*

All but one of the specific suggestions made for improving oversight and governance bodies related to RECs. In order to achieve greater consistency (especially of interpretations and decision-making), several respondents proposed that RECs should work to clear, national guidelines—or a simple, consistent set of frameworks and interpretations—that they can follow easily. Given the professional interests of most our respondents, it is unsurprising that their concerns were focused on 'genetic' research. Thus, one University/Hospital-based Clinical Scientist who worked on small-scale clinical research advocated introducing a 'generic genetics application' and REC approval process (R14). Another mentioned formulating clear national guidelines 'particularly for family history and genetic databases', that should be 'checked by some other professional bodies as well' (R47), and a third advocated a centralised REC guideline for use in genetic studies. The aim here was to avoid researchers having to make repeated REC applications—for example, to authorise research using the same materials but looking at different genes—and to alleviate associated administrative burdens.

More generally, one Bioinformatician suggested having a 'central registry' to give ethical approvals for international collaborations (R39: Repository). This was seen as a means to overcome problems for multinational collaborations due to different countries having varying ethical standards, Another respondent, a Clinical Scientist, urged that RECs should have their own source of legal advice (R47: Uni/Hos), rather than requiring applicants to obtain and furnish legal advice when questions about the law arise.

Finally, in relation to other types of oversight bodies, two respondents each suggested having some method by which researchers could 'establish a relationship with a suitable body', such as a 'steering committee' R1: Bioinformatician, Uni). One idea behind this was to give practitioners a body to consult, or to which to submit queries for advice, on a rolling basis during the lifetime of projects. Another idea was to have a body that could make policy decisions, and decide such matters as when ethical approvals needed to be sought (R1; R23: Research Scientist, Pharma). It was unclear whether these respondents had in mind a project-specific internal committee, or some external body—perhaps a national bioethics council or committee, such as exists in virtually all Western European and in many other countries around the world[6]—to perform such functions. Clearly, the nature and constitution of such an oversight system might need to differ drastically for different kinds of research, and within different settings (and some researchers may already have this kind of oversight in place).[7] The key implicit point seemed to be the ability to have direct access to, and to build up a working relationship with, a 'suitable' body, based on that body having some knowledge and understanding of the project.

### *Guidance Sources and Dissemination*

Only three respondents, all working in University/Hospital settings, made concrete suggestions about alternative guidance sources, and how information about the law best could be disseminated to practitioners. One Clinical Scientist suggested that each Primary Care Trust should have a 'genetic lawyer' (R28). Part of that person's role would be to 'sift' through all of the laws and 'mountains of information' that come in (mostly 'from outside'); provide relevant information, or summaries of relevant governance requirements, to staff; be a source of legal advice; and produce a

[6] D Dickenson, 'Should the UK have a National Ethics Committee?' *BioNews* 479 (6 October 2008), <http://www.bionews.org.uk/commentary.lasso?storyid=4023>.

[7] See further SMC Gibbons, 'Regulating Biobanks: A Twelve-point Typological Tool' (2009) 17(3) *Medical Law Review* 313–46.

'crib sheet' for common situations (R28). A second Clinical Scientist more vaguely proposed a similar refinement process:

> [W]ith the current rules on databases and on sample collections [...] there are an enormous number of rules and more and more seem to be coming out. And I think somebody needs to take a look at them and refine them in some way. (R9)

A third respondent, an Epidemiologist, proposed making information about key aspects of the HTAct easier to locate and understand—for example, by having a website to 'break it down into perhaps more easy to access information', about such matters as 'exactly who to approach, where to get information from, what sort of ethical approval we would need' (R46).

### *Alternatives to Governance*

Finally, a few respondents felt that governance reform was not the answer—or, at least, not the complete answer. In the view of one Clinical Scientist, government-imposed regulation 'always backfires' (R17: Uni/Hos). Accordingly, it is better to rely on peer review: 'leave it to the professionals' (R17):

> [T]here's probably enough governance in place from my point of view. And [...] I guess the idea of governance structures for me is minimalist and you know we need the ethics sort of control. But the more [...] government tries to micro-manage research or tries to say we should only be doing research on [...] these applied questions or whatever it always backfires in every country I've seen them do it. [...] It doesn't work, people subvert it [...] governance I would like to be as minimal as possible. (R17)

Another suggestion was to use education—rather than governance driven by genetics being 'over-hyped' (R30: Bioinformatician, Uni), scandals, and ungrounded speculation over supposed public concerns—to change the culture, correct public misunderstandings, and allay unfounded fears. One idea was to have a full-scale, public education campaign about genetics and biobanking.

## Analysis and Discussion

These findings are instructive and useful at multiple levels. At a practical level, they help to illuminate further the topics of special concern expressed by our sample of respondents. They address some of the key areas where regulatory gaps either exist, or are thought to exist. They draw on and reflect practitioners' own experiences, insights and knowledge. Many of the suggestions made have relevance and application to particular regulatory actors. While some are vaguer or more broad brush, many suggestions are quite specific—and, potentially, suitable for being translated into workable measures.

In more analytical terms, while many respondents did put forward concrete suggestions, it is interesting to note the high ratio of complaints and concerns to actual concrete proposals. Both the volume of complaints and concerns reported in the previous chapter and in this chapter, and the range of topics covered, clearly outstrip the concrete suggestions made for improvement. This is, perhaps, partly a reflection of the fact that biobanking is a complex area, in which there are no obvious, easy, quick fixes or solutions. It may well be a function, too, of the nature of our sample—which, for the most part, comprised professionals engaged in biobanking practice (in a wide range of contexts) who did not describe themselves as having had any prior personal involvement or expertise in contributing to processes of designing regulations or guidelines or in acting as regulators. But it also is consistent with our findings that some respondents were not always fully aware of the current laws and governance framework, and/or were not overtly investigating what the laws or guidance say.

Reflecting on the spectrum of suggestions that were made, their breadth is noteworthy—from the political and policy level (often from those involved in grander-scale projects), down to the 'hands on', day-to-day practice level. That the recommendations are more heavily weighted towards the practical, implementation end of the spectrum is also a valuable finding. Specifically in relation to oversight and governance bodies, it is striking—although, not surprising given the amount of negative feedback reported previously—that almost all suggested improvements concerned RECs. Moreover, all but one of our respondents implicitly simply took for granted the appropriateness, inevitability or ubiquity of the REC system. Despite the catalogue of complaints about RECs reported in Chapter Seven and above, not one respondent suggested doing away with, replacing or fundamentally altering, the REC system. Their suggestions focused exclusively on possible ways to improve it. Indeed, only one respondent stated explicitly that all biobanks *should* go through the REC process (with improvements). This could imply that all others presumptively took it as 'a given'.

Also worth highlighting is the fact that two suggestions made for improving guidance sources and dissemination—namely, the call for NHS 'genetic lawyers', and for more accessible information breaking down the HTAct—suggest that the respondents concerned were unaware of resources already available. For researchers and others who are associated with the NHS, the Service has in-house legal departments. They may not act as information 'sifters' or conduits in the manner suggested by our respondent, or include genetics law specialists. It may be worth the NHS considering these additional, or complementary, roles for its legal teams. However, such departments are, at least, available to provide legal advice to NHS staff. Equally, an array of internet-accessible, 'broken-down' information about the HTAct, specifically designed for practitioners, is already available—notably, on the Human Tissue Authority and Department of Health websites. These observations reflect our findings in Chapter Five, about respondents' generally low awareness of guidance on offer. They also suggest that, alongside any governance reforms, an educational campaign to keep practitioners properly informed of the best places to go for authoritative information, guidance or advice may well be beneficial.

## CONCLUSIONS

This chapter has presented the third, and final, tranche of research findings covering respondents' attitudes, beliefs, assumptions, perceptions, and concerns regarding governance and guidance on biobanks and related activities in England and Wales. The data here covered two substantive topics. These were: (1) expressions of a positive desire for governance, and the underlying rationales; and (2) concrete suggestions for improving how biobanks currently are governed, together with supporting reasons.

As was foreshadowed in Chapters Six and Seven, the 'positive', welcoming attitudes towards governance explored in this chapter contrast markedly with the more 'negative', and often strongly hostile, attitudes that we encountered in the previous chapters. It is for this reason in particular that the data and findings from all three chapters must be read together, so that an accurate picture is portrayed. Overall, as the trilogy of chapters has unfolded, we have seen many illuminating and instructive findings emerge from our data analysis, including some that may raise concern for those charged with responsibility for governance. Notably, analysing our data on attitudes has helped to reveal where existing laws, guidance instruments, actors, and processes—or the lack thereof (actual or perceived)—are causing the greatest concern for, or impact on, working practices. Discussion of these issues also has served to highlight potential 'mismatches' between the existing regulatory regime, and the understandings, perceptions, experiences, concerns and behaviour of professionals engaged in biobanking activities.

Drawing together all of the data, findings, and analyses from Chapter Six, Seven and Eight, together with insights drawn from previous chapters, we might conclude that, broadly speaking, four principal themes emerge relating to practitioners' attitudes towards governance, and the impact of those attitudes on day-to-day biobanking practice. Those four themes relate to: (1) core attitudes, beliefs, and assumptions, and how these appear to influence practitioners' responses to governance (both implicitly and explicitly); (2) significant areas of concern, the consequences of these for governance and practice, and suggestions for possible improvements; (3) apparent or potential mismatches between current governance on the one hand, and practitioners' attitudes, perceptions, expectations, and experiences, and the realities of working practice on the other; and (4) what practitioners themselves see as being the fundamental purposes (or value) of governance, the kinds of governance they positively desire, and why. Working through each of these four, broad themes in turn affords a convenient framework for synthesising and encapsulating our key findings, cross-cutting themes, and major conclusions from Chapter Six, Seven and Eight—and for highlighting important implications and lessons to be learned from them.

## Influences on Outlooks of Practice

The first thematically arranged set of findings comprises insights into how respondents' attitudes, beliefs, assumptions, and perceptions about governance (heavily) informed their decision-making and, ultimately, shaped their practice. Key findings may be summarised as follows.

In addition to 'explicit' or 'extant' governance and guidance sources, our data analysis uncovered a powerful underlying role played by three distinct forms of 'implicit' guidance. These stemmed from respondents' professional backgrounds, practical experience, and their thinking about other stakeholders and actors (notably, participants, potential participants, and peers). We have argued that such elements subtlely, yet materially, shaped respondents' attitudes, influenced their reflections on the biobanking governance landscape, and either positively motivated or negatively constrained their behaviour. Especially noteworthy was the pervasive influence of norms and assumptions embedded within 'professional cultures'. Key aspects here included inculcated values and principles; shared practice within professions or working communities; common or cumulative experience or knowledge; and shared presuppositions, conventions, understandings, concepts, or 'etiquette' about what is legally, ethically or scientifically appropriate.

The current biobanking governance system depends very heavily both on explicit or extant governance, guidance and oversight structures—particularly those that may be classed as 'informal', rather than 'formal' or 'official', in nature—and the various forms of 'implicit' governance. Thus, the principal governance structures at work are: (1) the crucial 'gatekeeping' role played by RECs; (2) entry controls, policy preferences, conditions and other requirements imposed by leading biomedical research funding bodies; (3) internal, institutional oversight systems; (4) biobanks' or projects' own steering, scientific, ethical or other such committees or boards; (5) the widespread role played by individuals, informally, privately, and ad hoc, through the operation of professional and social networks; (6) the de facto power of 'trailblazers' to become, by default, norm-setters or standard-setters, especially in the absence of clear rules or regulations on point; (7) the pervasive influence of forms of 'implicit' guidance in shaping practitioners' attitudes, and positively motivating or negatively constraining their behaviour; and (8) the concomitant power of those actors (both institutions and individuals) involved in educating or training practitioners, through their capacity to control, dictate, inculcate, perpetuate, reinforce, and/or modify key aspects of 'professional culture'.

In addition to the three forms of 'implicit' guidance, respondents also characteristically brought five kinds of implicit beliefs or attitudes to bear on the process of engaging with, interpreting, and deciding how to enact (or not) governance in practice. Once again, many of these core beliefs, assumptions, attitudes, expectations, or presumptions derived from aspects of 'professional culture'—notably, disciplinary or professional background, training, expertise, personal experience of practice, and anecdotal knowledge or beliefs. Extrapolating from these beliefs and assumptions, we constructed a generalised, somewhat caricatured, 'ideal type'[8] of the outlook of our respondents—and, potentially, practitioners more generally—towards governance: (1) experienced practitioners 'already know' what to do; (2) they do not need to look at laws or governance to find this out; (3) laws and governance presumably just mirror what they 'already know' to be right in any event; (4) practitioners only wish to 'do good'; (5) their motives, intentions, interpretations and judgments all are trustworthy, honourable and accurate; (6) therefore, they should be trusted, and left alone to act as they see fit. We do not claim that all of our respondents overtly expressed all of these views (although many certainly subscribed to at least some of them), but rather that there was a pervasive sense of self-confidence, self-belief and self-regard that informed a leaning toward self-regulation.

As an 'ideal type', this outlook rests on a high degree of self-confidence in an ability to judge three things correctly: (1) what is 'good'; (2) when (and what form of) governance is required or appropriate in relation to one's own practices or projects; and (3) what other stakeholders (notably patients, participants and the 'public') really want, expect or need. Against this model of a positive self-image we found a common, negative perception that governance systems—including laws (notably, the HTAct and DPA), and regulatory actors (notably, 'out-of-touch' RECs that lack relevant specialist knowledge or expertise)—routinely misjudge each of these three things. Where respondents demonstrated a high self-confidence about 'knowing' what participants really want, expect or need—and in being able to 'speak for' them in this regard—this was accompanied by a pronounced tendency (rightly or wrongly) to elide, align or equate respondents' own self-interests, especially

[8] M Weber, *The Methodology of the Social Sciences*, trans and ed EA Shils and HA Finch (Glencoe, IL, Free Press, 1997 (1903–1917)) 88.

in conducting research unhindered, with participants' interests, expectations and needs. For some respondents, participants' (perceived) 'real' interests, needs, and expectations were the 'bottom line' benchmark for deciding how to behave, rather than 'out-of-touch' laws or regulatory actors.

The ideal typical positive self-image contrasts with several interrelated, negative beliefs or perceptions about the current biomedical and biobanking governance framework. Once again, using a generalised, somewhat caricatured, ideal type to simplify our findings, this outlook may be summarised as follows. The governance framework is premised on distrust and hostility towards practitioners and biobanking-related activities (especially research). It erroneously and excessively focuses on preventing potential (unlikely) harms or abuse, rather than on fostering and facilitating beneficial research. It is frequently unclear, inefficient, incompetent, inconsistent, obstructive and unduly burdensome. Consequently, key laws (especially the HTAct, and its genesis), regulatory actors (especially RECs), governance systems (notably, within the NHS) and the prevailing 'culture'—especially around 'sensitive' topics such as biobanking, human tissue and genetics—can be excessively cautious, over-sensitised, overly protective, driven by anxiety or fear and seek to protect against 'imaginary' harms. Thus, current governance measures frequently operate to stifle or 'hamper' scientific progress. They are themselves unwelcome and unwarranted 'obstacles' or 'barriers'. Again, we are not claiming that all of our respondents overtly expressed all of these views (although many certainly subscribed to at least some of them), but rather that this characterisation reflects the pervasive sense of criticality toward governance.

We argue that this broad, but complex and multifaceted, underlying attitudinal frameworks informed the enactment of governance across all substantive areas and topics addressed in Chapters Six, Seven and Eight. Not least, it underpinned the decisions that some respondents' made both to engage in various forms of selective compliance (for example, 'pick-and-mix'; creative interpretation), and to side-step, evade, avoid or simply ignore relevant laws, guidance or regulatory actors (for example, where multi-jurisdictional complexity arose; where regulatory actors were not highly esteemed; where laws conflicted with one another, or with what respondents felt they should be free to do; or where governance was considered bizarre, excessively cautious, over-protective, uncertain or too difficult or costly to access or assimilate into practice). Based on their responses given in the context of the interview, a proportion of respondents appear to operate, day-to-day, on the basis that: (1) laws and governance simply corresponded with, and mirrored, what they 'already knew' and already were doing; and/or (2) that selective compliance (or avoidance) techniques were a legitimate and sufficient response to governance. Yet, from a legal perspective, both attitudes may be quite unsafe—thereby exposing practitioners (and their institutions) to potential liability.

These insights and findings have important implications for how future governance systems might (best) need to be designed, introduced, implemented, managed and enforced. This is especially so if policymakers, lawmakers and regulatory actors wish the biobanking governance framework to be as effective as possible, and to command widespread support, respect acceptance and, ultimately, compliance from practitioners. Notably, this first set of insights suggests a need to be alive to, to understand, and to factor into any governance arrangements: (1) the evidently significant role of professional cultures; and (2) the profound (and undoubtedly inescapable) influence of practitioners' various implicit sources of guidance, beliefs, assumptions, attitudes and perceptions. If our findings are generalisable more widely, then it appears that these outlooks may well be having a substantial, pervasive and widespread impact over how governance initiatives are enacted (or not) by professionals engaged in biobanking activities in their day-to-day practice. If so, then our analysis also may be helpful in illuminating key potential flashpoints, or reasons why practitioners might choose to avoid, subvert, undervalue or simply disregard governance measures, based on their own attitudes, beliefs and self-interests. These may well warrant closer attention (including further tailored empirical research).

## Concerns of Practitioners

The second broad theme to emerge from our data analysis in this series of chapters on attitudes relates to significant areas of concern identified by respondents relating to governance, the consequences of these for governance and working practice, and suggestions for possible improvements. As noted above, the findings here help to reveal where existing laws, guidance instruments, actors and processes—or the lack thereof (actual or perceived)—appear to be particularly problematic or worrying. Highlights and major cross-cutting themes include the following points.

While some respondents expressed praise for certain laws or regulatory actors, overwhelmingly their comments were negative. Indeed, many respondents were highly critical. An enormous number—and range—of general and specific concerns, problems and difficulties were reported. These touched on various specific laws (notably, the HTAct and IPR); guidance (such as from funders); regulatory actors (predominantly RECs); and governance systems (within the NHS, especially). Overall, however, comparatively few of the leading laws, guidance sources and regulatory actors that populate the regulatory space (as were identified in Chapter Three) were discussed in any detail. Notably, even the DPA—which, as Chapter Five demonstrated, respondents generally were well aware of—did not feature as a forefront consideration.

The majority of concerns expressed about governance were instrumental in nature. They related to practical problems encountered in particular working contexts, including implementation issues. Illustrative major fields of concern included: difficulty locating or accessing relevant governance sources; difficultly understanding laws and guidance (notably, key aspects of the HTAct); guidance being inadequate (notably, from funders); bureaucratic problems, wastefulness, inconsistency and competency issues (particularly affecting RECs and NHS systems); costs and resource implications (notably, HTAct licences; retroactive policy changes; IPR); and a host of problems associated with multiple governance systems (including domestically, multinationally, and stemming from jurisdictional difference). Other concerns rested on more normative grounds (and to some extent spanned working contexts). Most notable here were problems in principle with specific laws. Leading examples included: profound concern, confusion, and criticism over the scope of 'relevant material' under the HTAct (exclusion of DNA; lack of differentiation between arguably 'different' kinds of material—both implicating a more general, also widely shared, concern over whether or not the law should embody 'genetic exceptionalism'); the deleterious nature (and practical impact) of the culture change ushered in by the HTAct (anxiety; difficulty obtaining samples, especially post-mortem); and objections in principle to the IPR and patents systems.

Many respondents described concrete examples of detrimental consequences that they attributed to perceived flaws or gaps in the governance framework. As well as a host of difficulties for working practice (many examples of which are encapsulated immediately above), some psychological effects also were evident. These included unease, uncertainty, fear (including of unintentional violation), and often marked resentment or hostility towards certain laws and/or regulatory actors. Detrimental consequences for the governance system as a whole also emerged from our data analysis. These included both instances of subversion or non-compliance, and the perceived loss of social good, through potentially beneficial research and scientific progress not happening. Notable examples were: avoidance strategies; exclusionary techniques (including avoiding working with certain studies, research activities, collaborators or settings); administrative wastage and duplication (including through (probably) superfluous governance steps sometimes being taken, such as obtaining REC approvals); the undermining of practitioners' trust, faith, support and, ultimately, adherence; and a corresponding undermining of the efficacy, authority, status, operation and successful implementation of governance.

Respondents put forward a broad range of concrete suggestions for possible improvement. These touched on six areas: (1) political and policy steps; (2) new governance frameworks (notably, a specific, 'generic' biobanking governance framework, plus detailed, step-by-step implementation 'frameworks' or guides); (3) specific law changes; (4) changes to oversight and governance bodies, and their operations; (5) guidance sources and dissemination; and (6) alternatives to governance. Overall, the recommendations were weighted towards the practical, implementation end of the spectrum.

Overall, this second group of findings affords two kinds of valuable lessons. The first relates to problems or difficulties associated with specific laws, guidance sources and regulatory bodies. Each may well warrant individual attention (starting with further consideration of whether the views of our sample are more widely held). Our findings—both as to the problems and concerns reported, and concrete suggestions for possible improvements—could provide helpful insights, to trigger and inform appropriate and beneficial remedial action. Many of the suggestions made have relevance to particular regulatory actors. Potentially, many may be suitable for translation into workable governance measures. Alternatively or additionally to being useful instrumentally (for improving existing governance systems and guidance), these suggestions may shed light on key areas that warrant closer attention and further empirical research work (including to explore the extent of concerns, or their impacts on practice, in specific working contexts).

The second batch of lessons from this second broad theme concerns more generic, systemic matters. Three leading examples are worth drawing out. First, our findings and analysis suggest that the historical, de facto, heavy dependence on RECs to govern biobanks—and the ways in which the REC system currently operates and is constituted (both domestically and internationally)—are unsatisfactory. We argue, however, that RECs are not well suited to, nor adequately equipped for, governing many types of biobanks or multinational collaborations, particularly those that are organisationally complex, geographically fragmented, and/or that share materials widely. Secondly, policymakers, lawmakers, and regulatory actors may wish to consider carefully: (1) the quality of the practical, step-by-step, implementation guidance that they provide for practitioners; (2) how they go about formulating, introducing, publicising and disseminating new guidance and governance reforms (especially additional or retroactive requirements; consultation; the involvement of practitioners; educational campaigns); and (3) the nature and extent of opportunities that should be permitted for regulatory bodies and practitioners to interact, and to establish ongoing working relationships. Thirdly, prioritising the investigation of the extent and impacts of the problems reported herein relating to multiplicity (such as multi-jurisdictional issues, and multiple governance actors), and potential measures to overcome them—such as through streamlining, centralising, co-ordinating, rationalising and/or harmonising biobanking governance laws, bodies and systems (both nationally and internationally)—would seem to be urgently required.

## Potential 'Mismatches'

The third theme concerns potential 'mismatches' between current governance on the one hand, and practitioners' expectations and experiences, and/or the day-to-day logistical realities of biobanking practice, on the other. Key findings and insights are as follows.

Numerous apparent mismatches uncovered by our data analysis relate to specific governance laws, actors, and systems, or flow from (perceived or actual) gaps in the existing regulatory framework. We have encapsulated many examples already above. Further potential mismatches appear to flow from deeper conceptual tensions. Two prominent examples canvassed by our respondents relate to: (1) whether—and, if so, how—the law should distinguish between 'biological material' and 'data'; and (2) how data are regulated. In respect of the intersection between 'biological material' and 'data', our findings exposed a diverse mixture of individual opinions about the essential nature of the two; the 'divide' (if any) between them; and how they should be defined, conceptualised, classified and regulated under the law. Respondents' varying viewpoints were underpinned by a range of (potentially conflicting) forms of practical and ethical reasoning. These included: logical or rational arguments; practical or pragmatic grounds (including in relation to working contexts); and more ontological or emotional reflections. No clear consensus views emerged. Genetic materials and genetic data were a noteworthy blurry or 'grey' area. Again, this reflected respondents' more general opinions over the notion of 'genetic exceptionalism', and how far (if at all) special rules for genetics should be enshrined within the law, or in ethical guidelines.

Respondents' varying perceptions, understandings, expectations and experiences contrast markedly with the currently dual, bifurcated, and (especially) conceptually sharply differentiated, approach to regulating 'biological materials' and 'data'. Our findings suggest that this 'legalistic' governance approach may well be premised on an artificial—or, at least, artificially 'bright-line'—distinction, that does not accord with practitioners' perceptions, or with the scientific realities of biobanking or genetic/genomic practice. A widespread concurrence among our respondents that there needs to be (greater) harmonisation between tissue and data regulation further strengthens this conclusion. Not least, it reflects the general impression that emerged from our findings that, for many respondents (and, perhaps, other practitioners more generally), 'tissue', 'data', 'samples', DNA, 'genetic' materials and so forth lie very much along a 'continuum', or are 'nested', rather than existing in separate boxes (conceptual, physical or virtual). In respect of data regulation mismatches, three dominant areas of concern emerged. These were: (1) calls for greater legal flexibility, to enable confidentiality to be negotiated or balance against other, competing interests; (2) difficulties caused where what is expected, desired, or assumed to be possible (including by non-IT practitioners, and by governance requirements) exceeds actual techno-scientific capabilities; and (3) complaints that data currently are over-protected under the law (raising, again, concerns over governance being obstructive).

Overall, mismatches reported in both areas suggest that, in some instances at least, current governance arrangements for regulating tissues and data fail to meet, align or accord fully or adequately with, the perceptions, expectations, experiences, and needs of practitioners and, potentially, other stakeholders. They also appear (at least, partially and/or in some contexts) to be somewhat inconsistent or incompatible with the logistical realities of working practice (including scientific and IT).

Such mismatches potentially carry significant ramifications for governing activities relating to biobanking. Some could result in practitioners (inadvertently) violating governance requirements, including civil and/or criminal laws. Some mismatches arguably render existing forms of governance at least partially unworkable or unfit for purpose—thereby undermining their efficacy, utility, authority and acceptability. Meanwhile, the potential mismatches identified immediately above suggest that, in the opinions of some of our respondents, the two major, existing, parallel legal regimes used to regulate key 'raw materials' used in biobanking—namely, biosamples (especially via the HTAct) and data (especially under the DPA)—are neither mutually consistent nor adequately fit for purpose. As well as mismatches specific to each regime (including the lack of coverage for genetic materials), their very duality itself is called into question by our data—premised, as it is, on a legalistic, sharp, conceptual differentiation; and applying, as it does, separate and differing rules. Policymakers, lawmakers, and relevant regulatory actors may well wish to pay careful attention to our findings and conclusions, and further explore and address the various problems, concerns and mismatches identified.

## Governance Needs

Finally, the fourth theme covers what respondents themselves saw as being the fundamental purposes (or value) of governance, and the types of governance they positively desired, expected or welcomed. Major findings and insights may be summarised as follows.

A significant proportion of respondents actively supported the need for (appropriate) governance. This included a widespread, common desire to be governed personally in some way. The very fact that so many respondents positively desired governance is highly instructive. This is particularly so given that, at first glance, it might appear somewhat inconsistent with our other findings, as summarised above—notably, respondents' widespread criticisms, condemnation, and, occasionally, outright rejection of existing governance; attitudes about 'already knowing' what to do, and wishing to be left alone simply to get on with it; selective compliance and other manipulation or avoidance techniques; and perceptions that many governance mechanisms 'stifle' or 'hamper' research and should be 'swept away'.

Part of the explanation for this apparent disjuncture lies in our respondents' very particular vision of what biobanking and biomedical governance should be designed (and used) to achieve. As a broad characterisation, our respondents tended to express views that 'governance' should perform two fundamental (and intersecting) purposes or functions. Its primary, twofold purpose (and value) is: (1) to confirm practitioners' (perhaps) pre-existing knowledge or belief that their projects, purposes, and practices are 'appropriate' and 'legitimate' (normatively, ethically, scientifically and so forth); and (2) to demonstrate such appropriateness and legitimacy by conveying a 'message' to other key stakeholders. The second, subsidiary purpose (and value) of governance is simply to spell out for practitioners, in fully detailed, practical, readily accessible and legally compliant terms, precisely what they need to do.

Seen in this light, the positive desire for governance among some respondents in fact accords with our earlier findings. Not least, it reflects their attitudes and beliefs that, while some practitioners may express the view that they do not *need* to be governed per se, governance is in part a necessary and crucial tool for 'reassuring' others that all is well. In addition to providing a marker of legitimacy, a number of other significant reasons were apparent in our respondents' desire for governance. Notable additional reasons given for desiring governance included: wanting professional 'comfort' or support (especially to assuage anxiety due to governance gaps, or legal uncertainty or confusion); confirming 'appropriateness' at a normative level (alignment with 'public sentiment'); and various instrumental or practical benefits (including fostering participation and 'trust', and securing funding). These five kinds of reasons for wanting governance interlink with an important cross-cutting theme within our data—namely, a widely held perception among some respondents (and, doubtless too, other practitioners, stakeholders and regulatory actors) that biobanking is a particularly 'sensitive' area. This perception was linked especially to recent scandals or controversies (such as those that precipitated the HTAct, high-profile data protection lapses and the police's National DNA Database); heightened public sensitivity over genetics in general; and the 'newness' of biobanking as a scientific field of endeavour.

Once again, these findings offer valuable insights that, arguably, warrant taking into account—and, perhaps, usefully may be harnessed by policymakers and lawmakers—when considering the future governance of biobanking in England and Wales. In particular, and assuming that they can be shown (through further research or consultation) to be widely generalisable, then they further illuminate the general kinds of underlying schemas of attitudes and beliefs and outlooks that practitioners realistically might be expected to bring to bear when they encounter governance mechanisms—and which materially appear to influence their decisions about how (or whether) to engage with, and enact, governance in practice. These findings also highlight key expectations and needs relating to what kinds of governance ought to be provided. Notably, these potentially may include not only an appropriately tailored, specific, high-level governance framework for biobanking activities, but also sufficiently detailed, accessible, practical guidance for implementing governance framework requirements in working practice.

# 9

# Enacting Governance—The Case of Access

*Catherine Heeney and Andrew Smart*

## INTRODUCTION

IN THE FACE of a regulatory framework which is so extensive and dynamic that it is only partially comprehended by those working in the field, one of the goals of the GGD project was to gain insight into the *practices that constitute governance*. We wanted to expose the ways in which governance is constituted in and through working practices; to examine how governance is produced, reproduced and changed in everyday activities. We thus asked questions of our respondents that aimed to reveal the routine ways in which governance is 'enacted' by actors and organisations through examples of reasoning, activities and technologies. In this chapter, the ways in which governance is constituted and enacted in practice will be discussed and outlined in relation to one issue in particular: access to data and/ or samples. In essence, access refers to matters of sharing: who is allowed and/or supposed to share, what are they allowed and/or supposed to share; and when they do share it, where does the sharing happen, and how is it facilitated or prevented?

One reason for focusing on access was its prominence in our data. The majority of interviewees (all but two) offered experiences of negotiating access or data sharing and/or opinions about what should and should not happen in the future. Interviewees voiced a range of different opinions, sometimes fairly strong opinions, about access. Ten interviewees took either a strong position for or against moves towards so-called 'open-access' or more liberal approaches to sharing. These divergences suggested to us that access was a topic which was currently contested and in flux. Latour has described controversy as a methodological tool; he argues that Actor Network Theory (ANT) 'claims to be able to find order much better *after* having let the actors deploy the full range of controversies in which they are immersed'.[1] The controversy associated with access therefore made it a pertinent focus for sociological analysis of how governance is enacted.

To consider how interviewees were enacting governance arrangements, this chapter focuses on the range of factors described as contributing to decisions to allow access or not. In what follows we adopt a number of ideas from Hilgartner and Brandt-Rauf,[2] who reject the notion that people either have or do not have (or give) access; rather there are various degrees of access, limitations on access and different ways of allowing access, which become evident when consideration is given to how actors take or defend positions about what to share, with whom, when and how. These ideas are supported by recent empirical research on biobanking in France, which shows the plurality of access practices.[3] We thus accept that access and its governance is influenced by numerous factors, among which the law does not necessarily occupy a privileged position. Following this lead, we explore the factors that our interviewees said they considered in their access decisions and reveal the role played by organisational, technical, social, ethical as well as regulatory factors in shaping the norms which informed access decisions. Our analysis reveals aspects of governance that occupy the regulatory space without expressly being part of the official regulatory framework.

## ENACTING ACCESS: KEY FINDINGS

In this section we analyse a range of issues that influenced the practices surrounding access. We identify some common variations that are relevant for thinking about governance, and the ways in which these are embedded in the social contexts in which practice occurs. This data analysis is organised under five key questions: what is shared with whom; how does sharing occur; who decides on access and on what authority; how are applications for access judged; and how is access restricted?

### What is Shared With Whom?

All of our interviewees shared something with somebody and some were involved in multiple projects with a range of access arrangements. As was described in Chapter Three our interviewees were usually involved in a web of overlapping projects, research groups and studies. They were thus

[1] B Latour, *Reassembling the Social: An Introduction to Actor Network Theory* (Oxford, Oxford University Press, 2005) 11.

[2] S Hilgartner and SI Brandt-Rauf 'Data Access, Ownership, and Control: Toward Empirical Studies of Access Practices' (1994) 15 *Science Communication* 355–72.

[3] F Milanovic, D Pontille and A Cambon-Thomsen 'Biobanking and Data Sharing: A Plurality of Exchange Regimes' (2007) 3(1) *Genomics, Society and Policy* 17–30.

parties in complex and variable 'exchange relationships': social interactions between groups or individuals formed around the practices and processes of sharing materials. We identified three cross-cutting dimensions for analysing these relationships: the *actors* with whom sharing occurred; the *artefacts* that were shared; and, the *forms* in which these were shared. Here we simply wish to illustrate that exchange relationships are broadly analysable using these dimensions, rather than comprehensively report the various combinations which were evident in our data.

An important characteristic of *actors* in exchange relationships was the perception of their status as 'internal' or 'external'. Other actors within a research team were usually the most 'internal' and the wider scientific community or public the most 'external', with individuals or groups within consortia and collaborations existing somewhere in-between. In respect to sharing different *artefacts*, the most obvious division that was made was between samples and data. In addition, judgments were made about exactly which samples and data, or which aspects of these, were shared (for example, specific parts of a database or all of it). This revealed the existence of finer-level distinctions about the content and quality of materials. One Clinical Scientist involved in a well-established longitudinal study explained a change in access policies had meant that 'the more detailed geno-coding and occupation coding is no longer available as a routine' (R15: Uni/Hos), while a University-based Research Scientist reported that shared data needed to have satisfied 'various performance criteria' (R3). Such restrictions on sharing related to the value and/or risks that were perceived to pertain to the specified artefact (in these examples, data of questionable quality, or high sensitivity).

The positioning of actors and artefacts was cross-cut by conceptualisations of data and/or samples as existing in different *forms*, for example, respondents used terms such as 'identified', 'anonymised', 'unpublished', 'published', 'unfiltered', 'aggregated' or 'summary'. This Bioinformatician at a repository lodging genomic data, whose job entailed making information available to others, spoke of sharing 'the data collections we have ... without restriction to the scientific community', although it transpired that what the research team were sharing was 'a sort of simplified summary' (R29). Meanwhile, a Clinical Scientist distinguished between his collaborators who 'can see all the data that's ever been generated' and the 'summary data on the web' (R31: Uni/Hos). The 'summary data' in this extract refers to aggregated and/or anonymised data.[4]

[4] Across our sample, data that could be readily traced back to study participants was only shared within a given research team. When such 'identified' data was shared, two justifications were offered: 1) commonly, a need to link to records for a specific research purpose (to follow individuals in the progress of a disease, or to enable linkage to health records at a later date) and/or 2) for oversight (to check that consent procedures had been followed).

**How Does Sharing Occur?**

We identified a variety of technical arrangements that facilitated access to data and samples. There were, of course, differences between the ways in which samples and data were shared, as the former are material objects, while the latter could be accessed remotely (for example, via databases on the internet). When interviewees discussed methods for sharing, they often involved IT systems, including internet portals and purpose built IT 'platforms' which in some cases actually underpinned collaborative networks. Eleven interviewees were involved in projects which were specifically planned to facilitate data-sharing. Two of these projects aimed to establish IT 'platforms' to support collaborative sharing of data about particular conditions or disease areas, including one which was set up to gather data and samples from NHS clinics that could subsequently be shared with a wider research community, including the private sector.

Collaboration was a primary mode of sharing. Over half of our interviewees shared and received data and/or samples through collaborations. A variety of practices were included under the concept of collaboration. One respondent viewed a database *itself* as collaboration, as it embodied a way of working together that would not otherwise exist, as well as capturing the results of the collaboration. More usually, however, collaborations were conceptualised as research relationships between individuals or groups (with a consortium being the most formalised of these). Such collaborations were seen as a reciprocal (and mutually beneficial) relationship. This Clinical Scientist involved in a well-established longitudinal study described the additional research benefits arising for those who collaborated with 'people outside', and noted that 'one way in for the people outside is to collaborate with members of the study team' (R15: Uni/Hos). Although no respondent overtly stated that they would deny access to non-collaborators, their descriptions of the possibilities of access suggest that anyone who was not providing some 'value' in return would face resistance or at least reticence with regard to some artefacts or forms of these artefacts. As such, collaborative exchange relationships were a means of facilitating access, but sometimes 'full' access was only available to those who could offer of other valuable resources in return (a point we return to below when we discuss motivations for and against sharing).

**Who Decides on Access and on What Authority?**

We found variety in the loci of responsibility for decisions-making and the extent to which this was part of a formal structure. Five of the projects with which one or more of our interviewees were involved were described as having a designated or formal committee or steering group to make decisions about access and collaboration. Projects which had such a committee ranged from medium to very large in size.[5] A Clinical Scientist who was involved in one such project explained that 'the data is only available to collaborators and they have to be approved by the steering group of the study which is the PIs" [Principle Investigators] (R31: Uni/Hos), while in another, experienced individuals were given the authority to vet access proposals and advised the access committee. In small and some medium-sized research projects, PIs also had the significant role in making decisions about what was shared, how and with whom. The authority for decisions about access can thus be seen to be located in established organisational or bureaucratic practices, and/or the expertise and status of key individuals.

Taking a different viewpoint, five respondents discussed how clinicians take the initiative to find recipients for data or samples through informal networks of colleagues or, in some instances, on an ad hoc basis. For example, this Clinician explained:

> I'll see a patient with something rare, you might have a look on the internet to see who last published something, who discovered the gene for this rare condition and then you find out their email address and you send a kind of "You don't know me but I've just seen a patient with your disease, are you interested in their DNA?" And they usually email back and say yes great, send it. (R13: NHS)

In these practices two powerful source of authority were apparent: the norms of the biomedical science community and the reported desires of patients/research participants. One Clinical Scientist involved in small-scale clinical studies explained:

> I mean generally speaking anonymised DNA samples can be sent around and that's acknowledged as a way of behaving that's acceptable and I tell patients that I might send your DNA to somebody in Australia if they're doing some research I think might give us an answer and people are very happy about it, in fact they like it because they think well they know they're going to get, the whole world is now thinking about their problem not just Dr X. (R14: Uni/Hos)

Informed consent agreements and documentation (which could define the nature and scope of sharing) were consistently referred to as significant elements in decisions about access. As a source of authority, informed consent could be seen to embody both the regulatory power of ethics committees and the wishes of research subjects (this issue is discussed further in the subsection on normative motivations below).

### How are Applications for Access Judged?

A 'good scientific proposal' was a primary basis for judgement (R20: Bioinformatician, Repository). The criteria for (and processes of) assessing the

[5] See Appendices for explanation of size of 'genetic databases'.

quality of scientific proposals were, however, not described with particular precision. The following two extracts indicate that the criteria for judging scientific quality can be grounded in specific knowledge or expertise of some kind, but that they can also involve a relatively flexible approach to seeking further advice. This Clinical Scientist argued that questions about scientific quality could be judged against the original goals of the study, but if proposals were submitted in which 'it wasn't clear that scientifically it was the right thing to do I think we would have to set something up' (R47: Uni/Hos). A Bioinformatician based at a newly established repository, who foresaw limitations in their ability to make scientific judgments (because they did not possess the requisite knowledge), concluded that 'if there was something that made us think that research wasn't worthily [sic] we would probably take scientific advice' (R36). These two responses also reveal that initial judgments might be required to establish if an application needs further consideration, although the grounds on which such judgments are made were not made explicit (suggesting a reactive approach). Furthermore, there is a degree of uncertainty about what should happen if there were concerns (indicated by statements such as 'I think we would have to set something up' and 'we would probably take scientific advice') which demonstrate some aspects of decision-making about access were not captured by formalised processes. However, neither of these interviewees expressed particular concern at the current lack of standardised procedures for decisions about seeking specialised scientific advice, implying that a 'wait and see' approach was actually preferred.

Professional status was also used when assessing suitability. This was judged by factors such as professional interests, institutional affiliation and publishing record. In the words of this repository-based Bioinformatician: 'you have to be a bona fide [*name of disease*] researcher, associated with either a university or a research organisation' (R20). Aspects of status rest on professional standing, but as the following Clinical Scientist involved a web-based data sharing platform indicates, social networks could also be important:

> [Y]ou are only allowed to join the consortium if you're sort of you know a member of the professional regulatory body in your country and you are working in an academic or genetic centre and an existing member of the consortium can personally vouch, sort of regulate it in that way (R11: Uni/Hos).

Knowing 'a member of professional regulatory body' and having someone to 'vouch' for you are forms of social capital that (at least in the views of this respondent) contribute to judgments about the suitability of would be collaborators. Two explicit reasons were given for making judgments about professional status. The first was maintaining the quality of the database; the Clinical Scientist quoted above argued that 'in order to keep it of a high quality you need to very carefully manage the people who are allowed enter data' (R11). The second reason was ethics. This Bioinformatician at a newly established repository referred to the norms of his project to facilitate access to clinical and genetic data, adding: 'anyone who wants to be a member of the confederation has to sign up to those things' (R36). Another Bioinformatician who was analysing data on behalf of a large collaborative project described the responsibility of those working with data more particularly in relation to protecting 'the confidentiality of subjects' (R30: Uni).

From the perspective of those hoping to *gain* access, reference was made to the use of negotiation or persuasion. The following Repository-based Research Scientist explained that, despite working for a large and high profile programme of sample and data collection, securing data on participants from other existing sources required an act of persuasion as they 'wanted to be convinced that they were not unhappy about {allowing access}' (R8). Factors that appeared to influence success in these negotiations included existing personal and professional relationships, and issues of proximity or status within an organisation. This Clinical Scientist who worked on small-scale clinical studies for example, explains:

> If I wanted access to the laboratory database as long as the question I was asking was in some way related to my work then I think in my position as a consultant within the department I'd probably get access to those (R48: Uni/Hos).

This extract also illustrates, however, that such a request would need to relate 'to my work' thus implying that this respondent would expect those receiving the request to make a judgment about whether he was 'bona fide' (that is to say the justification and authority of the request would be assessed).

### How is Access Restricted?

Respondents reported variation in the extent to which they, and others, had formalised access procedures. This Clinical Scientist, for example, reported:

> Well sometimes there's quite a stringent procedure in place and you will send a written consent to release, release information and/or a sample. And sometimes there are less stringent procedures in place (R2: Uni/Hos).

A Repository-based Research Scientist provided further detail on the range of access procedures used by his study.

> We have a full range from just putting it on the web with pretty much a gentleman's agreement that people should not publish the things that we are working on, to datasets where we make them available but people have to make themselves known as a bona fide researcher and provide a letter that they will not redistribute the data, in which case we give them access through a secured website mechanism by issuing them a password (R37).

One kind of formal restriction (evident in both of these extracts) was contractual, whereby licenses or equivalent permissions were granted (most commonly these allowed access on the understanding that shared materials could not copied or further distributed). Other kinds of restrictive practice involved imposing physical and/or technological constraints on usage. Two respondents who worked within two different medium-sized, well-established[6] longitudinal studies reported that 'if somebody wants to analyse genetic data then it has to be done here in-house' (R5 Research Scientist, Uni), or 'people come here and work' (R24: Epidemiologist, Uni). On these projects physical restrictions were accompanied by the use of technological constraints, whereby data is analysed on 'a non-network machine' (R5), or 'on isolated networks that aren't part of any wider area network' (R24), although technological restrictions were also evident in isolation from physical ones (such as the password 'secured website' mentioned by R37 above).

Decisions about restricting access, or managing the risks of unauthorised access, were also manifest in descriptions of security. Interviewees, in very different contexts and in relation to a variety of artefacts, readily reported on their security practices. Routine examples for Clinical Scientist's involved in small-scale clinical studies were paper questionnaires being secured in a 'locked filing cabinet' and 'password protected databases' (R16: Uni), and biological samples being 'given a coded number' (R14: Uni/Hos). Nevertheless, one Research Scientist who managed a collection of data and samples in the private sector, talked of confusion about key terms, like anonymisation, which could have different meanings for different people, and explained that in order to facilitate clarity 'we have actually laid down this is what we mean by all of those' (R22: Pharma). In some cases more complex security arrangements were described. Over half of the interviewees referred to there being a 'key holder'—a person or organisation holding the information that would be required for putting together 'identifying' information with biomedical information; and for fifteen of our respondents, the organisation holding the key was the NHS.

Such restrictive practices, however, have the potential to conflict with ideas about collaboration and sharing, a dilemma noted explicitly by five interviewees. Managing access was thus reported as a 'problem' by this Epidemiologist, who worked for a well-established longitudinal study:

> [T]hat's been a constant tension between potential solutions to this problem so [...] we've now developed special licenses where organisations as well as individuals sign up to these licenses and if people break the rules then the organisation as well as the individual can suffer' (R24: Uni).

It is apparent that, at least in this case, different options for restrictive practices have been developing over time, and moreover there is a perceived

[6] Both studies had been running for around 20 years or more.

need for sanctions to enforce compliance. It should also be noted that outside of the above-noted formal access arrangements four interviewees also drew attention to routine technical or organisational constraints that they thought prevented them from linking administrative health records with research data in ways that they thought would be beneficial.

## AN INITIAL ANALYTICAL REFLECTION

Hilgartner and Brandt-Rauf[7] use the concept of the 'data stream' to characterise the fluidity of data which can change state from 'raw' to 'pre-processed' and so on. They argue that sharing data is a complex and multi-faceted process: there are various degrees of access, numerous physical and technological limitations on access and different ways of allowing access:

> Access to a data-stream could be provided in many ways. At one extreme, a laboratory could use the Internet and Federal Express to provide daily access to its entire data-stream. At the opposite extreme, a laboratory could withhold the entire data-stream and even keep secret the fact that it is looking for the gene.[8]

Access practices are thus presented as highly diverse, and numerous social, legal, technical and other factors contribute to decisions about what is are shared and how it is shared. Using ethnographic research on French biobanks, Milanovic *et al* report three different kinds of exchange regimes, which they characterise as 'cooperative', 'gift' and 'sub-contracting'.[9] Each of these regimes entails associated forms of social relationship and they are articulated in different forms of documents and agreements. Importantly, in their study these regimes of exchange co-existed in single biobanks and they argue that this reflects 'the different ways in which biobankers conceive their biological resources and circulate them in various networks.'[10]

Our analysis is consistent with these ideas about the fluidity of the artefacts generated by scientific research and the variability in access practices. The three dimensions of 'exchange relationships' captured how our respondents shared various artefacts with different actors in a variety of forms. Our notion of exchange *relationships* denotes the importance of social aspects of access practices. The centrality of collaboration as a mode of sharing highlighted that access is rooted in and sustained by social interaction and negotiation, based on a need to develop relationships that could (ideally) become reciprocal and mutually beneficial. If we accept collaboration is a primary, and sometimes the only, mode of gaining preferential access to materials, establishing such social relationships becomes a foundational

[7] Hilgartner and Brandt-Rauf 'Data Access, Ownership, and Control', above n 2.
[8] *Ibid*, 34.
[9] Milanovic *et al*, 'Biobanking and Data Sharing', above n 3.
[10] *Ibid*, 28.

component to understanding how access is governed. The overt sociality of collaboration is, however, just a part of the picture; the influence of social relationships was evident in more subtle ways. Variations in contexts and circumstances—ie social factors—can help to explain why we found different models of decision-making about access. Social networks were intrinsic to aspects of judgments about professional standing and could be used to aide negotiation when trying to secure access. The physical, technological and contractual restrictions to access revealed ongoing social processes—imprecise definitions, changes in policy, negotiation and the subtle placing of obstacles in an attempt to control. These reveal an interpretative and interactive quality to practice; places in which social relations, norms and values enter into judgments and activities about access.

The kinds of variations in decision-making we have charted indicate a degree of flexibility even within the boundaries of particular organisational, project or study contexts. Even things which were seen as 'standard' by most interviewees, such as security procedures, on closer consideration have interpretative and interactive dimensions, where personal and social norms and values can enter into judgement and practices. Although these kinds of findings could prompt questions about the legitimacy of decision-making powers and processes, the extent to which we could say that there are 'problems' with the current mechanisms for governing access would depend upon us, as analysts, adopting a normative position in relation to the practices that we have described. The existence of a social dimension in the governance of access may have ethical implications, but our aim in this chapter (following Latour[11]) is to reveal and map these, rather than judge their probity.

## MOTIVATIONS FOR AND AGAINST SHARING: KEY FINDINGS

In the following section we consider respondents' motivations to share or withhold access. We begin by considering important changes in the wider context of the scientific research community in which decisions are made, and then look at the practical and ethical motivations and disincentives for allowing or broadening access.

### Changing Models, Changing Etiquette

It was reported that an alternative culture, with new associated models of practice, was emerging in the scientific field and threatening to supplant previously dominant 'ways of doing things'. The 'open-access' model, and the

[11] Latour, *Reassembling the Social*, above n 1.

emblematic Fort Lauderdale Agreement,[12] which refers to existing collections of data or samples as 'community resources', has altered the social, political and economic context for data sharing; it is effectively a new set of norms and practices regarding the governance of access to samples and data.

> I think the real tension that is arising is between a model which is built around the, I think, Fort Lauderdale agreement on the free and potent sharing of genetic sequence data and a model which is much more rooted in the clinical trials and long term follow up studies which is concerned with protecting the identification and confidentiality of individual [study participants]. (R15: Uni/Hos)

The tension referred to by this Clinical Scientist involved in a well-established longitudinal study related to the norms underlying or implied by different methodologies; and particularly a difference between those norms embedded in contemporary collections of materials designed to be 'common resources', versus those embedded in existing collections of materials that had been accumulated during the process of answering set research questions. Accordingly, some interviewees felt the rules, norms or etiquette according to which they had conducted their scientific careers, and managed their projects and their relationships with colleagues and research subjects, were being challenged or redefined, as indeed were the collections of research materials they had assembled. For some interviewees there was a sense that an alternative model was being imposed and tensions arose where there was resistance to it.

This new model and its associated culture and etiquette was supported and enacted by a number of high profile actors, research studies, research organisations and research funders. The Human Genome Project (HGP) was mentioned by five interviewees in connection with open access, and was argued by this repository-based Bioinformatician to be 'one of the best examples' whereby 'open data release' provided a resource which had encouraged 'genuine scientific research, really innovative scientific research' (R10). The challenge that the HGP posed to the accepted norms and practices in the field of biomedical research is demonstrated by the following repository-based Research Scientist who describes the adoption of the open-access model as a selfless, almost heroic, act of independently minded scientists:

> Now the thing that changed that was really the genome project where a group of scientists, not anyone else but scientists themselves, decided that they were going to release the data directly into the public domain, completely unfiltered, even if it meant that other people could gain competitive advantage over them (R25).

The Wellcome Trust Case Control Consortium was mentioned as a programme of work which was an heir to the norms and practices which

[12] The Wellcome Trust, 'Sharing Data from Large-scale Biological Research Projects: A System of Tripartite Responsibility' (2003) <http://www. genome.gov/Pages/Research/WellcomeRe-port0303.pdf> (accessed 7 March 2011).

the HGP had shown to be successful. Organisational actors that were recognised as championing sharing included the Sanger Centre, which was 'proactive in the community to get people to make their data available' (R37: Research Scientist, Repository), and funding bodies. For example, one Clinical Scientist suggested that in 'large scale genetic studies [...] quite a lot of the funders have been the driving force behind early data release' (R12: Uni/Hos). The Wellcome Trust in particular was identified by interviewees, including the following Clinical Scientist, as encouraging data sharing and open access: 'at the moment, the major players in terms of voice are Wellcome Trust because they know what they want' (R15: Uni/Hos). The Trust's prominence in the data is in part explained by the fact that over half of our interviewees cited it as a current or past funder of their research, although as the last extract implies there was also a sense that it had a particularly strong policy stance towards open access. One way in which the influence of the Trust was experienced was through conditions of funding. As this Bioinformatician explained, it was 'a stipulation that as well as putting your publication into *Pub Med Central* or some equivalent, you have to make the data available' (R30: Uni).

There was, nevertheless, evidence that the emergent culture and practices of sharing were a source of tension, or viewed as a negative pressure. The following Bioinformatician, who worked at a repository lodging genomic data, was ambivalent about publishing in open-access journals because of the implications for researchers with smaller groups with fewer professional, institutional or financial resources 'I do actually think it's a good idea but it can be really costly for small groups to publish in those papers' (R38). A range of disincentives to sharing will be reported below, including issues of reward in scientific careers, and concerns about the social and ethical relationships that exist between researchers and research subjects. The existence of these tensions was illustrated by a Bioinformatician at the same Repository who claimed that those who want to resist sharing for reasons of self-interest sometimes used the 'ethics dimension' to 'restrict the... data to a chosen set of researchers for their own research aspect' (R10). This demonstrates that the attempt to introduce the new culture and practices of open access was a matter of conflict and controversy, where positions were often defended via moral arguments. As one Research Scientist with considerable insight and influence within the open-access debate pointed out, however, it should be recognised that, a range of 'release' practices (including for unpublished data) has existed historically and this variability continues:

> [B]ut the extent to which raw data are released has been hugely variable over the years, some people release a lot of stuff and other people don't and it varies with fields as well, the different fields of science behave differently (R25: Repository).

This may be interpreted as evidence that it is overly simplistic to portray open access or the 'community resources' model as entirely supplanting the 'traditional' norms and practices. It could, however, also be regarded as an attempt to counter the claims of detractors of open access, by playing down the extent of change to existing norms and practices.

### Practical Motivations

The most common motivation for sharing was a need or a desire to access something for which an exchange was necessary. Some interviewees referred directly to this, while others merely implied it, as we noted how potential collaborators may be judged on the quality of the data that they offer.[13] Such motivations varied depending on the interests and circumstances of the interviewee. For example, a Clinical Scientist who conducted small-scale clinical studies allowed other researchers access to his patients' DNA to find 'a genetic answer to [the patient's] problem' (R14: Uni/Hos). Two interviewees who worked for different pharmaceutical companies both noted the importance of collaborating with individuals working within the NHS because of the potential to recruit patients into clinical trials. Other than data and samples, desired 'objects' of exchange included technology, knowledge, finance and kudos or status. Some academic sector organisations, such as The Sanger Centre, held a particularly esteemed position because of particular knowledge and technology (which draws researchers without the resources to analyse large datasets into collaborations and sharing relationships). The Sanger Centre, according to some interviewees, was also regarded as having organisational stability (because of the way it was 'funded'). Life sciences companies were viewed by some respondents as a source of financial backing; for example, one Research Scientist, argued that it was unfeasible to undertake clinical trials for interventions into rare disorders 'if you don't have company involvement' (R43: Repository). Such commercial involvement, however, was also regarded as an ethical disincentive (a point we return to in when we discuss practical disincentives to data sharing below). Finally, kudos or status could result from being involved in successful research; for example, this Clinician who had little opportunity to be directly involved in research explained: 'you get your name on their paper because you've contributed to clinical information' (R13: NHS).

Three projects in our sample were motivated to share by their very raison d'etre; they had been established as repositories to promote or facilitate sharing, either with pre-specified collaborators or more generally with the wider scientific research community[14]. Sharing was thus intrinsic to the core business of such projects: 'we are collecting that data and these samples and

[13] See above, section 'How are Applications for Access Judged?'.
[14] See above, section on 'How Does Sharing Occur?'.

bringing them into a central location where it can be made available to the scientific community for research' (R20: Bioinformatician, Repository). This aim infuses the entire organisational and technical arrangements, including for example employing staff to develop and maintain the ICT that would enable sharing.

A third motivation for sharing was (more accurately) a reason not to oppose sharing if or when asked to do so. This Bioinformatician speculated that a relaxed attitude to sharing among the research team stemmed from the fact it had considerable resources in terms of data and samples, technology and funding. This confidence, moreover, was arguably linked to the fact that the individual was based within the robust organisational structure of a high-ranking university:

> [I]t might also be to do with the size of the organisation because we're quite a big group so perhaps we're not threatened by other people having the data, it doesn't bother us (R30: Uni).

The implication of this is that this respondent does not perceive how sharing could challenge (threaten) the standing of the group, perhaps because of its high status or capacity to compete with other groups who had access to the same materials. Conversely, this implies that smaller, less well-established and less well-resourced individuals or teams might be less happy with a culture of sharing, if they do not have the capacity to exploit their own or other available materials quickly or effectively.

## Normative Motivations

The imperative to share with particular groups and individuals were (in part) based upon trust in the 'good' intentions of others. Three respondents (all of whom occupied roles that were not primarily dependent on the exploitation of the samples and/or data in their keeping) expressed the view that people with inappropriate intentions, or without the relevant expertise or qualifications, would be unlikely to approach them for access. When asked 'who wants access to the data' one of these respondents replied 'the bona fide researchers, the sort of people who *should*, who *could* have the data' (R30: Bioinformatician, Uni *emphasis added*), and another commented 'people at this level [...] don't ask for pieces of information which are not necessary' (R2: Clinical Scientist, Uni/Hos).

A second explanation of why people *should* share materials was rooted in the expectations, or views, of patients or study participants, at least as these were invoked by our interviewees. Respondents' claims to know what patients or study participants 'want' were justified in relation to three (potentially overlapping) rationales: the existence of formal consent agreements, whereby: 'you know patients are willing for that information to be shareable'

(R11: Clinical Scientist, Uni/Hos); a desire 'to maintain, to uphold the original intention of the donor' (R35: IT Specialist, Repository); or, simply on the claim that 'what they [patients who donate samples] want is the best possible science that gives them the best possible chance for survival and getting better' (R34: Research Scientist, Pharma). From different professional backgrounds, these respondents all mobilised the purported desires of study participants or patients using the supposition of shared health goals, as a stand-in for actual consent in the latter two cases. This last interviewee, however, suggested that such invocations of patients or study participants desires could mask the self-interests of the researcher:

> [A]nd most of them at the same time also have the academic ego in addition to that [...which] extends not only to giving them the ability to treat their patients better but also, you know, to publish. And if you have those two interests come together, then we don't have a problem (R34).

This opinion suggests that claims of the alignment of the interests of scientists and those of patients or study participants it is not always unproblematic, and highlights that such justifications may be treated with scepticism within the scientific community.

A third motivation for sharing was 'social good', especially maximising the use of research resources in light of the investment of public money. For example, this Clinical Scientist who had built up a large number of research samples relating to a particular disease, argued that:

> Given the amount of money that has been, that is spent doing all this work it would be a shame if, especially the healthy, you know if the controlled population could not be shared (R45: Uni/Hos).

Another aspect of the social good argument was more abstract; as this Bioinformatician at a repository lodging genomic data expressed it: 'the high flowing kind of moral arguments about this being mankind's property and so on and so forth' (R29). However, as the tone of this extract implies, this motivation for sharing was considered by this respondent to be rather idealistic. Indeed, another Bioinformatician at the same repository acknowledged 'genuine ethical concern' (R10) about potential threats to confidentiality, but suspected that, in practice, sharing was obstructed by a professional reward system that means individuals must seek 'credit' for their contributions to research as an IT Specialist at a different repository put it (R35). Nevertheless, the respondents in our sample who talked most readily about these barriers generally supported, or were involved in, projects which (to some extent at least) promoted sharing. These proponents of broadening access tended to use moral arguments against those who opposed greater sharing. From this view, those who were unwilling to share were impeding the social good by obstructing the benefits to science and patients and failing to maximise the potential of existing resources or public money. For example, a University/Hospital-based Research Scientist whose research required large numbers of data and samples described public sector organisations that lacked infrastructures to enable sharing as 'out of touch' (R41).

**Practical Disincentives**

The depletion of finite resources was given in support of a parsimonious approach to sharing samples by interviewees from six different studies. This highlights an important difference in the considerations relating to sharing data and samples, as the following repository-based Research Scientist explains: 'it's right in a sense to focus on use of the depletable aspects of the resource, ie the samples; the data are in effect inexhaustible' (R8). None of the interviewees reported that resource depletion would wholly preclude access, but the view that any sharing would have to be an 'efficient use' (R5: Research Scientist, Uni) of materials was fairly common. Even the following Clinical Scientist, who had previously expressed support for an open approach to access and sharing, viewed depletion as a significant influence on decisions about sharing:

> It's still finite and we don't want to do every bright idea that people want to come up with. What we'd prefer to do is develop a sort of a list, and we've essentially started doing that, of things that people think might be interesting and as I say some of them, by the time we get to the end may no longer be interesting and the question may have been answered one way or the other elsewhere (R49: Uni).

This extract equally demonstrates the overlapping considerations in access decisions, such that judgments about what was 'interesting' at a particular time to the person or people making decisions about access intersects with practical considerations about the nature of the materials that are being shared.

The need to have systems for dealing with access requests, including a formal attempt to 'undertake some sort of curation' (R1: Uni), was seen as a practical disincentive to sharing by this Bioinformatician, whose web-based data-sharing platform meant he was often in the position of meeting requests for access. The time and effort involved in allowing access to individuals outside of the project or organisation lead this *Epidemiologist* in a well-established longitudinal study to suggest a fee would be charged for access to the collection he was involved with, 'not to put big barriers in the way, but to encourage activities but at the same time to make it manageable' (R6: Uni). Part of this respondent's desire to 'make it manageable' is the expectation that the costs of enabling access will not be met by his own organisation but rather by the person or organisation requesting the materials. He acknowledges, nevertheless, that this cost could act as a barrier to sharing, albeit one that he regards as small.

Protecting a research niche or domain of expertise/interest, including potential accompanying financial aspects, was another practical disincentive for sharing. Broadening access to research materials was described by this IT Specialist at a newly established Repository, who was attempting to gain access to existing data sources held within other organisations, as raising 'issues of territorialism' (R35). Restricting access is a way for people to 'protect' their work or particular areas of knowledge from those who were (or who could become) direct competitors. This Bioinformatician at a repository lodging genomic data, despite being a supporter of open access, acknowledged the risk that other scientists could exploit data produced by others, and in the worst case could eventually be 'shutting you out for future developments' perhaps due to patenting and the enclosure of areas of research (R39). Efforts to protect the potential financial benefits that might stem from research gave rise to some fairly complex confidentiality arrangements, which included trying to ensure that other scientists could not guess which part of the genome or which gene a team was working on. It also led to some complex everyday practices designed to maintain secrecy:

> So we have to mirror the public domain databases internally and update them the whole time so that if I want to say here's my top secret gene I can do this experiment internally without actually sending out over the wires. It's really weird stuff that we would never have thought about (R12: Uni/Hos).

In this case, the need for this secrecy and the methods for achieving it were imparted by the private sector collaborators of this public sector Clinical Scientist, who had argued that they wanted to protect their investment and any potential Intellectual Property (IP) to be gained from use of the data. Indeed, the desire to protect potential financial benefits arising from the research was especially a concern for individuals based within collaborations involving the private sector, whereas in other cases it was reportedly more a matter of the protection of a time investment and the negative career implications of losing control over the fruits of past labour.

### Normative Disincentives

A primary set of concerns related to equity, which captures the perceived problems of 'free-riders' and adequate recognition or credit for peoples' efforts. This Clinician explained that 'ensuring that you get the credit for your input' (R13: NHS) was an expectation which preceded the decision to share. A Repository-based Bioinformatician again acknowledged that this was a problem for researchers: 'you put stuff in the public domain but you've then got no control over anybody else exploiting it' (R39). This Research Scientist, who was subject to a great many requests to access data and belonged to a team which was responsible for the building and maintaining a well-established longitudinal study, offered a practical solution to the issue of credit suggesting embargoes on the use of materials, to allow the originators of a study 'time to get recognition' (R5: Uni). For example, a period of restricted use was favoured by this Bioinformatician: 'you still want to leave it six or twelve months to let the locals [the research team which collected the data and/or samples] have a go at it' (R30: Uni).

Despite respondents mostly sharing anonymised data and samples for research, some interviewees' worried that sharing may result in the inappropriate disclosure of information relating to individual research subjects. Various interviewees, especially those who had built relationships with study participants over time (for example, by working on longitudinal studies) and those who recruit research subjects though clinical work, expressed concern that broadening access could threaten the confidentiality of research subjects. Sometimes this was simply because of 'the practicalities about the way we run our studies, you know, certain contextual ways of doing things' (R6: Uni), as this Epidemiologist in a well-established longitudinal study put it. Alternatively it could be because of the nature of the information itself; as this Bioinformatician at a repository lodging genomic data expressed it there was 'a major headache about releasing that data in an entirely, truly public—anybody can do anything they like with that data—because it genuinely can't protect the individuals' (R10). This concern also extended to the potential for collaborators to identify research subjects as a result of combining information sources, as this private sector Research Scientist noted: 'that can happen if a statistician who's worked on a project looks at our data and knows their data' (R23: Pharma).

Beyond simply protecting confidentiality, some interviewees talked about the need to protect the relationships of responsibility embedded in the 'original context' of their research. For example, this Clinical Scientist involved in a longitudinal study worried that processes that involved 'linking' datasets would mean that the original collectors of the samples and data would 'lose that element of supervision' (R15: Uni/Hos). In particular, the 'original context' was important to interviewees who were involved in well-established longitudinal studies, which had required a great deal of effort to initiate and sustain. From their perspective it was difficult for outsiders to understand the efforts and relationships which both gave rise to—and importantly, sustained—the original collection of samples and data. Similarly, this Clinical Scientist, involved in a longitudinal study claimed that arguments to broaden access came from those who did not appreciate the socio-ethical parameters of the original context: 'that there is a trust relationship between the study subjects and the people who actually physically take the blood, talk to them, ask them for consent' (R4: Uni). Indeed, consent was a key touchstone in the accounts of those of our interviewees involved in studies where the original contact with data subjects was made in a clinical setting.

A Clinical Scientist who had recruited research participants from among his/her patients claimed that the original terms of consent reflected what these people had agreed to and claimed that if those terms were violated, the study participants would be 'pretty mad' (ie angry) (R31: Uni/Hos). There was also, however, evidence of the above-mentioned respondents who worked in (different) longitudinal studies seeing the limitations of adopting a rigid position towards the maintenance of the original context. For example, one argued that it can be possible to share by finding 'common ground where useful work can be done without violating the original terms of consent' (R15). This was illustrated by an example whereby the original terms of consent were used to support a case for widened access, by allowing other scientists to work on data and samples while requiring that: 'what they discover is all fed back [into the study database] for the common good of the [study participants]' (R4).

A fourth normative concern related specifically to private sector involvement. In some instances this intersected with the aforementioned issue of 'original context', whereby data/samples had been collected under the assumption that they would not be shared with commercial organisations. The negative associations with the private sector included questions about its motivations for carrying out the research (ie the profit motive); for example one Research Scientist, who had received private sector funding for research, nevertheless expressed distaste for research outcomes being used 'for their marketing exercises to actually sell their drugs' (R43: Repository). Private sector involvement was also viewed by some as restrictive. The following two respondents were employed in the public sector but worked in collaboration with commercial organisations. A Clinical Scientist who worked on small-scale clinical studies, while not couching it explicitly in negative terms, explained that they operated 'under fairly strict, well very strict confidentiality rules' as a precautionary measure intended to retain the opportunity for future profitability (R12: Uni/Hos), while a Bioinformatician based at a repository lodging genomic data complained that a commercial partner blocked the release of gene sequence information that might be more widely useful if widely disseminated: 'a restriction that I think is quite bad' (R38:). There was also concern that commercialisation could mean that the scientific community, research participants or both might be less willing to donate materials for research. This Bioinformatician who worked at the same repository, that promoted the open-access model, suggested that if the data-sharing project he worked on started to make a profit, scientists would become less willing to allow his group access to their datasets:

> [I]t's not really our data, we're just custodians of those data on behalf of the scientific community and I actually think that if this turned into a commercial operation it would, it would completely change the whole, the whole ethos in the sense that scientists would say well I'm not giving those buggers my data (R29).

**An Initial Analytical Reflection**

An emergent culture of practice, which broadens access to materials at every level, is being exemplified and promoted by key actors within the biomedical research community. As might be expected, where existing cultures of practice differ in significant ways to the emerging norms and practices there was evidence of resistance, illustrated by perceptions that broadening access could be damaging, or inappropriate given certain contextual factors (such as existing agreements and understandings between scientists and research participants, or the particular financial and organisational conditions of a given research team). Discord appears to arise from, on the one hand, the perception of some actors that there is a lack of respect for well-established professional norms, etiquette and practices, or on the other hand the perception of other actors that scientific advance and the public good are being hindered by narrow self-interest or a professional culture that is characterised by 'territorialism'. Territorialism is indeed an interesting term to be used in this context, in that it captures the ideas of demarcation and defence of boundaries, which are, at least in part, tied to the particular criteria for rewarding scientific endeavour, we shall return to this point in the discussion. What is most notable at this point is that actors' opinions about this discordance, and their positions relative to the stance of powerful social actors, can shape their choices and decisions about access.

The above-reported incentives and disincentives for sharing imply a network of exchange relationships, which is based on perceived needs and desires for data and/or samples but also equipment, knowledge, finance and kudos. Expressions of such needs or desires related to interests and circumstances; for example, we noted the potential 'security' that came from having the capacity (technological, financial, knowledge etc) to be able to extract 'value' from data and/or samples and how this could dispose the interviewee positively or negatively towards sharing. The significance of this finding relates to the differential distribution of resources, power and influence among organisations, research teams or individuals, which can introduce tension into exchange relationships between those with different status. Those that lack status can find themselves in a position where they need or desire the resources of others, but are insecure about sharing their own for fear that these might be more effectively exploited by others, thus further entrenching their disadvantaged position. It is hardly an analytical revelation that those in powerful positions have disproportionate and potentially cumulative control over resources and influence. For example, such key organisations, research teams or individuals can be regarded as having gained the status associated with 'obligatory passage points'[15]

[15] M Callon 'Some Elements of a Sociology of Translation: Domestication of the Scallops and the Fishermen of St Brieuc Bay' in J Law (ed) *Power, Action and Belief*, Sociological Review Monograph 32 (London, Routledge and Kegan Paul, 1985).

in socio-technical networks. Nevertheless, the ways in which formal and informal patterns of exchange, and the governance regimes that accompany them, are influenced by the dynamics of these relationship is highly relevant when considering negotiations and decisions around access.

Our findings about the normative motivations to share included broad ideas about maximising social good from common (biological and financial) resources. One intriguing sociological issue here was the ways in which these working scientists mobilised notions of the desires and expectations of patients or study participants to add moral weight to the particular position they had assumed in relation to access. In some instances these supposed desires and expectations were embodied in formal consent agreements, but in others they were apparently assumptions about what others 'want'. This is not to suggest, however, that respondents were erroneous in their perceptions or intentionally seeking to manipulate the situation by citing the needs and desires of absent patients and study participants. What is notable is the process of 'translation' in which scientists strengthen their standpoint and actions by strategically using the rhetoric of 'what patients or study participants wanted' in the process of 'enrollment'.[16] This active process of alignment reveals the social and moral positioning involved constituting the governance of access arrangements, which is simultaneously a type of 'boundary work'.[17]

Normative motivations or considerations were further evident in the findings relating to original consent agreements. Some respondents had concerns about the relationships of responsibility embedded in the original context. These are reminiscent of the notion of 'contextual integrity'.[18] The original context of data collecting contains tacit understandings about what the various parties are signing up to. In some instances, however, the original consent agreement appeared to be viewed less as a fixed object, and more as a kind of negotiating platform that might offer the originators of a study an opportunity to maintain control and influence over the basis on which their 'resources' were (or were not) shared. The wording of the original consent could be offered as a way of finding 'common ground' when sharing was considered desirable, or as a defence for not sharing even if that decision was motivated by other considerations (such as fears about appropriate 'credit', or a lack of trust in potential collaborators). These variations in outlook and the potential for negotiation again illustrate the ways in which access and its governance are constituted and enacted in everyday practices and decision-making.

[16] *Ibid.*

[17] T Gieryn, 'Policing STS: A Boundary-Work Souvenir from the Smithsonian Exhibition on "Science in American Life"' (1996) 21(1) *Science, Technology, & Human Values* 100–15.

[18] H Nissenbaum, 'Protecting Privacy in an Information Age: The Problem of Privacy in Public' (1998) 17 *Law and Philosophy* 559–96.

## DISCUSSION

One of the goals of the GGD project was to gain insight into the practices that constitute governance. The aim of this chapter has been to look beyond official, formalised, governance frameworks, such as laws and guidance to consider how governance is both constructed and enacted in and through a variety of social practices and contextually located motivations. When considering how the governance of 'access' is enacted we have been partially guided by the questions posed by Hilgartner and Brandt-Rauf,[19] including identifying 'the multiple interactions among access practices, law, and the political economy of research', and asking how participants use elements of an official governance framework to 'attain protection'.

### The Contingency of Enacted Access and Its Implications for Formal Governance

The notion of the 'data-stream'[20] aptly characterises the plasticity of the artefacts that are generated during the scientific research process and the variability in practices of sharing these. Our analysis further revealed how interviewees' 'exchange relationships' were based on multiple considerations, which interacted differently depending on a variety of contextual factors. Individuals and groups faced immediate choices that were shaped by a range of motivations or disincentives to allow or restrict access. Furthermore these actors operated in broader working contexts or cultures that also shaped their opinions and activities, as well as their range of choices. The factors that could influence decisions to share or withhold materials were multiple, intersecting and, potentially conflicting. We found evidence for all of the following social considerations: perceived needs or desires to share materials for research and develop professional working relationships; ethical concerns about consent, the public good and protecting participant privacy; tacit judgments about the scientific merit of projects and the suitability of the people making a request; organisational imperatives, possibilities and obstacles; and, the norms, practices and etiquette of the scientific community (including its reward system). This range of factors highlights that exchange relationships are dependent on the standpoints of actors, which are entwined with the social contexts in which they operate.

It would be tempting to conceive of a formal model of the '*governance of access*', in which the variety of factors we encountered in our interviewees' descriptions could be accounted for and where their effects on a

[19] Hilgartner and Brandt-Rauf, 'Data Access, Ownership, and Control', above n 2, 368.
[20] *Ibid.*

final outcome of sharing or not sharing could be judged in a unified way. This would, however, misrepresent the situation we encountered. Not only did we find a variety of motivations, interests and values, but also there was potential for a variety of interactions between these factors that were context specific and where any given context was itself dynamic and evolving, as described in Chapter Four. For example, exchange relationships were influenced by the type of resource the interviewees had themselves or wanted to access, what agreements were already in place, which determined what type of thing could or must be shared and in what form, how the scientific merit of proposals or collaborators were judged and what social or economic relationships pre-existed requests for access.

The capacity for formal governance frameworks to capture the social and socially contextualised aspects of enacted governance is questionable (as indeed is this need for this). One reason for this is that scientific developments can run ahead of legal and regulatory systems, and indeed law and policy makers frequently need scientists to explain to them what it is that needs to be regulated.[21] We could add, however, that it is not only the science that can be in advance of formalised governance. The social, cultural and economic contexts in which science is carried out, and consequently the concerns of those who might be (or become) 'stakeholders', also develops over time. This said, however, the importance and influence of the dynamic social and contextual factors needs to be recognised in the policy debates which shape and formalise governance frameworks. If they are not, the resulting governance models will be at odds with the practices of the intended objects of governance (ie biomedical researchers and the 'genetic databases' or biobanks which they create and maintain).

## Discordance and Controversy: Emergent Normative Issues

By taking Latour's[22] advice to follow controversy we have explored the discordance surrounding movements to broaden access. In the section above entitled 'An Initial Analytical Reflection', we noted that we did not aim to make ethical judgments about the practices and viewpoints we found in this chapter (although an ethical reflection is provided in Chapter Ten). Instead we have attempted 'to trace connections *between* the controversies themselves rather than try to decide how to settle any controversy'.[23] For this reason we are circumspect about drawing conclusions about how our finding in this chapter might translate into formalised governance, and prefer instead to draw attention to points of tension within the scientific

[21] *Ibid.*
[22] Latour, *Reassembling the Social*, above n 1.
[23] *Ibid*, 23.

community that may raise dilemmas for law and policy makers seeking to regulate the practices of sharing research materials.

The first of discordance relates to ethical issues. Adopting a sociological viewpoint, we claimed that some researchers who are asked to share materials used the original terms of an informed consent agreement as a 'tool' for maintaining some control and influence over 'their' data and/or samples. This analysis should not, however, detract from the ethical duties and practical concerns that face researchers in respect to discharging their duties to protect research subjects. According to interviewees who had concerns about confidentiality that arise from broadening access, in some circumstances their duty of protection may not be entirely met by procedures designed to protect anonymity. This claim, however, was itself a matter of contention. Some other researchers interviewed in this study, namely the proponents of broadening access, tended to characterise concern about confidentiality as a 'smoke-screen' employed by researchers who did not want to share their resources for the 'public good'. The use of 'ethics' in this controversy is reminiscent of Latour's[24] account of the use of 'facts' in scientific argumentation and Callon's work on translation.[25] Moral arguments alone or in conjunction with appeals to actors who are said to hold particular views, are marshalled by the proponents of different viewpoints in order to support their case. As such, while ethical issues like confidentiality and the 'public good' are significant and need to be addressed by official governance regimes, it seems unlikely that these will be amenable to straightforward resolution.

A second controversy is the distribution of the 'benefits' that can arise from generating and sharing materials. Achievement in science is judged (at least in part) by one's peers, and is rewarded with increasing prestige that can translate materially, into improved salary, resources or working environments,[26] or in a less materialistic interpretation, into professional legitimacy and self-esteem.[27] Arguably, therefore, 'scientists derive much of their motivation and personal satisfaction from their emotional involvement in their work',[28] although (presumably) these emotions may be partially linked to desires for recognition and reward. While each of these characterisations of scientific endeavour may be somewhat idealised, [29] they highlight

[24] B Latour, *Science in Action. How to Follow Scientists and Engineers Through Society* (Cambridge MA, Harvard University Press, 1987).

[25] Callon, 'Some Elements of a Sociology of Translation', above n 15.

[26] Eg, RK Merton, 'Priorities in Scientific Discovery: A Chapter in the Sociology of Science' (1957) 22 *American Sociological Review* 635–59; S Cole and JR Cole, 'Scientific Output and Recognition: A Study in the Operation of the Reward System in Science' (1967) 32 *American Sociological Review* 377–90.

[27] WO Hagstrom, *The Scientific Community* (New York, Basic Books, 1965).

[28] MJ Mahoney, 'Psychology of the Scientist: An Evaluative Review' (1979) 9 *Social Studies of Science* 364.

[29] DJ Hess, *Science Studies: an advanced introduction* (New York, New York University Press, 1997) 56.

an intimate relationship between scientists and the scientific community. Indeed, 'communism' is one of the 'Mertonian norms' of science, such that (again as an ideal): 'the findings of science constitute a common heritage to be shared with the whole community, with recognition and esteem the sole property right of scientists.'[30] Scientists, however, also display independence, individualism and autonomy,[31] and competition is recognised as playing central role of in the organisation of science.[32] This is manifested in struggles to win the so-called 'priority race',[33] and the fear of being 'scooped',[34] which creates a tension between individual and community:

> [T]he reward system sets up an immediate tension between cooperative compliance within the norm of full disclosure (to assist oneself and colleagues in the communal search for knowledge), and the individualistic competitive urge to win priority races.[35]

Mahoney[36] argues that competition, and its links to personal advancement and reward, can (partially at least) explain evidence of the sometimes aggressive nature of scientific disputes that centre on priority claims and personal credit. In public sector science, reward and recognition rests on displays of ownership and 'contribution', while in the private sector the importance of attribution is further entrenched by the need to protect and indeed profit from financial investment.

As the value of material generated in scientific work is tied-up with its exchange within social systems, it should be clear that long-entrenched reward systems are unlikely to easily align with new ideas about liberalising access. Respondents' fears about not fully or satisfactorily exploiting the value of their research efforts arose in a variety of ways including concerns about kudos, credit, status, career and financial reward. One particularly important concern reported by respondents who were critical of, or ambivalent toward, broadening access was that scientists may actually cease to invest in the creation of the artefacts which have now become so sought after. It would, however, be disingenuous to proffer selfishness as the primary explanation for resistance to liberalising access. A sea-change in norms and practices has created a working environment that threatens long-standing ways of working; those who were critical or ambivalent toward the broadening of access still displayed attachments to communal professional or social values; and, solutions to the 'problem' of credit were

[30] *Ibid*, 56.
[31] Hagstrom, *The Scientific Community*, above n 27.
[32] P Dasgupta and PA David, 'Toward a new Economics of Science' (1994) 23 *Research Policy* 498–99.
[33] Mahoney, 'Psychology of the Scientist', above n 28, 362.
[34] Hagstrom, *The Scientific Community*, above n 27.
[35] Dasgupta and David, 'Toward a New Economics of Science', above n 32, 500.
[36] Mahoney, 'Psychology of the Scientist', above n 28.

forthcoming. In such a context 'self-protection' may be a fairer and more accurate portrayal of evidence of resistance to sharing. Nevertheless, there was also evidence of ill-feeling and resistance toward policies (in funding and publishing) that have already shifted toward broadening access. In our view, the movement to liberalise access requires a reconsideration of the relationship between scientists and the scientific community, including an appraisal of appropriate systems for reward and recognition. This represents a significant and urgent challenge. Moreover, any solutions need to address not only career advancement and material reward, but also the less materialistic issues such as emotional investment and professional self-legitimacy (and not simply reducing the latter to the former).

It also seems likely that the differential distribution of power and resources that we have described will have an impact on exchange relationships, and accompanying governance regimes. Some individuals, teams and organisations are in a weaker position than others (both in terms of exploiting the research materials they have gathered and benefiting from open access to other materials). This power imbalance seems likely to affect formal and informal patterns of exchange. Moreover, these power relationships may also influence how ethical concerns relating to sharing materials are communicated and evaluated. Recognising and accounting for differences in power will be important if inequity and exploitation are to be avoided in, or at least not further entrenched by, official governance regimes.

## Changing Cultures of Practice

Star and Griesemer[37] view working contexts as frameworks that enable and constrain a given course of action in both immediate and subtle ways. At any one time there are numerous (and potentially competing) possible methods[38] or 'ways of doing things', and the dominance of some 'ways' over others is not an accident but instead embodies (to some extent) the interests, values and requirements of certain actors. These dominant cultures of practice include 'obligatory passage points', which are gateways that must be traversed to in order to reach a desired set of goals (eg procedures that must be adhered to).[39] Establishing and maintaining the dominant cultures of practice is a process that requires agency, and when there are attempts to change these it is possible to see the conflicting interests

[37] SL Star and J Griesemer, 'Institutional Ecology, "Translations" and Boundary Objects: Amateurs and Professionals in Berkeley's Museum of Vertebrate Zoology, 1907–39' (1989) 19(3) *Social Studies of Science* 387–420.

[38] SL Star, 'This is Not a Boundary Object: Reflections on the Origins of a Concept' (2010) 35(5) *Science, Technology and Human Values* 602–17.

[39] Callon, 'Some Elements of a Sociology of Translation', above n 15.

of different actors. As Star and Griesemer put it, once an obligatory point of passage is established, 'the job then becomes to defend it'[40] against the constant challenges from new methods, residual categories and attempts to standardise practices.[41]

As we have seen in our findings, certain powerful actors have been constructing and implementing new 'ways of doing things' (such as key funding agencies and scientific organisation). This process has included the redefinition of existing objects (whereby existing 'collections' of data and samples have been relabelled as 'common resources'). Arguably, this process has even involved the discrediting of previously accepted practice (whereby failing to make wider use of research materials is portrayed as being unethical). Researchers who continued to adhere to the old 'ways of doing things' risk being viewed as the 'defenders' of what was previously an accepted set of values and practices. We conclude that, while it is likely that there will continue to be diverse 'ways of doing' sharing, the imbalance of power between actors espousing differing cultures of practice will result in disharmony that takes a certain form. 'Resistance' to policies that aim to broaden access will be met with accusations of self-interest or territorialism. This stand-off seems likely to remain until other aspects of the working culture of science, particularly its reward system, are aligned to the goal of increased sharing.

[40] Star and Griesemer, 'Institutional Ecology, "Translations" and Boundary Objects', above n 37, 391.

[41] Star, 'This is Not a Boundary Object', above n 38, 613.

# Part III

# Reflections, Ramifications and Recommendations

# 10

# Reflections on Practice and Governance

*Jane Kaye*

THE RELATIONSHIP BETWEEN practice and governance is an under-researched area in the field of biobanking and genetics but one where there is growing interest and scholarship. This chapter will draw on the analysis carried out in the mapping of the regulatory space in Chapter Three and the findings from the empirical research in Chapters Four, Five, Six, Seven and Eight. The purpose of this chapter is to understand the relationship between the current governance frameworks, and the experience of practitioners who are involved in the establishment and management of biobanks in the UK. In doing so, it will provide an insight into the discords and convergences between the legal and regulatory requirements and current practice in regard to biobanking. This will provide an understanding of the ways that laws and the institutions with regulatory powers (both formal and informal) affect and have relevance for current biobanking practice. This chapter will provide an overview of the current governance framework and compare this with the experience of our respondents in navigating the regulatory space. From this synthesis, this chapter will identify the areas where practitioners involved in biobanking have encountered problems with the system that was in place at this time. This will have relevance for thinking about the regulation of medical research in general, and will provide a basis from which to make further policy recommendations in our concluding chapter.

## THE NATURE OF THE GOVERNANCE FRAMEWORK

The analysis of the regulatory space in Chapter Three pointed to some deficiencies in the legal framework and the governance mechanisms that are in place for biobanks. This is because the current framework for medical research is patchwork of legal instruments and informal and formal regulatory bodies, which have developed over time, each applying to different aspects of the research process and no one being responsible for oversight of all. As detailed in Chapter Three, the system is heavily dependent upon self-regulation, self-policing, trust, voluntary compliance, social networks and social sanctions, the moral agency of individual professionals and practitioners, and a willingness (and ability) on their part to follow 'good' or 'best' practice. Subsequently, there is no clear normative orientation, policy direction, or explicit solid theoretical foundation for the system as a whole. The result is a highly complex governance structure that can be inconsistent and incoherent and difficult for researchers to navigate. The relatively new area of biobanking and genetics has struggled to be accommodated within this patchwork governance framework for medical research. This raises significant issues as to how best ethico-legal practice in new areas of medical research is being developed but more importantly it questions the effectiveness of the current governance system for medical research as a whole.

## The Legal Framework

It is not possible to understand the legal framework for biobanks without understanding the general framework for medical research in the UK. This is because like many other jurisdictions within Europe, there is no purpose-designed legal framework or dedicated legal instrument that applies to biobanks in the UK. Instead, biobanks come under the general legal framework for medical research. However, unlike other jurisdictions within Europe,[1] the UK does not have legislation that applies to all kinds of medical research on human beings. While there are the Clinical Trials Regulations[2] which relate specifically to clinical trials, the tendency in the UK has been to include references to medical research in other more general legislation. For example, the Human Tissue Act 2004 in regulating of the use and storage of human tissue also refers to how this should apply for research purposes. The Mental Capacity Act covers the protections that should be in place for those lacking capacity to consent on their own behalf and details how these should apply in relation to medical research. Similarly, general law, such as the Data Protection Act 1998, the Human Rights Act 1998 and the legislation that applies to patents also have implications for medical research. This patchwork coverage does not necessarily cover all the dimensions of medical research governance in a comprehensive way, and each piece of legislation must be read in conjunction with each other in order to develop a real understanding of how the law applies to medical research. To understand how these might apply to biobanking requires expert legal knowledge.

[1] Noteably Norway, Iceland and Sweden.

[2] The Medicines for Human Use (Clinical Trials) Regulations, Statutory Instrument 2004 No 1031.

However, there are also various statutory and non-statutory codes of practice and guidelines issued by the Department of Health, Medical Research Council (MRC), Human Tissue Authority (HTA) and Information Commissioner's Office (ICO) that have relevance in biobanking and are easier to access. Informal sources that were identified in Chapter Three are well in excess of 30 codes of practice, a host of guidelines, circulars, letters, best practice guidance documents, fact sheets and statements of ethical principles—all of which have a direct, or at least potential bearing on governing biobanks or the professionals who establish and use them. In the absence of specific provision, biobanks are instead subject to a vast overlapping, yet uncoordinated, patchwork of legal instruments and policy guidance, which are both formally and informally constituted. When practitioners were asked about the laws that applied to their work, all 49 respondents gave different answers, reflecting not only diversity in our group of respondents but also the diversity of law that has application in this area but also the amount of guidance that has been generated for medical research.

The way that the respondents in our study dealt with this complexity of information and the law that they identified as relevant for their research is revealing. Firstly, while our sample of respondents as a whole referred to many of the instruments reported as relevant in our analysis of the regulatory space, there was no one source of guidance, whether formal or informal, that was identified and used by all respondents and no clear ranking of sources. The 'formal' governance instruments of statute, regulations, statutory codes of practice and common law doctrines were prominent sources of guidance for practitioners but so too were informal sources of guidance which in some instances appeared to be considered more relevant. Although the HTA and the DPA were the statutes mentioned most frequently, our analysis of the way that these were discussed during our interviews suggest that, generally speaking, people did not appear to have an in-depth knowledge of the requirements of these pieces of legislation and their claimed knowledge was sometimes inaccurate. Respondents were largely unaware of the guidance that was provided by these regulatory bodies as to how these statutes should be implemented. Sources of guidance such as 'good practice' or 'best practice' principles were cited as key sources of information, such as the policy guidance of the Department of Health, Medical Research Council (MRC), the Wellcome Trust, the US National Institutes of Health (NIH), and the Good Lab Practice Monitoring Authority. It is argued that some such guidelines and websites could be expected to have great significance for our respondents (perhaps in some instance even greater significance than statute) because of their focus on biobanking in particular. Nevertheless, this only serves to highlight the importance of informally constituted bodies, other than bodies formally constituted by the state, as regulators in this field.

The analysis of the interview data (in Chapters Five to Eight) supports the idea that the laws and legal frameworks at the time lacked clarity and was open to interpretation, and this was compounded where multiple sources of law, legal actors or jurisdictional requirements were involved. Another factor here is the lack of consistent and clear definitions of much-used biobanking concepts and terms in legal instruments or guidance. Where there was room for interpretation, some respondents talked about implementing laws and guidance to the best of their ability. As the analysis in Chapters Five to Eight argues, however, this can mean that law or guidance is interpreted in such a way that allows individual practitioners to do what they believe to be appropriate (albeit within the framework of their professional culture and routines of professional practice). However, the analysis of the interviews also revealed that at least some practitioners had concerns or anxieties about whether they were working in accordance with the law, or whether they were appropriately differentiating between the relative weight, or authority, that should be given to governance instruments.

The work in this project thus highlights some areas that may give rise to concerns for those charged with regulating the field. On the one hand, practitioners are sometimes placed in a situation where they are left to decide which of the governance possibilities on offer best fits their needs. On the other hand, this is a situation in which they appear to be relatively unaided when trying to decide why some guidance should be followed in preference to others. Furthermore, this process of identifying and applying governance is time-consuming, since sorting through the law and guidance produced by bodies in the UK, Europe and at an international level that may be potentially applicable requires effort and expertise. One difficulty, for example, would be deciding which one has the greater authority or applicability. For example, if the Information Commissioner Office has not issued any relevant guidance—is it better to follow guidance provided by a funding body and hope that they are offering sound advice? Without clear direction from Parliament or a delegated authority it is very difficult for biomedical practitioners to make these judgements. The absence of one clear authority and a legal framework specifically for medical research has resulted in uncertainty about what best practice in biobanking should be—but it has also contributed to a diversity of practice.

## Regulatory and Governing Bodies

The analysis of the regulatory space for medical research carried out in Chapter Three demonstrated that there exist vast numbers of bodies that have issued guidelines on medical research. However, there are only four bodies in this domain that actually derive their authority from statute or regulations sanctioned by Parliament and the democratic legitimacy it endows. These bodies have the authority to enforce sanctions, such as withdrawal of licences and the imposition of fines, if the standards that they set are not complied with. The key bodies constituted under statute are the HTA, the ICO, RECs and the General Medical Council (GMC). These bodies are given powers to write guidelines and codes of practice, with the exception of the RECs, who are guided by SOPs (Standard Operating Procedures) that are written by the NREC. The powers that are conferred on each of these bodies give them the authority to control the various stages in the research process and to act as gatekeepers. For example, the HTA has the power to issue licences for the use of tissue; the ICO requires that a data controller is registered in order to process data; REC approval is required before any research using NHS patients, premise or resources is started; and the GMC sets the requirements and standards for medical practice. The powers of these bodies appear to apply to the key areas of activity that are important to medical research by ensuring that professionals are appropriately qualified; tissue is licensed before use and data usage is monitored; and research applications are scrutinised before proceeding.

However, when we consider how these regulatory bodies may apply to biobanks, a different picture emerges. Firstly, the HTA powers and licensing requirements do not apply to extracted DNA a fact which effectively excludes much of the activity carried out in genetics research from the scope of the HTA's authority. Secondly, the ICO must regulate data processing for the whole of the UK and so it largely relies on complaints to be made before an investigation will be made because it does not have sufficient resources to do otherwise. The implications for medical research and biobanks are that essentially as long as there are no scandals, the ICO will take an information-providing role. It relies on those regulatory bodies specifically developed for medical research oversight to ensure adherence to the law, rather than using its own resources to investigate practice. The ICO is also dependent upon registered data controllers in the medical sphere to make decisions as to when data processing is legitimate. In terms of biobanking regulation, it is only the broad principles of data protection that apply, and how these might apply is left largely undefined. The result is that issues such as whether genetic information is covered by the Data Protection Act are not stipulated and practitioners in the UK rely on guidance by the European bodies such as the Article 29 Working Party. The GMC only applies to registered practitioners. Increasingly those involved in biobanking and research are derived from a number of disciplinary backgrounds, of which doctors may only be a small number. RECs are therefore the only formally constituted bodies that appear to have some direct authority over biobanks and have a regulatory influence.

However, even in the case of RECs, the scope and appropriateness of their decision-making appears to be mixed. Firstly, there is anecdotal evidence that genetic research and its implications are not adequately understood by the vast majority of RECs.[3] In addition, RECs only have

[3] R Ashcroft, AJ Newson, PMW Benn, 'Reforming Research Ethics Committees (editorial) (2005) 331 *British Medical Journal* 587–88.

the authority to approve research applications; their on-going powers of oversight or review are requirements for self-reporting once the research has commenced. Biobanks that develop as a result of a number of research projects are therefore subject to scrutiny for each new research project and to some extent reuse of existing samples and data. *De novo* biobanks such as population biobanks, are infrastructures designed to continue for many years, and do not necessarily involve just one research project, but are likely to involve many research projects that are not possible to describe, or determine at the time of the establishment of the biobank. Therefore, while the RECs can oversee each new project, they do not have the powers, authority or resources to ensure that these biobanks are governed efficiently or effectively. Furthermore, the introduction of the NRES SOPs that now give biobanks authority to approve research conducted by researchers[4] on the biobank has the effect of side-stepping the oversight that RECs may have previously exercised.[5] Amongst other things, SOPs altered the arrangements for ethical review of 'research tissue banks' by NHS RECs partly to reflect the new Human Tissue Act 2004, and partly to streamline and reduce the level of ethical review of tissue research. This now means that a REC can give approval for generic consent for the use of samples in a biobank. This has further placed biobanks outside of the scope of formal regulatory oversight bodies. Our analysis demonstrates that regulatory bodies constituted under statute, have little or no oversight responsibilities for the activities of biobanks. Although these bodies should be incredibly influential, as their powers are sanctioned by law, in practice, this is not the case.

The analysis reported in Chapter Five detailed a number of other bodies that do not have formal regulatory powers but which were considered by our respondents to be key sources of guidance or bodies that influenced biobanking activities. The RECs and funding bodies featured most commonly, but so did the NHS and professional bodies, such as the Royal Colleges and the BMA. The sources of authority for professional bodies are not always mandated through the force of law, but rather stem from their possession and control over other key regulatory resources, notably information, wealth, and organisational capacity and the role they have in providing professional training and information. These bodies have considerable power and influence in the governance structure as they represent practitioners and therefore have legitimacy for their activities. However in the case of biobanks, their influence and role in developing guidelines has not been evident, in contrast to the role that RECs and funding bodies have played.

Funders are extremely influential, as they can provide or withdraw funding and in doing so can influence policy and activities. Yet at the same time, they

[4] See ch 3 above for more detail.

[5] Standard Operating Procedures for Research Ethics Committees in the United Kingdom ('SOPs'), version 3.3 (April 2007).

also rely very much on researchers and a system of peer review to validate and decide on important research initiatives and strategies. The influence of funding bodies has been significant in terms of genetic research, as funding bodies have sponsored and initiated the current infrastructure of sample and data collections that have emerged over the past ten years. Important initiatives have been the Wellcome Trust Case Control Consortium, the extensive investment in biobanks by the MRC,[6] as well as the funding of UK Biobank by the MRC, the Wellcome Trust and the Department of Health. Almost in recognition of the lack of appropriate regulatory bodies and structures for biobanks within the medical research context, UK Biobank has established its own structures of oversight with an Ethics and Governance Council. Access committees have also been established for the Wellcome Trust Case Control Consortium and other biobanks, as internal mechanisms to control access to the resource, but also to supplement the lack of capacity by other bodies such as RECs to carry out this governance role. Funders in the UK have also been responsible for the funding and the accompanying policy that has been required for these new infrastructure and resources in genomics established to share sequence data.. Funders play an important role in governing research, but as this role is largely unaccredited and they are not subject to the same accountability mechanisms as other formally constituted regulators.[7] This important role in research governance is a largely a 'de facto' effect of their funding activities.

## The Interdependency of Organisations

The picture that our legal analysis provides is of a complex system of governance for medical research that has grown up over time. The absence of specifically tailored and comprehensive, formal legal provisions for medical research, or a single or coordinating body that would be formally charged with oversight, has lead to a highly complex and confusing web of actors, interrelationships and interdependencies. Such complexity and confusion are paralleled in the formal and informal documents that currently exist for medical research. The uncoordinated roles of different formal regulators have created a context in which different bodies, such as funders, step in to breach the gap. This has lead to a plethora of guidelines and other documents that frequently overlap and at times can be mutually inconsistent. In addition, bodies that have been charged with authority to regulate in this domain, such as the ICO and the RECs often do not have

[6] J Kaye and P Martin, 'Safeguards for Research Using Large Scale DNA Collections' (2000) 321 *British Medical Journal* 1146–49.

[7] Such as the Regulators' Compliance Code: Statutory Code of Practice for Regulators (17 December 2007).

sufficient resources to take a more proactive role. This has allowed other actors, such as funders to exercise their powers by influencing the direction and development of policy and the research agenda. This system is largely self-regulating as the state depends upon a combination of formal and informal actors in the medical research domain, to regulate with little external oversight or interference by bodies outside the medical field. The result has been the development of a complex interdependency between different kinds of bodies who have different and sometimes conflicting approaches, interests and powers.

The influence and regulatory role of formally constituted regulatory bodies and the legal instruments developed by Parliament, such as the Data Protection Act, appear from our respondents to have little direct bearing on how biobanks are governed. This is not to say that formally constituted regulatory bodies have no bearing on the regulation of medical research, but rather that it is not the detailed 'hands-on' regulation that exists in other jurisdictions such as Iceland. The regulatory space that biobanks are embedded in, is not a domain that is dominated by state intervention, but rather is a negotiated space, where a number of different bodies have authority, power and control, even though the basis for authority and influence may vary. Within this governance system, a number of powerful bodies co-exist but also at times compete, as there are not always clear areas of control and responsibility. The status and legitimacy of an organisation, whether it is mandated by statute or not, will depend upon knowledge, resources and the people to run the organisation; the expertise to design rules; and, the powers to monitor and police enforcement of those rules. These powers and the ability to enforce them can vary enormously. It is only in rare occasions that this distribution of power is challenged through the courts, or the authority of different bodies is challenged by other regulators.

The way that regulators perceive their role as regulators is part of a complex negotiation with other regulatory bodies but also with practitioners. The bodies that have delegated responsibilities under statute, such as the HTA, still negotiate the content of their guidance to reflect current practice while remaining within the requirements of the law. The regulatory attitude of the HTA is not to be a remote and authoritarian regulatory body as, according to the HTA's website, 'collaboration' is one of its founding principles. The HTA seeks to engage closely with professionals and public stakeholders in developing its policies and regulatory arrangements, to create and foster 'a robust and proportionate regulatory system' under which 'professionals support and understand the regulatory framework they must follow'.[8] This approach means that the HTA, one of the key bodies responsible for the regulation of

[8] http://www.hta.gov.uk/about_hta/how_we_work.cfm.

medical research, can work closely with practitioners, which also enables its decision-making to be influenced by, and in accordance with, the realities of practice. However, in the case of biobanks involving genetic material, the HTA has refrained from providing specific advice because research on DNA is outside its legislative remit.

Another example of the interdependency between organisations is found in the approval of research and the pivotal role, which RECs have in this process. As discussed in Chapter Three, RECs are relied upon to keep biobanks and biobanking activities in check by the principal UK policymakers, lawmakers, formal regulators and the governance framework. However, the powers of RECs are not comprehensive, with scrutiny being focused at the beginning of the research process and with few enforcement powers once the research has commenced. As pointed out in Chapter Three, due to these deficiencies and the fact that no other body is charged with this task, other agencies must provide the missing 'teeth' to ensure that research is adequately supervised. Therefore, RECs depend heavily on the existence of other bodies and mechanisms—both formal and informal—to ensure compliance, to enforce their ethical conditions, and penalise violators. Prior ethical approval is required by the Department of Health, NHS, GMC, professional associations and all leading research funding bodies (notably the MRC) for research that involves NHS staff, facilities and patients. If there are breaches of these requirements, disciplinary action by employers or funders may be taken, but not by RECs. Currently there is not a clear relationship between those who set the standards and requirements and those with the power to enforce them. This also means there is not a clear line of authority that is evident to researchers.

## NAVIGATING THE GOVERNANCE FRAMEWORK

The description above provides an insight into the complexity of the governance framework that builds on the analysis found in Chapter Three and the findings from the interviews reported in Chapters Five to Eight. It goes some way to explaining how regulatory bodies negotiate and operate within this framework and how our respondents perceived and navigated the governance structure as it existed at a particular point in time. The comparison of these two different analyses points to the complexity of the governance structure in this area and its polycentric and diffused nature. For example, according to a strict legal analysis we would expect to find that the regulatory bodies sanctioned by Parliament would have the greatest influence and authority. However, our respondents barely mentioned formal bodies, such as the ICO and the HTA which are legally significant and have formal legal status, powers of enforcement and responsibilities. Instead our respondents reported more active engagement with RECs, funders and other professional leaders in the field, as well as advisory bodies and steering committees.

## Sources of Advice

RECs and funders were named by more than half of our respondents as the important external bodies that they turned to for advice and guidance. The way that they were used by researchers can be attributed to the roles that they have in the research process. Our legal analysis showed that the RECs are one of the main gatekeepers in allowing research to proceed, as approval from a REC is required before researchers can carry out any kind of research on patients in the NHS. They provide initial ethical approvals for projects and their Chairs are on hand to provide additional opinions as research protocols develop. As part of this approval process, researchers are required to submit an application, to attend a REC meeting and to provide follow-up, progress or final reports. In our interviews, the second most commonly mentioned form of external oversight were the funding bodies, such as the MRC, Wellcome Trust, the US National Institutes of Health (NIH) and the Economic and Social Research Council (ESRC). The way that funders influenced the research process was varied. Some respondents described funders as being influential in helping to set the requirements for project management, rules of engagement, data handling, third-party access, and the content of participant consent forms. Funders therefore have the ability to set and shape the research agenda and the way that research practice around biobanking is carried out. The mechanisms that funders can use to influence research practice is by setting individual grant conditions but also by providing ethical guidelines, position statements and good practice guidance. A prime example of the effect of funder's policies on practice has been open-access policies that require that research data is made accessible to other researchers.

The internal oversight bodies and processes mentioned by respondents during interviews were not a feature in our 'on-the-books' analysis of the regulatory space. What becomes clear through the sociological work is that practitioners rely in part on personal and professional networks, as well as formal oversight and advisory systems that are in place within their own working contexts. Examples given of internal non-specialist structures within the public sector were general internal audit, monitoring and control systems, university lawyers, Caldicott Guardians and NHS Research and Development teams. In addition, committees or boards, set up especially for individual projects played important advisory, oversight, management or executive roles. These bodies were not uniform in composition or in the kind of role that they had, but, for many of our respondents, they were the first port of call when faced with a challenging situation. Other forms of internal oversight mentioned by numerous respondents were regular project group, team or partner meetings that helped to shape the direction of the research or biobank.

Another notable finding from the interviews with practitioners was the preference and reliance that was given by some to the wisdom and knowledge of other professionals in the field. Analytically, this was contrasted to a relative dearth of descriptions of respondents approaching external regulatory bodies for advice. It was reported in Chapter Six how some respondents spoke to people whom they already knew, such as immediate colleagues, project team members, external collaborators and specific individuals in funding bodies, or professional associations. Likewise, respondents based in the private sector relied on 'in-house' specialists but also looked to other pharmaceutical companies websites to ascertain their policy. Going to the right conferences, meetings or training courses and talking to people were also seen as a valuable way to get the latest ideas and information on emerging practice. Researchers or research projects that were considered to be 'path-breakers' or 'trailblazers' were used as models or examples to follow. This last example highlighted the importance of cutting-edge projects, as the standards that they develop become the precedent for other projects, which in turn become accepted practice. An example of this is UK Biobank as many of their procedures and ways of doing things have become accepted as best practice and have been implemented by other biobanks, in the UK and internationally.

The perspective of respondents lends a very different view of the bodies and ways that the establishment and operation of biobanks are governed than our legal analysis provided. There was also a marked difference in the way that those working in industry experienced governance compared to most of the practitioners working in public sector settings such as hospitals and universities. Private sector respondents navigated the regulatory system with support from in-house teams of lawyers and experts. Corporate structures were described as having clear vertical and horizontal divisions of labour, decision-making, specialisation and responsibility. Notably, traits of this model of divided responsibilities were also evident in the accounts of some respondents working at the (to a large extent) publicly funded organisations we described as 'Repositories'. The bodies that provided direct oversight of the activities of the respondents in our study can be divided into those that were external and internal, respectively, to their own organisations.

## Social Relationships and Networks

In addition to the explicit guidance provided by colleagues and internal and external bodies, our research revealed three additional factors that influenced respondents' reflections about the biobanking governance landscape. These were their professional background, practical experience and sensitivity towards the opinions of other stakeholders and actors. In particular, the combination of professional training reinforced by practical experience within an institution and the interaction with others were a source of implicit guidance for our respondents. Respondents who had different professional backgrounds or training, or who came from differently constituted working communities therefore had different values and approaches to their research.

Professional backgrounds can thus have a significant bearing on the course of action that individuals may choose and the values and principles that they identify to provide the basis for their action. Genetics research requires the involvement of people from a variety of different disciplines with expertise in recruitment of patients, research design, sample and data management, curation and analysis. Each person will bring specific disciplinary expertise that will have been developed through their training and professional experience. This can lead to different ideas about what is the most appropriate course of action in new situations which may challenge disciplinary or professional boundaries of what is considered appropriate behaviour. For example, clinicians with regular contact with patients may have quite a different perspective on what is considered an ethical course of action compared to statisticians or bioinformaticians who primarily work with data. While disciplinary background is important in guiding decision-making about the selection of research questions or methodologies, it also has a bearing on the norms and values inherent within one's practice, and when deciding what should be appropriate behaviour in a given situation.

Practice is a dynamic and evolving thing that it is based on the experience, disciplinary perspective and knowledge base that an individual brings to a situation, which may in turn be tempered by the approach that a group or institution may have. Our findings suggest that practice is continually enforced by 'doing' but also through interaction with other professionals both within an institution and externally. The culture of an institution will have an influence on the action and the development of best practice. Embedded practice can come to be seen as self-evidently correct, uncomplicated and unproblematic and therefore something that is unquestioned. The benefits of embedded practice are certainty, efficiency and stability. In addition, this shared understanding of practice and notion of socially shared conventions may be put to the test by individuals from different disciplinary backgrounds but also by the challenge of having to work together to establish new initiatives such as a biobank. This dynamic situation can be a basis for learning, and is part of the evolving shared knowledge within larger professional or working communities.

The importance of other people's views in directing, aligning and supporting decision-making was identified as a key source of advice by our respondents. It is also a means of developing and maintaining certain behaviours and practices. Some of our respondents in our study regarded the opinions of others—whether they were patients, their immediate colleagues, eminent members of relevant professions, senior staff within institutions or expert advisers as influential in their own decision-making. The possibility that they might be taken to task or publicly shamed for not doing the 'right thing'—which might lead to long-term reputational harm—was a consideration in decision-making. At the same time, being connected with others also held the potential of generating obligations. We reported instances where our respondents regarded themselves as the protectors or champions of their research participants' interests, which involved protecting them against harmful or unethical use of their data or samples. We also reported that in some respondents the desire to develop or follow appropriate governance practice was motivated by a wish to safeguard public trust in medical research, or by a concern about using biobanking resources as justly and as productively as possible. Therefore, in this myriad of ways, the views of 'others' played an important role in guiding behaviour and developing the norms that have become embedded in biobanking practice for those in our sample group.

## The Importance of Individual Decision-Making

Without clear external guidance on biobanking, our respondents had to exercise considerable judgement about the best sources of guidance to follow, who to contact as well as having to decide upon the most appropriate courses of action. In many cases, if a funder or REC did not clearly outline what should happen, respondents were forced to select from a wide range of sources to identify standards and practice that were relevant for biobanks. Even if a funder was prescriptive, there was still room to 'negotiate' how a resource would be managed. It appears for some of our respondents that individual action and professional culture were an important purveyor of norms, values and standards of good practice. Our respondents also reported on the importance of other professionals who provided mentoring and advice, as practice was developed through 'doing' and a process of trial and error. Such relationships are an important part of the overall structure and process of governing biobanks and genetic research.

Biobanking requires the long-term curation of a resource as well as making it accessible to a wide range of users. While the whole purpose of biobanking is to make samples and data available for others, the application of open-access policies to existing projects created a number of dilemmas for our respondents about how this role should be exercised and how it might transform the relationship with research participants. The curation of a resource and the managing of access by third parties are both activities where the custodian relationship is important. The respondents in our sample regarded that the execution of these activities was carried out with an awareness of how this might impact on the relationship with their research participants. In this capacity as a custodian of samples and data entrusted to them by research participants, our respondents described a number of tools that they used to regulate access to samples and data and exercise their custodian responsibilities. Examples included the authority to decline access, anonymisation of the samples and data, and the use of consent forms and various contracts to govern access.

Respondents had a key role in determining who would have access to a biobank, and on what basis, as part of their role as custodians. According to our respondents the key factors in assessing whether to enter collaborations or grant access to materials that were not widely available was on the basis of professional standing, the scientific quality of the research proposal, social networks and an ability to work to appropriate standards. Assessing professional standing involved determining the credentials of the researcher, or their professional legitimacy—such as their professional qualification, their university or a research organisation and their previous published research. This could also involve establishing with whom they had trained and who their current collaborators were. This focus on the 'personal' was an important criterion in this small, self-regulated community of genetic research. While the research planned was important, the focus is on the individual. This, in many ways, could become a personal assessment of the individual rather than an assessment of what the researcher is aiming to do. Therefore, for some of our respondents the personal and professional integrity of individuals embedded within this professional network are factors that have become part of the considerations in decision-making.

Generally in research, it is the custodians of samples and data who design the consent forms that participants are asked to sign as acknowledgement of their voluntary participation in research but also as a future record of what they agreed to. The wording of the original consent defines the nature of the obligations owed by custodians and can be used a basis for refusing or granting access to data and samples. The decision whether secondary research falls within the scope of the original consent is one that is made by the custodian, and if in doubt, they will refer this to a REC or the appropriate body attached to the biobank. The original consent is also used by RECs to determine whether proposed secondary research on the research material can be carried out. Therefore for our respondents, the informed consent form was an important mechanism to determine whether further research could happen. It could be used by the custodian as a basis for restricting access by other secondary researchers or specifying the conditions that should be attached to the planned research.

Standard practice is to use anonyminisation as a way to protect the confidentiality of the information and to keep the promises made to research participants that they will not be identifiable through the research process. The custodians may hold the keys or the 'identifiers'—the information that could re-identify research participants. Although anonymised data and samples are mostly shared, there may be situations where being able to re-identify research participants may be necessary to track individuals for research purposes or to feed back results to individual participants. Therefore, those who hold the keys to de-identify the individuals become the gatekeepers to further research. Our respondents identified a number of reasons why it might be justified to re-identify samples and data. Two reasons given by our respondents were the need to link to records for specific research purposes or to ensure that the consent is appropriate for the proposed research.

Key contractual documents were used by our respondents to stipulate the terms of access and sharing. Standard Operating Procedures (SOPs) as well as different kinds of written contracts or agreements were important mechanisms that shaped the way that biobanks were established, managed or accessed. Respondents reported using contracts such as material transfer agreements (MTAs), memoranda of understanding (MOUs) or similar agreements between collaborators, intellectual property rights permissions (including patent licences) and third-party access agreements. Some reported that negotiating the content of these bespoke documents in some cases slowed down the research process considerably as lawyers and other specialists had to be involved. It is evident that these were important mechanisms by which custodian obligations can be reinforced and limitations placed on how samples and data are used. In the absence of overall frameworks, these individual contracts are important mechanisms that may be used to fill or supplement the governance vacuum.

## PROBLEMS WITH THE CURRENT GOVERNANCE SYSTEM

The attitudes towards governance and laws in this area were diverse. Although some respondents expressed praise for certain laws or regulatory actors, there were some who were negative and some were highly critical (see Chapter Eight). These respondents saw the governance system that was in place at the time as doing little to encourage or promote research, patient care or scientific progress, but instead created barriers and obstacles that hamstrung or stifled basic research. Some of these comments could be attributed to the newly introduced Human Tissue Act, which at the time was in the process of being implemented in the UK. Some respondents reported spending a considerable amount of time and resources on navigating and negotiating legal and regulatory frameworks not designed for biobanks or genetic research. Our analysis suggests that best practice has been developed through trial and error, with support from colleagues in the research community but with little encouragement, sanction or endorsement by external regulatory authorities. Unsurprisingly, many of the problems encountered by our respondents could be directly related to the nature of the regulatory framework and the fact that it had little relevance for biobanking activities.

### A System that Lacks Relevance

Our legal analysis demonstrated that despite the diversity of regulatory bodies, statute and guidance that applies to medical research, very little of this has relevance for biobanking. This was supported by our interview findings as while our respondents could collectively identify many of the regulatory bodies in the regulatory space, the knowledge of individual respondents was very limited. This suggested that there was no one source of guidance that all respondents could point to for direction nor specific regulators who were used regularly by our respondents. Instead, our respondents had knowledge of different aspects of the regulatory space and the bodies that operated within it. There was no one body that all respondents identified as being able to provide clear authorative advice on what was considered good practice in the field of biobanking. Instead the key sources of guidance were funders, RECs, colleagues and informal advisory bodies, composed of experts in the field. Decisions about which body to approach for advice appeared to be influenced largely by the professional backgrounds, roles or responsibilities, as well as the social networks and institutional culture that an individual was anchored in.

### Failing to Meet the Requirements of Respondents

The governance system failed to meet many of the concerns and needs of our respondents who were trying to establish biobanks and wanted very practical advice. At the time of the interviews our legal analysis demonstrated that there was no clear, common regulatory framework for governing biobanks either at the national or international level.[9] Respondents wanted a framework that could provide higher-level principles but also detailed guidance on how to carry out day-to-day operations. There was a desire for a flexible 'road map', 'checklist', 'step-by-step guide' or 'wizard' that people could work through to ensure that they were compliant and had not missed any vital implementation steps or key issues. In particular, bioinformaticians wanted a clear, detailed framework for designing, setting up and running biobank IT systems that would set out the security

[9] This has since changed with the 2009 OECD Guidelines on Human Biobanks and Genetic Research Databases (HBGRDs).

standards and procedures for handling participant-identifiable data. If such a framework had existed it may have provided some certainty and clarity. Our legal analysis showed that such clarity did not exist within the law and regulatory system that was in place at the time.

There were key issues concerned with biobanking that were not covered by the existing legal framework nor could they be found in the general law as explained in Chapter Three. For example, the law did not set out clearly the normative principles that should apply to property rights and ownership in respect of human biological materials held in a biobank. Clarity was needed on who 'owned' samples or collections, especially where multiple institutions are involved. This was important in the case of buying, selling or transferring biobank resources but also to understand the roles and obligations of funders and the researchers who established biobanks. Another issue that was not covered by the law was benefit sharing and the requirements for feedback of individually relevant research results to research participants. Although feedback of individual results is addressed in the guidelines of the Medical Research Council (MRC),[10] these are not necessarily binding on researchers who are not funded by the MRC. As stated earlier, the law also failed to provide direction on terms and concepts that were essential for biobanking, such as the requirements for anonyminisation, informed consent and broad consent.

## Ambiguity in Interpretation

Without clear legal guidance, there were different interpretations of what was considered to be appropriate practice by regulators but also researchers had to justify their own choice and interpretation of particular guidance. Respondents reported being faced with the situation of different regulators interpreting the same guidelines differently as well as regarding different guidelines as more authorative. For example, there were differences between the RECs and what institutional Data Protection Officers wanted. In grey areas of interpretation, researchers may be referred to different bodies that may have different answers to the same issues. Our respondents were also put in the position of having to justify to bodies, such as RECs, why they had either followed specific guidance or why they had interpreted it in a certain way. One respondent described having to defend their interpretation of MRC guidelines with the REC. Within the UK, it is well known that different RECs may give different opinions on the same kind of research. In the case of new practices such as biobanking, this created problems for our respondents. It was suggested by a number of respondents

[10] Medical Research Council 2000 Personal Information in Medical Research.

that there should be clear consistent national guidelines for RECs that they can follow easily to eliminate some of the problems created by inconsistent decision-making.

## The Impact of the Law

At the time that interviews were carried out, the HTAct was being implemented in the UK and subsequently many of our respondent's concerns about laws were to do with the HTAct. Respondents were confused by the exclusion of DNA from the Act and found that it complicated rather than simplified their work. This corresponded with our legal analysis that showed that there was inconsistency, irrationality and gaps, in the way that DNA is treated under the Act. Even though the exclusion of DNA meant that respondents were relieved from the licensing requirements of the Act, some respondents would have preferred that DNA was included under the provisions of the HTAct. It was evident from our respondents that the HTAct had lead to a change in culture, which had resulted in people being more cautious and as a consequence tissue samples for research purposes had been harder to obtain.

Respondents' experience of Intellectual Property (IP) law and patents lead some respondents to hold the viewthat IP had a detrimental effect on practice. Patents were seen by some respondents as a purely 'defensive' mechanism that could be used to 'block' access to materials and thereby stymie potentially beneficial research. Some respondents' experience with the patents system had resulted in the protracted negotiation between the parties and the involvement of lawyers which led to high costs, delays and complexity. Another negative impact on research mentioned by some respondents was the high cost involved in establishing systems to implement new pieces of legislation, such as the HTAct and the EU Clinical Trials Directive 2001/20/EC.[11] One particular concern was the cost of HTA licences, which was seen as being something that would add to the costs of research with no perceivable benefit for actual research.

Interacting with colleagues from abroad, where there might be slightly different regulatory and legal frameworks, also created problems for some of our respondents working in international projects. Obtaining REC approval from nationally based RECs, could involve multiple REC approvals from different countries involving considerable time and resources. This could result in additional costs, bureaucracy and having to deal with inconsistent decision-making. One respondent had simply decided not to engage with the laws because of the prohibitive time and

[11] [2001] OJ L/121/34.

cost entailed in identifying and analysing the legal requirements. Some of the problems identified by respondents were the conflicts between laws, as well as restrictions on transfer between countries for samples. A solution that was used in large projects was to aim for the highest legal standard in a project, in the hope that this would exceed or capture the legal requirements of all participating jurisdictions. If this was too difficult to achieve, some collaborators had been left out of projects, if it were found that the requirements in their country were too onerous, or out of step with those in other countries.

## IN CONCLUSION

The conclusion from our analysis is that current biobanking governance framework in England and Wales depends very heavily on 'informal' oversight methods. The system appears to be dependent on self-regulation or 'soft' regulatory techniques based on our analysis of the governance system at this time. Reflecting on the regulation of medical research as a whole, what emerges is a predominantly self-regulatory system, heavily reliant on voluntary compliance. Key formal agencies do wield some authority, yet, they still depend fundamentally on the medical profession. For example, the HTA and ICO depend upon expert advice and information and RECs rely on the medical profession to supply their members. The benefit of this approach is that regulators are very familiar with current practice and rely on practitioners to be active in advising how medical research should be governed. However, this approach also raises the issue of regulatory capture and the difficulty of ensuring that governance structures are sufficiently independent to make legitimate decisions. Funding bodies, also rely on professionals, through systems of self-review to validate research findings, peer-review of research proposals, and consultation on important research initiatives and strategies.

This complex system has grown over time and allows a number of powerful bodies to co-exist, usually without extreme conflict, even though there are not always clear areas of control. It is 'decentered'—that is the state is not the centre of regulation—and it is 'polycentric'—there are multiple sites in which regulation occurs at sub-national, national and transnational levels.[12] The regulatory space that medical research is embedded in, is not a domain that is dominated by state intervention, but rather it is a negotiated space, where a number of different bodies have authority, power and control, even though the basis for authority and influence may vary. Even the bodies that have delegated responsibilities under statute still negotiate the

[12] J Black 'Constructing and Contesting Legitimacy and Accountability in Polycentric Regulatory Regimes' (2008) 2 *Regulation & Governance* 137–64.

content of their guidance to reflect current practice, while remaining within the requirements of the law. The status and legitimacy of an organisation, whether it is mandated by statute or not, will depend upon knowledge, the money and people to run the organisation, the expertise to design rules, and the powers to monitor and police enforcement. While such a system can be seen to allow representation, it can also mean that some constituencies, such as patients, may not be adequately represented if they do not have sufficient resources. This is an issue of concern, particularly as biobanks are dependent upon a high level of public trust in order to be operational. It also means that there may not be a targeted and uniform approach to new areas of research, such as biobanking, that raises issues that do not fit neatly within existing paradigms.

The governance system that was in place at the time of our interviews did not appear to be responding quickly to the needs, motivations and interests of all researchers or people who are responsible for establishing biobanks. It was evident that any new initiatives in biobanking were required to fit with the existing disperse regulatory framework that covered medical research in general. Since then there have been some initiatives such as the generic approval that was brought in for biobanks that have accommodated the specific issues that biobanks present. However despite such a concession, for some of our respondents, considerable time was spent on trying to re-invent the wheel and establish new systems to operate within a framework that is not designed for what they do. It could be argued that operating outside of the regulatory framework would appear to give people in biobanking considerable latitude and therefore it would be a welcome opportunity to be free of oversight mechanisms. However operating outside of the system actually makes things harder for practitioners as they must make up their own rules and protocols and must justify them to existing regulatory bodies. Because of this dilemma, a number of respondents even expressed a positive desire for more regulation, in the form of a purpose-designed system for biobanks.[13]

Our analysis shows that the experience of our respondents when engaging with the governance structure for medical research was diverse and context specific. The governance system in place at the time of our interviews was complex and dynamic, with significant changes to the law and the bodies that regulated medical research. Not all of these requirements had direct application to biobanks which at that time was a new and emerging field. Therefore, it was uncertain how far day-to-day biobanking and genetic research in England and Wales was actually in compliance with all of the legal, ethical and professional requirements that existed at that time. This raises questions as to how best provide guidance for new initiatives

[13] See ch 8 above.

such as biobanking, as an ineffective governance system has the effect of slowing down research leading to uncertainty and inefficiency. It also raises the possibility that any new initiatives unless properly governed may not be in conformity with ethical and legal requirements. In the worse possible scenario, this could lead to criminal or civil action and ultimately the undermining of public trust. Therefore, this analysis suggests that there are important reasons why attention to the development of a responsive governance system that can deal with new initiatives as they emerge is necessary for the progress and advancement of medical research.

# 11

# Ethics and the Governance of Biobanks

*Catherine Heeney and Michael Parker*

THE PRIMARY MOTIVATION for the research project underpinning this book was a concern that proposed moves towards the regulation and governance of biobanks might be insufficiently informed by the subtleties and variety of day-to-day practice in genetics research. Our particular concern was that models of governance appropriate to the development and use of large-scale biobanks designed to act as a community resource might not be appropriate, and might indeed be unethical, in relation to small-scale genetic 'databases' such as those built up by clinician-researchers working in the context of family studies on single gene disorders. The last few chapters have focused on the ways in which governance is understood by various research actors and the previous chapter investigated some of the ways in which governance is enacted in practice, using the sharing of data as a case study. In this chapter, we return to some of the themes first identified in Chapter Four regarding the variety and complexity of research practices in biobanks to investigate the implications of these aspects of practice for the ethics of genomic research and for the ethical governance of biobanks.

## THE EMERGENCE OF BIOBANKS

The emergence of biobanks is closely linked to recent developments in research in genomic medicine including: technological and scientific developments; policy initiatives; and, the strategies of major donors and funders. The key contemporary scientific and technological drivers of these developments are reasonably well established.[1] They begin with

[1] WG Feero, A Guttmacher and FS Collins, 'Genomic Medicine: An Updated Primer' (2010) 362(21) *New England Journal of Medicine* 2001–11; J Witte, 'Genome-wide Association Studies and Beyond' (2010) 31 *Annual Review of Public Health* 9–20.

the completion of the sequencing of the human genome, which first made it possible to contemplate the realistic possibility of whole-genome approaches to medical research.[2] This was followed by the detection and validation of more than ten million single nucleotide polymorphisms (SNPs) which together characterise common variation across the human genome—a SNP is a variant occurring in more than 1 per cent of the population. The subsequent finding that the occurrence of SNPs is often closely correlated with neighbouring SNPs (the degree to which this is the case is called 'linkage disequilibrium') and the implication of this that much of human genomic variation could be captured by means of a smaller number of tag SNPs—combined with the work of the HapMap Project to characterise the nature of linkage disequilibrium in different populations—made genome-wide approaches to medical research on the relationships between genomic variation and disease in humans a practical reality. These scientific developments have been interdependent with rapid technological progress in array-based genotyping which has made it possible to simultaneously measure very large number of SNPs (currently arrays can genotype up to 2 million SNPs), and in the speed and reduced cost of whole genome sequencing. These developments in New Generation Sequencing technologies continue and strengthen this trend.[3]

These scientific and technological developments have led to the emergence of a number of new forms of research in genomic medicine which, broadly speaking, might be said to fall into four main types.[4] The first and probably the most high-profile of these is the Genome-wide Association Study (GWAS) which uses the genotyping approaches described above, in conjunction with epidemiological methods, to compare the mapped common genomic variants (SNPs) in very large numbers of patient cohorts to those in large numbers of unaffected controls to establish whether an association exists between the disease and particular genomic variants. Over the past five or six years, a large number of well-validated associations have been established through GWAS.[5] The second main type of research uses genomic techniques—both array-based genotyping and whole-exome sequencing—to investigate the relationships between *insertions and deletions* (including *copy-number variations*) and other structural rearrangements in the genome and disease: for example, the relationships between deletions in particular parts of the human genome and the occurrence of

[2] ES Lander *et al*, 'Initial Sequencing and Analysis of the Human Genome' (2001) 409 *Nature* 806–921; FS Collins *et al*, 'The Human Genome Project: Lessons from Large Scale Biology' (2003) 300 *Science* 286–90.

[3] ML Metzker, 'Sequencing Technologies: The Next Generation' (2010) 11 *Nature Reviews Genetics* 31–46.

[4] See n 1.

[5] See, eg, the National Human Genome Research Institute at <http://www.genome.gov>.

developmental disorders which may not be explicable in terms of variation at the level of SNPs.[6] While these studies do not generally require unaffected controls, they do call for the collection of large numbers of samples and clinical data from affected patients and relatives. Third, genomic methods are also being used by clinically focused research laboratories to focus on more narrowly defined areas of the human genome to develop new diagnostic and predictive tests for and greater understanding of *single-gene disorders*.[7] Such studies generally take the form of research involving disease-focused cohorts of patients and their families.

Fourthly, and finally, a form of genomic research which is of relevance to medical research, even if not in itself primarily medical in focus, is what might perhaps be called the *genomic diversity study*—studies which aim to better characterise the forms of genomic variation in and between particular populations in ways which will enhance the tools available to researchers using other approaches such as GWAS. Research into diversity is important for a number of reasons in this context: one of these is a concern that there is the potential for variants common to particular sub-populations,even if rare elsewhere, to be important factors in susceptibility to disease, eg malaria; and another is the potential for a better understanding of diversity to enhance the statistical power of genomic studies more broadly.[8] Whilst these types of study involve the collection and storage of biological samples they do not generally involve the collection of samples from people who are affected with a disease and so may not be associated with clinical data.

Although different in other important respects, each of these forms of genomic research requires the collection and storage of biological samples and well-characterised phenotypes, ie clinical and/or other relevant information collected and recorded according to shared standards and protocols. These samples and data are collected in two main ways. In some cases, samples and clinical data are collected prospectively with the explicit aim of developing a biobank. In others, perhaps the majority, in order to avoid the delays and costs involved in generating large sample and data collections from scratch, researchers will draw upon existing sample and data collections, eg cohort studies and archived sample collections associated with other research projects. This requirement for, sometimes very large, collections of samples and data combined with the necessity of access to cutting-edge sequencing, genotyping facilities, expensive research artefacts such as 'knock

[6] See, eg, Decipher at <http://decipher.sanger.ac.uk>.

[7] S Fokstuen *et al*, 'Rapid Detection of Genetic Variants in Hypertrophic Cardiomyopathy by Custom Dna Resequencing Array in Clinical Practice' (2011) *Journal of Medical Genetics* doi: 10.1136/jmg.2010.083345.

[8] YY Teo *et al*, 'Genome-wide Comparisons of Variation in Linkage Disequilibrium' (2009) 19 *Genome Research* 1849–60.

out mice' and teams of analysts only available in a limited number of settings has led to the emergence of often very large, multi-partner, international, collaborative research networks bringing together health professionals, cutting-edge sequencing facilities and researchers from a range of different disciplines including: bioinformatics, statistics, epidemiology and so on. It is against this backdrop that the phenomenon of the 'biobank' as a new form of scientific collaboration has emerged. For, whilst it might be argued that something akin to the biobank has had a long history in medical practice and research, eg the collection and systematic storage of Guthrie cards, in the post-genomic era biobanks and the research collaboration associated with them are taking forms which are both quantitatively and qualitatively different from these earlier manifestations.

In the interviews, we encountered researchers involved in each of the forms of research outlined above—ranging from those involved in small-scale family-based studies investigating the nature and effects of a particular mutation all the way through to those involved in very large epidemiological studies including several hundred thousand participants. As will be apparent from the discussion above, different types of research will tend to generate rather different kinds of biobank both with respect to data and samples and in relation to the nature of the research collaborations they call into existence. And given the increasing tendency for biobanks to draw upon archived samples and data from multiple sources, this suggests the potential for interesting and morally significant interplay between the various research purposes associated with the samples and data at the time of collection and other significant features of the historic database. Given the potential for morally significant differences to emerge between such practices and the various features of the historic database, it is perhaps worth briefly revisiting some of the different practices identified and explored in Chapter Four in the light of the variety of scientific research purposes outlined above.

In addition to the different types of purpose or method outlined above, key variables identified in the Chapter Four included: history, as captured by the feature of age; size; contents; types of participants and modes of recruitment; institutional setting; and, finally, change and multiplicity,[9] which were an important characteristic of almost all the databases we encountered and, therefore, less suited to serve as a way of typing databases. How do these features interact with those associated with the four main scientific purposes outlined above and what sorts of actual or potential ethical questions does this raise?

[9] We have slightly renamed these categories for the purposes of focusing on factors relevant to ethics.

## History

Most of the collections created and managed by the researchers interviewed for this study used at least some and often a great many samples and data originally collected for quite different types of projects. Whilst some biobanks have been or are being purpose-built and samples and data collected prospectively, even these may draw on existing sources of data, such as health records. It is not uncommon for the data and samples used for a genome wide association study, or sequenced using NGS technologies to have their origins in a number of different smaller studies and to draw upon a wide range of different projects each of which had quite different purposes—for example, samples and data related to the genetics of heart disease might be brought together with those collected as a part of a longitudinal birth cohort study. In addition to the diversity of the origins of the sample and data used in a genomics research project, the history of the database might be significant in other important respects, for example, in many cases, the samples will have been collected in an era when the possibility of their being used in genomic research or incorporated into biobanks could not have been envisaged. It became apparent in our interviews, for example, that some of the sample and data collections in these contemporary 'biobanks' were up to 30 years old whilst others were very much more recent. The possibility of explaining technology-enabled uses of data and samples to research participants would therefore have been impossible, as they could not have been foreseen.

## Size

The 'size' of biobanks also varies—both in terms of the number of samples included and with regard to the amount of data collected in relation to or produced from each sample. A biobank might include data and samples from anywhere between 100 participants—in a genetic diversity study or a family-based single gene study—and 1 million. To complicate things still further, the database might include anything from one sample per participant up to 50 and each sample might itself be associated with very few or a very large number of 'data-points'. In a large genomic epidemiology study for example, each sample might generate data relating to more than two million SNPs. Similarly, in relation to clinical or phenotype data, the amount of information collected can vary significantly. There is inevitably a relationship between the history of a biobank and its size. As new samples and data are added and new technological tools are applied the database can change both quantitatively and qualitatively. For example, the introduction of whole exome sequencing has the potential to both dramatically increasing the number of data-points per sample and to change the nature of the information contained within a database.

**Contents**

Whilst most of the biobanks encountered empirically contain both data and biological samples, the kinds of each vary. Biological samples can include: blood, urine, faeces, hair, tumours, saliva and healthy tissue; and the kinds of data might include: medical records, genomic data, demographic data of various kinds, age, sex, educational level, physical measurements, ethnicity, disease status, histopathology reports, imaging, pathology reports, data on birth defects, family pedigrees and so on. Moreover, the relationships between the data and samples and the types of either is not always easily predicted from, for example, the type of work in which the researchers managing the data collection process are involved. As was discussed in Chapter Four, scientists may engage in speculative data and sample collection, anticipating a time when these materials may be relevant and useful in research.

**Types of Participants and Modes of Recruitment**

Participants whose data and samples are collected in a biobank might have been recruited in a number of different ways and in a range of settings. In the vast majority of cases in our study, participants were recruited by clinician-researchers in the context of clinical interactions in a hospital or primary care setting. Participants in cohort studies can also be recruited through hospital channels. Recruitment by clinician-researchers might take place in a public or a private healthcare setting. In other cases, participants will have been recruited outside of the clinical setting, eg as participants in clinical trials or perhaps as a result of recruitment from the general population. The term 'recruited' is perhaps not the correct one to use here, for in some cases, where archived samples and data are brought together from previous studies, although the research has been given ethics approval and the associated consent judged appropriate to allow the sample and data to be used in a new study, the participant will not have been explicitly recruited to this particular study—they will have become participants as a consequence of their enrolment in another kind of study. In this sense it might perhaps be more accurate in the context of biobanks to speak of routes or pathways to participation rather than recruitment. In addition to the routes by which they become participants, the types of participants also vary. They might include: members of a population with a phenotype associated with a particular mutation, members of a population with a mutation predisposing to a particular condition (but not yet affected), carriers of a mutation of interest, people with a particular disease irrespective of mutational status, people with a family history, cohorts of people born at a particular time in a particular geographic location, participants in clinical trials, healthy controls and members of the general population.

### Institutional Setting

Biobanks can be located in a number of different institutional settings, including: public healthcare provider systems such as the NHS, private health institutions, commercial companies, private or public research institutes, and universities. In practice, biobanks are often complex multi-institutional entities bringing together diverse partners in different countries and institutional affiliations. In some cases the samples and data will themselves be distributed across or have originated in several different institutions, thus constituting a potential database ready to be called into existence around particular projects as appropriate.

### Change and Multiplicity

In combination with the research reported in previous chapters the discussion above shows that biobanks are complicated, changing entities characterised by a great deal of internal heterogeneity. They may be temporary in character, dependent as they are on winning and maintaining support of relevant collaborators, a quality captured by the notion of the Boundary Object.[10] Change can happen in a number of different ways. Some of these include: the addition of new samples and data, the development and application of new technologies and techniques and the development of new research questions leading to the production of more or different types of data on existing samples. Change might also happen through the development of better understanding of existing data. Techniques in use today, such as whole exome sequencing and high density genotyping are producing vast amounts of data the implications of which are currently not understood but which may be interpretable at some point in the not-too-distant future with the potential for the significance of data to change without the need for it to be subjected to further tests or research.

Biobanks are also characterised by diversity and are highly internally heterogeneous. The multiple histories of the samples and data by which biobanks are constituted and the diverse purposes for which they were collected mean that the types of samples and data relating to individual participants may vary significantly across a database. For example, a participant recruited prospectively for a study on the genetics of heart disease because they are affected by early onset coronary artery disease may have different types of data and sample associated with them, including consent information, to someone who was recruited as a member of a birth-cohort. Biobanks are characterised by a range of overlapping and multiple material

[10] SL Star, 'This is not a Boundary Object' (2010) 35 *Science Technology Human Values* 601.

practices and histories and individuals and their associated samples and data can be members of a number of different and potentially overlapping studies and databases at the same time. Moreover, the different communities being brought together around biobanks may have multiple understandings and experiences of what it is or what it is for.[11] What this suggests is that they are inevitably complex, diverse and heterogeneous objects taking different overlapping forms at different times with different connections and relationships and forms coming into life in the context of different research projects and as new technologies and research questions and collaborative relationships come in to being—and these changes and forms of heterogeneity can be of real moral significance.

## Biobanks as Objects of Ethical Concern

It has been the founding premise of this book that the development of successful and appropriate models for the governance of biobanks requires attention to be paid to the complexity and variety of day-to-day practice in genetics research and clinical practice and for overly simple conceptions of the 'genetic database' and the 'biobank' to be resisted. As with 'governance', so too with ethics: the development of these complicated and shifting forms of research collaboration and their associated biological samples and data is leading to the parallel emergence of ethical considerations not previously encountered in combination. Whilst there has been a great deal of discussion of these ethical issues and there is now a significant literature in this area, there has been a tendency for the biobank to be conceptualised as a relatively unproblematic and stable object. However, this not the case and indeed they are not always experienced as one unified and enduring *thing* by those working with them. Very little serious attention has as yet been paid to exploration of the implications of the important but subtle heterogeneity not only between but also within the various forms of biobank. It is important to pay attention to these features of biobanks, which have ethical as well as sociological significance. It is highly likely that at least some, if not most, of the differences between different types and settings of research and the life histories of the samples and data which comprise the material resources contained within them will be of moral significance and have implications for the development of appropriate forms of governance.

In what remains of this chapter, we begin to investigate some of the main ethical issues arising in the context of biobanks highlighting the potential

[11] SL Star and JR Griesemer, 'Institutional Ecology, "Translations" and Boundary Objects: Amateurs and Professionals in Berkeley's Museum of Vertebrate Zoology, 1907–39' (1989) 3 *Social Studies of Science* 387–420; AM Mol, *The Body Multiple: Ontology in Medical Practice* (Durham, NC, Duke University Press, 2002).

impact and relevance of the complexities explored earlier in this chapter and in Chapter Four. For the sake of convenience these issues are discussed under the following familiar bioethical headings: valid consent; benefits and harms; duties of care; and, data-sharing and data-release.

## Valid Consent

The obtaining of valid consent from participants is a key benchmark of ethical research. For consent to be valid, it must be informed, voluntary and competently given.[12] The kinds of research associated with the development and use of biobanks can present challenges to each of these aspects of the validity of consent. An important concern is the question of what might count as being adequately informed or as having sufficient understanding to provide valid consent for participation in genomic research. The concern arises, partly, from the highly technical nature of such research and the complexity of key concepts, as well, arguably, as the distance the data and perhaps the samples will travel from the original context in which they were collected. Genomic research brings together a wide range of complementary forms of expertise, eg epidemiology, statistics, bioinformatics, and those of health professionals etc in collaborative research networks in which the various forms of expertise involved may not be fully understood by many of the participating scientists. For example, the bioinformaticians involved in a particular research collaboration may not be aware of the clinical considerations relevant to the recording of a valid phenotype. Against this background, the answer to the question of what constitutes being adequately informed is not straightforward. For, although the achievement of an appropriate level of understanding and knowledge is an important aspect of any account of valid consent, what is to count as appropriate or adequate in this context remains unclear?

It may be particularly difficult for participants to adequately grasp the implications of the fact that biobanks are embedded in a range of other scientific practices associated with the sharing of data with the wider scientific community, as was discussed in Chapter Four. It has become standard practice in genomics (as in many other forms of research) for major funders to require researchers to make their data available to the wider scientific community very quickly after its production and this has led to the emergence of a wide and complex range of data-sharing practices. The existence of these practices presents important questions about what is to constitute adequately informed consent from participants. What is it essential for

[12] Exceptions not addressed in this chapter might include research in emergency situations, research involving people lacking decision-making capacity, and research involving children.

participants to know and understand about the ways in which genomics data will be released and shared in the future? Until very recently, considerations related to the achievement of valid consent to medical research envisaged a process by which participants were informed about the aims, practices and implications of participating in quite specific research 'projects'. This might have been expected to include information about the research question being investigated; why it was important; what would be involved for the participant if they took part, eg the potential harms to the participant; and ensuring that participants were aware that they could withdraw from the study at any point without this affecting their care and so on. Moreover, it has been argued that participants in research draw on the setting in which they provide data to provide clues as to what will happen to their data.[13] This understanding of the processes and purpose of consent is difficult to square with the research associated with biobanks which is intrinsically future-oriented, open, collaborative and inherently uncertain in important respects.[14]

One response to concerns about the validity of broad consent and about the open-endedness of research practices in genomics has been to explore the potential for consent to be given, not to specific research uses, but to agreed forms of governance or oversight of the future uses of data and samples, for example, participants might be asked to consent to their samples and data being contributed to a database whose research future uses are to be overseen by a suitably constituted data-access committee. Whilst this is a solution which has been adopted by some large genomic networks—usually in combination with a requirement for researchers to sign up to a data-access agreement specifying the limits of acceptable future research uses—it presents practice ethical challenges of its own such as: should the governing body be 'representative', for example, should there be research participants representation, or community member involvement and if so, what might count as appropriate representation and acceptable limits to the decision-making powers and remit of such bodies? Moreover, there are practical considerations relating to the constitution of such committees in that depending on the volume of requests for data, the vetting process may become onerous perhaps requiring dedicated personnel with relevant expertise to oversee and manage requests for data. It also raises the question of whether valid consent can be given to governing bodies that may themselves be subject to change, as what counts as good practice and the relevant law changes?

Thus far, this discussion has taken as its focus consent for *future* uses. However, given that many of the large genomic studies currently in existence

[13] O O'Neill, 'Some Limits of Informed Consent' (2003) 29 *Journal of Medical Ethics* 4–7.

[14] M Sheehan, 'Can Broad Consent Be Informed Consent?' (2011) *Public Health Ethics* doi: 10.1093/phe/phr020.

are formed through the bringing together of diverse established sample collections and data, it is also important to consider when, if ever, it is acceptable to use samples and data collected in the past. To some extent this is going to come down to a question about whether the consent originally obtained for a particular sample might reasonably be said to include the proposed use and in many cases this will be straightforward. A person will have given consent to their participation in, say, genomic research on heart disease and the proposed study will be of exactly this type. However, in many cases this will be less clear and the question of the applicability of consent will be one requiring judgment and interpretation. For example, if previous consent has been given to 'research on heart disease' but this was obtained before the genomic era is it reasonable to assume that the participant would have been happy for the sample to be used for 'genomic research on heart disease' or, even, perhaps, research to 'develop statistical models with the potential to be of use in the understanding of heart disease (and other diseases)'. It is sometimes argued that research participants became involved in research in order for their samples and data to be exploited to the maximum.[15] However, given it is difficult to know how far to extrapolate from this original act of participation this alone cannot provide a basis for ethical decision making around future use of data and samples. In particular cases, the task of interpretation will be more or less difficult and this raises the question of how such decisions should be made and by whom? And, these questions will be made more difficult still in situations in which documentation of consent is either non-existent or poor. The status of the consents obtained for the use of samples in a particular database is likely to vary.

Against this background, one possible response would be to adopt the position that there should be no re-use of samples without re-consent. But whilst this would have the merits of taking consent very seriously, it is an approach which is itself not free of ethical difficulty. For it suggests that in situations where re-consent is not possible the collection of new samples would be required, entailing both the financial costs and other implications of collecting new samples from very large numbers of, potentially vulnerable, people. In some cases, where the harms of the research are considered to be low, it seems likely that these considerations, ie the implications of obtaining new samples will outweigh those around the concerns about the precise validity of the original consent. And if this is in principle justified in some cases, the question arises here too of how such judgments are to be made, and by whom?

[15] L Brown, M Parker and M Dixon Woods, 'Whose Interest? British Newspaper Reporting of Use of Medical Records for Research' (2008) 13 *Journal of Health Services Research and Policy* 140–45.

The issues discussed above suggest that the question of validity of consent in the context of research using biobanks is complicated and that whilst it is important for broad principles and models of good practice to be agreed, it is also going to be important for local or perhaps even project-specific solutions to be developed, involving the making of moral judgments about how these principles and models of good practice might be applied in the context of particular studies.

## Benefits and Harms

It seems clear that genomic research of the kinds described earlier in this chapter has the potential to make a major contribution to the understanding of and ultimately the treatment of disease but what are the harms associated with research on biobanks and how might these be managed?

The potential harms of the kinds of genomic research outlined above fall into three categories: harms to individuals; harms to populations; and harms arising out of potential loss of trust in research and researchers. The foreseeable harms to individual participants are of a different kind and degree to those arising in many other forms of research. It is unlikely, for example, that participants will suffer significant physical harms as a consequence of their participation. Any harm to individuals arising out of this research is more likely to relate to the possible uses and misuses of personal or identifying information. These include concerns about confidentiality, ie the possibility that information provided to researchers or health professionals is used in ways which go beyond those that might have been reasonably expected at the time of consent. They also include concerns about privacy, ie concerns that the uses of information in research, even if used in ways which do not strictly constitute a breach of confidentiality, might reveal private information about individuals in ways which would constitute a harm to the individual or their family, for example through the revelation of misattributed paternity. The potential for breaches of confidentiality and for infringements of privacy are increased with the storage of very large amounts of personal, identifiable data.[16] This calls for particular attention to be paid to security and for honesty with participants at the time of consent about the limits of confidentiality and of security. Research participants will inevitably take different views about whether such risks are acceptable given the likely benefits of the research. There is some evidence in the context of research on very rare inherited diseases that participants who have a very strong interest in maintaining

[16] C Heeney, N Hawkins, J de Vries, P Boddington and J Kaye, 'Assessing the Privacy Risks of Data Sharing in Genomics' (2010) 14(1) *Public Health Genomics* 17–25.

confidentiality are nonetheless willing to trade some risk of a breach of this against the potential benefits of research.[17] However, this cannot be assumed and it is likely that views on this will vary between different people and research contexts. One way in which genomic epidemiology differs from much other research is in its implications for populations and it is an implication of this that it has the ability to produce information with the potential for harms to populations as well as to individuals. Some of these potential harms relate to the implications for how members of different populations, eg ethnic groups, will be perceived and treated by others as a consequence of the research. What, for example, are the implications for employment, insurance, marriagability, and forms of discrimination and so on of identified differences in susceptibility to disease between different populations? It seems likely that these harms will be greater in the context of research on some diseases than in others, eg mental health research or HIV. In some cases, the potential harms will be less to do with health status than with the possibility that genomics may question deeply held cultural beliefs such as those about ancestry.[18]

In addition to harms to individual participants and to populations, there is the potential for less direct harms resulting from a decline in trust in medical research where, for example, there is an increased perception that personal data are being misused, that security is less effective than expected, that the governance of research uses is insufficiently rigorous and so on with a consequent decline in willingness to participate in research and decrease in research of potential value. What this suggests is that in addition to the intrinsic importance of high standards of consent and the minimisation of harms to individuals and populations, these are also important in relation advances in medical understanding for the public benefit.

Broadly speaking, there have been two complementary ways in which concerns about harms of the kinds outlined above have been responded to in practice. The first of these has been to combine the removal of all identifiers from data with the maintenance of very high standards of security. The second has been to introduce governance mechanisms of various kinds to manage access to genomic and phenotypic data by researchers. These governance mechanisms have typically combined a formal process of application to an independent data-access committee with a requirement to sign a legally binding data-access agreement requiring researchers to commit not to make any attempt to identify individuals and to make reasonable efforts to avoid using data in ways which have the potential to harm people because of their membership of particular populations.

[17] M Parker, RE Ashcroft, A Wilkie, A Kent, 'Ethical Review of Research into Rare Genetic Disorders' (2004) *British Medical Journal* 288–89.

[18] O Bathe and A McGuire, 'The Ethical Uses of Existing Samples for Genome Research' (2009) 11(10) *Genetics in Medicine* 712–15.

**Duties of Care**

The diffuse and uncertain nature of the possible harms of genomic research, combined with those raised about the possibility of valid consent, generate questions about the duties of care of researchers and/or clinicians in the context of genomic research. The empirical research reported elsewhere in this book, has highlighted the many very different ways in which samples and data can come to be part of a biobank and the various pathways by which people can come to be participants in genomic research. One of the implications of this is that there is the scope for the existence of quite different kinds and degrees of obligations in relation to different participants in a database—and even potentially to those within a single study. For example, whilst some data and samples in a database will have been obtained by a health professional from affected patients in a clinical setting, others will have been obtained by academic researchers from the wider public as a result of an open call for participants in a community resource such as a national biobank. In some cases, participants will have been recruited for research on a particular condition, whereas others will have been recruited to a cohort study including all people born in a particular year or in a particular geographic area. And in each of these cases, researchers or clinicians may see themselves or be seen as having quite different—and sometimes conflicting—obligations in relation to 'their' participants. For example, those researchers who are responsible for establishing and maintaining cohort studies over many years and who have had a long term relationship with the members of the cohort may feel very differently about data and samples from the cohort being made available to third-party researchers through open-access agreements than do the researchers who have established biobanks for that very purpose. What this suggests is that the nature of the obligations to research participants in a particular study is likely to vary depending upon the nature of the relationship between the researcher and the participant, eg between researchers who see themselves primarily as clinicians and those who do not, and upon the nature and broad purpose of the project to which participants were originally recruited.

In addition to these kinds of uncertainties relating to the origin and provenance of the samples and data and to their possible future uses, there are other ways in which questions about the duty of care of researchers to participants can arise. One of the most hotly debated of these is the question of whether there is an obligation to feedback research results to participants.[19]

[19] Kaye *et al*, 'Ethical Implications of the Use of Whole Genome Methods in Medical Research' (2010) 18 *European Journal of Human Genetics* 398–403.

There is a degree of consensus that there is a legitimate public and participant interest in the conduct of genomic research—and in research more broadly—because there is a legitimate public interest in activities of this scale and impact and because they are frequently supported through public funding. In this context, it is widely agreed that public engagement and involvement in the oversight and management of biobanks and effective communication with participants and the wider publics about the aims, progress and outcomes of the research is a key feature of good research practice. What is less clear is the nature and scope of the obligations of researchers to feed back findings pertinent to individual participants. It is likely that the nature and scope of the obligations to feedback will vary dependent upon the nature of the study and the nature of the relationships between the researchers and participants. Two further background factors which are likely to be of pertinence to the decision about whether there is an obligation to feedback findings are, the fact that genomic research does not take place in laboratories accredited for clinical testing and hence may not achieve the standards of reliability required in the clinical context, and the important contribution of environmental and socio-economic factors which mean that assessment of the relevance of the findings of genomic research to personal risk is very far from straightforward. Having said this, there is the possibility—indeed the certainty in very large collections—that genomic techniques will sometimes identify features of a person's genome with immediate and clear-cut relevance such as, for example, an unexpected breast cancer mutation for which screening or surgery may be indicated as a risk-reducing intervention. What are the obligations to feedback to individual participants in these very different scenarios?

There are a number of possible different ways of managing the findings of genomic research. One route would be to make it clear to participants and health professionals at the time of consent that there will be no feedback of research results. There are a number of arguments supporting this including its potential for greater clarity about consent and about the distinction between research and clinical care and the fact that feedback assumes some sort of infrastructure in which the connection with participants is maintained to the extent that they can still be contacted and told to seek medical advice, for example. However, this approach has increasingly come to be seen as untenable in the context of research which has the potential, even if rarely, to produce very clear evidence of a serious harm which might be avoided by an easily available intervention and where there exists something akin to a duty of easy rescue.

Given that some obligation to feedback under certain circumstances is now increasingly recognised, the question becomes: which information and by whom? One possible way forward might be to take the view that only clinically relevant information should be fed back, ie information that would be fed back in the clinical setting were it to be produced in that context.

Were this position adopted, it might seem sensible for the decision about what to feedback to be made by a suitably qualified health professional and for the feeding back also to be done by a clinician. There are, however, a number of difficulties with this demarcation of duties of care. The data produced by genomic research are both extensive and complex and will require interpretation for the health professional to be in a position to interpret it for the patient. This means it will not be possible for the research group to simply ship the data over to the health professional in the expectation that they will thereby be in a position to make this judgment. This suggests that the research group, who are otherwise unlikely to be involved in clinical care or the making of clinical judgments about what is a pertinent risk for a particular patient are nonetheless going to be involved in—and perhaps ultimately responsible for—the identification of areas of a person's genome pertinent to risk and to work together with health professionals to establish the way in which such judgments will be made and this will bring with it not only duties of care but also associated models of good practice.

What this suggests is that the obligations associated with feedback cannot be the preserve of health professionals alone but are in practice going to be distributed across the scientific collaboration and include researchers with quite distant relationships with the participant in other respects. This implies that there is going to be a need to develop procedures, principles and practices for making decisions about feedback both in broad terms, eg through the establishment of a list of reportable findings agreed by clinicians and researchers, and mechanism for the making of value judgments in particular cases. Such processes and principles will also need to address the nature and scope of the obligations to feedback. For example, to what extent do researchers who access data through data-release mechanisms have obligations to feedback and if so, how should this be managed?[20]

## Data-sharing and Data-release

Discussion about the sharing of data in science is not new even if within the context of genomic research open-access models of data-release and data sharing—which have their origins in the Human Genome Project, the Fort Lauderdale Agreement and the Bermuda Principles—have recently become much more commonplace. The majority of major funders of genomic research now require a commitment to the depositing of sequence data in a central repository as a condition of funding. These requirements and practices reflect a wide-spread view that the sharing of data will promote the production of scientific knowledge through a combination of increased

[20] Parker *et al*, 'Ethical Review of Research', above n 17.

scientific use of data and the avoidance of duplication of effort. However, whilst arguments such as these emphasising the value of promoting the wider use of the results of genomic research by the scientific community and its potential to lead to important public benefits, moves towards open access have also generated debate about the compatibility of open-access with other important ethical principles and values.[21] The range of ethical issues identified is extensive.[22] It includes, as discussed above, concerns about privacy[23], about whether anonymity can be guaranteed,[24] about security, [25] about the implications of collecting and storing vast amounts of data and about its uncertain future use,[26] about the implications of data-release for populations,[27] and for family members of participants,[28] about the need to strike a proper balance between research and protection,[29] about the development of appropriate governance mechanisms,[30] about the implications for trust, consent and autonomy,[31] about commercialisation,[32] and about the ethical importance of the sustainability of databases.[33]

Of particular relevance are concerns related to the potential uses of data by third-party researchers who access the data through a data-access mechanism. These include concerns that the data might be put to inappropriate uses, for example, that attempts might be made to identify individuals or that data might be used for research incompatible with the scope of the original consent. This does not necessarily depend upon there being any malign intention on the part of the researcher. For there may

[21] W Lowrence and FS Collins, 'Identifiability in Genomic Research' (2007) 317 *Science* 600–02.

[22] Parker *et al*, 'Ethical Data Release in Genome-Wide Association Studies in Developing Countries' (2009) 6 *PLoS Medicine* 11.

[23] T Caulfield, AL McGuire, M Cho, JA Buchanan *et al*, 'Research Ethics Recommendations for Whole-genome Research: Consensus Statement' (2008) 6(3) *PLoS Biology* 430–35.

[24] Lowrence and Collins, 'Identifiability in Genomic Research', above n 21.

[25] Lowrence and Collins, 'Identifiability in Genomic Research', above n 21.

[26] Lowrence and Collins, 'Identifiability in Genomic Research', above n 21.

[27] GTH Ellison and IR Jones, 'Social Identities and the "New Genetics": Scientific and Social Consequences' (2002) 12 *Critical Public Health* 265–82; J Lunshof, R Chadwick, DB Vorhaus and GM Church, 'From Genetic Privacy to Open Consent' (2008) 9 *Nature Reviews Genetics* 406–10.

[28] Lowrence and Collins, 'Identifiability in Genomic Research', above n 21.

[29] Kaye *et al*, 'Ethical Implications', above n 19.

[30] A Cambon-Thomsen, E Rial-Sebbag and B Knoppers, 'Trends in Ethical and Legal Frameworks for the Use of Human Biobanks' (2007) 30(2) *European Respiratory Journal* 373–82; Sheehan, 'Can Broad Consent Be Informed Consent?', above n 14.

[31] AL McGuire and RA Gibbs, 'No Longer De-identified' (2006) 312 *Science* 370–71; AL McGuire, T Caulfield and M Cho, 'Research Ethics and the Challenge of Whole-Genome Sequencing' (2008) 9 *Nature Reviews Genetics* 152–56.

[32] G Haddow, G Laurie, S Cunningham-Burley and KG Hunter, 'Tackling Community Concerns About Commercialisation and Genetic Research: a Modest Interdisciplinary Proposal' (2007) 64 *Social Science and Medicine* 272–82.

[33] P Arzberger, P Schroeder, A Beaulieu, G Bowker *et al*, 'An International Framework to Promote Access to Data' (2004) 303 *Science* 1777–78.

be genuine uncertainty or legitimate difference of opinion about the scope of the consent and about the kinds of research it covers. For example, does consent to 'research on malaria' include research on 'statistical techniques to identify and take account of population structure'. Does it include research on the relationship between malaria and susceptibility to HIV infection? There is an additional concern that these judgments may be made differently by those researchers who access the data but do not have the same contextual knowledge or awareness of pertinent obligations as the researchers and clinicians who gathered the data and obtained consent.

In addition to concerns arising out of the potential for research uses to move beyond those envisaged at the time of consent, the release of data also presents ethical issues in relation to scientific collaboration and the sustainability of research arising out of the potential for tensions between the interests of data-producers and those of data-users. Even though in many cases, data-producers will also be data-users, in many cases there will be the potential for conflict between these two sets of activities. This is recognised in the Fort Lauderdale Agreement which emphasises the importance of scientific etiquette and ties the success of open-access as a policy to the need for data-users to respect the legitimate scientific interests of those who produce and release the data and not to try to scoop them. This can be particularly important in the context of research involving partners in developing countries where there is the potential for richer research groups in developed countries to undermine emergent scientific capacity in developing countries by analysing and publishing data much more quickly.

Because of concerns such as those above, and others discussed earlier in the chapter, there have recently been moves away from open access models of immediate release of data to more managed approaches which, whilst remaining committed to the sharing of scientific data attempt to do this through a managed process designed to allay some of the fears and to clarify the obligations of researchers using data. Whilst this offers a possible way forward it does inevitably raise issues of its own such as how in practice should decisions be made? How should the membership of the data-access committee be decided? How should it be held to account for the decisions it makes—not only those of inappropriate release but also those in which scientists feel they are being unfairly denied access.

## Developing Ethical Solutions in Contexts of Complexity

Biobanks are complicated, changing, heterogeneous objects associated with a diverse range of interconnected and overlapping practices. Some of this variety arises out of the different origins of the samples and data within the database and the different pathways by which people become participants.

Some of this variety arises out of the range of different forms of expertise and scientific practice associated with biobanks and genomic research and their different requirements for different types of samples and data. It has become apparent as this chapter has progressed, that biobanks are productive of different interconnected sets of ethical concerns and obligations. What has emerged in effect is the idea of the biobank as a complicated or *multiple* object of ethical concern. As was argued in Chapter Four, an important set of challenges arise in relation to the task of developing a governance framework for the management of biobanks, as a consequence of the fact that these are entities constructed in heterogeneous ways from a variety of collaborations, physical sample collections and numerous data-sharing arrangements.

None of this is very helpful to the medical researcher who is looking for an answer to the question of what is to constitute ethical practice in biobank practice. Against this background, researchers and funders might be tempted to turn to the law and to research ethics guidelines for assistance. However, as was seen in Chapter Three, there is not much help to be found there. The regulatory space within which biobanks operate is itself characterised by complexity, ambiguity and uncertainty. Whilst within ethical discussions it is acknowledged that there are 'trends' that challenge accepted research ethics frameworks, for example the still used but perhaps outdated model of the individual as a research subject able to consent, when the benefits and perhaps harms of genetic research are collective in nature.[34] As argued in Chapter Four, the lack of clear boundaries between biobanks and their rhizomatic[35] rather than nested organisational character further challenges the suitability of this model of consent as this arguably works best within a highly defined and bounded study.

Against this backdrop, how ought researchers and those interested in improving research practice to respond? Whilst there is certainly a real need to work towards greater coherence in governance mechanisms, policies and models, it is also clear from the empirical evidence gathered for this research and from the discussion in this chapter, that much of the complexity and morally significant contextual variation around biobanks and the research carried out on them is going to remain an enduring feature of research practice in genomics even if agreement is reached about some overarching principles of good practice. What this suggests is that it is going to continue to be important for collaborative research groups and those responsible for the oversight of biobanks to develop rigorous and transparent procedures to inform the movement from agreement at the level of principle to practice

[34] R Chadwick and BM Knoppers, 'Human Genetic Research: Emerging Trends in Ethics' (2005) 6(1) *Nature Reviews Genetics* 75–79.

[35] See, eg, G Deleuze and F Guattari, *A Thousand Plateaus* (Minneapolis, University of Minnesota, 1987).

when determining best practice in conducting research in particular cases. That is, researchers and others are going to need to continue to develop methods and models for the development and implementation of workable solutions in particular settings that are both justified and reasonable. To some extent this is going to be a matter of having purpose-built governance structures and procedures in place for addressing these concerns. But, it is also going to be important for researchers and those responsible for the governance of biobanks to have easy access to the kinds of ethics support and training necessary for them to be in a position to identify and analyse ethical issues arising in both in practice and in the development of policies and processes informed by these analyses.

## CONCLUSIONS

Building upon the discussion in previous chapters, this chapter has explored the ethical issues arising in the development and governance of biobanks. It has become increasingly clear that biobanks are complicated and heterogeneous objects of ethical concern. Rather than falling neatly within the boundaries of any one profession, organisation or sector these entities span boundaries, which in turn undermines attempts to simply apply the ethical frameworks accepted by a given profession or organisation. Moreover, the way in which boundaries are spanned and biobanks evolve is not always predictable as this evolution may depend upon the behaviour of the disease, new scientific and technological developments and changes within the wider policy context. Whilst it has been possible to map out some broad areas of agreement about what might constitute good practice, they resist simple or uniform solutions. The ethical complexity of this area of research practice is matched by the complexity and ambiguity of the governance contexts within which it takes place. This chapter has explored some of the implications of this complexity and variety for key areas of ethical practice: consent; duties of care; harms and benefits; and data sharing and data-release. The chapter concluded by arguing that this complexity and fluidity has implications for what is to count as good practice. It requires of researchers and those who are involved in the governance of biobanks, that they develop processes for the achievement of practical ethical solutions in particular contexts which they are able to justify in ways that are reasonable, transparent and accountable. And, this requires a range of forms of ethics support, advice, education and research. The issues raised in earlier chapters and treated here suggest a need for such support to be responsive to the specific goals and activities of a given biobank, which in turn requires ongoing cooperation between practitioners and other relevant experts.

# 12

# Conclusions

*Jane Kaye, Susan MC Gibbons, Catherine Heeney, Michael Parker and Andrew Smart*

THIS BOOK PROVIDES a case study of governance within the biomedical research field, over a period of time—2005 to 2009—when medical research in England and Wales was governed by a particular constellation of bodies and legal instruments, whose development had been influenced in part by a significant series of events leading to public concern about the uses of data and biological materials in medical research. It documents the experiences of researchers and others who were responsible for establishing, managing and running biobanks, as they encountered and engaged with this governance structure, and as they sought to implement this new approach to research—or tool for research—in practice. Our observations and findings relate to how medical research was governed during this period, and the issues that emerged as new tools were introduced and integrated into regular research practice. Our previous chapters have shown that biobanks are part of (and often are themselves) complex research networks;[1] and also that the governance framework within which they are situated is an example of a multifaceted web of interactions and interdependencies.[2] We assess the governance framework based on the analysis of interview data from the project and from the better regulation principles outlined in Chapter One to determine, first, the strengths and deficiencies of the governance structure for medical research in England and Wales and, secondly, how the new field of biobanking was accommodated within and interacted with this structure. In this concluding chapter therefore, we draw upon previous chapters to outline some improvements that could be made to the governance system for medical research, both domestically and more widely, based on our analysis of the views and experiences described by our respondents and our research findings.

[1] See ch 4 above.
[2] See ch 3 above.

## DYNAMIC NETWORKS OF PRACTICE

It is evident from our research that it is not possible to talk about all 'genetic databases' or 'biobanks' collectively, as if they comprised single, discrete entities. Significant differences exist, as biobanks come into existence through, are maintained by, and are embedded within complex sets of relationships and networks. We did not find 'the *ideal* database with one Principle Investigator (PI) based in one institution, who worked exclusively with one 'genetic database' with a dedicated IT structure, which had one clear source of data and samples, collected at roughly one point in time and connected with only one institution'.[3] Even those projects that were established as part of a dedicated sample and data collection were not self-sufficient for all of their data and/or sample requirements. Most of the biobanks that were discussed by our respondents were developed incrementally through different research projects rather than being established *de novo* as stand-alone entities. The 'genetic databases' managed by our respondents may best be described as fluid or 'rhizomatic'.[4] They typically contained a composite of samples and information gathered from a variety of sources through the process of carrying out a number of research projects over many years, and had been funded by different funding bodies, under different protocols and standards for research, as well as being assembled through collaborations with colleagues from many different institutions and localities around the world.

Accordingly, the collections of samples and data that emerged out of this activity could be 'constituted in multiple ways at multiple times, by multiple individuals or networks, and have any number of parts'.[5] They were not one closed *thing* with clear boundaries, but rather 'multiple'[6]—or capable of being' multiple'[7]—that is, described and conceived of in different ways by different people. Therefore, these entities were not experienced as complete, unified and enduring, but rather as something that could be utilised and accessed through various different networks and relationships. 'Where it was visible as one thing the behaviour of any given "genetic database" was that of a "boundary object" in that its use was not necessarily a settled thing but was constantly open to renegotiation and required maintenance'.[8] These entities embodied a certain set of social relationships that had brought them into being and helped to maintain them in addition to having a material element whereby the samples and data were stored and

[3] Ch 4, p 96.

[4] G Deleuze and F Guattari, *A Thousand Plateaus* (Minneapolis, University of Minnesota, 1987).

[5] Ch 4, p 96.

[6] AM Mol, *The Body Multiple: Ontology in Medical Practice* (Durham, NC, Duke University Press, 2002).

[7] *Ibid.*

[8] Ch 4, p 113; and see also below ch 11 as a whole.

accumulated over time. The nature of their existence, combined with their increasing value for research, meant that the perceptions of what they *were*, as well as their social and material compositions, were constantly reconfigured and reconstituted as part of new collaborations and research projects.

A significant characteristic of this field was the fact that the respondents recruited into our study had different disciplinary backgrounds and training. These differences materially affected both the types of involvement that they had with biobanks, and the professional cultures, expectations and practices that they would enact. Some respondents who had particular skills mixes were involved in multiple collaborations that involved different kinds of relationships and roles. This could be as the 'geneticist' or 'bioinformatician' in a large consortium bringing individual knowledge and/or the expertise of a team; as an expert on an Advisory Board; or as a PI responsible for a number of projects. 'Individuals, therefore, moved between different sectors and organisations applying their experience and expertise in new projects and different contexts'.[9] The leadership and involvement of such a variety of disciplines has had an important effect on the development of practice and the selection of governance mechanisms and their application in this field. It can also determine the regulatory authority and standards to which an individual may have to adhere. For example, a clinician who also carries out research will have to comply with the General Medical Council (GMC) requirements, or perhaps those of a Royal College, and those of their NHS employer—whereas a bioinformatician will not.

All of this has implications for how those who are charged with governing and/or regulating these activities and the entities that emerge from this enterprise carry out their tasks. And this requires us to return to our original research question—namely, whether the large biobanks, such as UK Biobank, that have been well funded, well publicised, and which have bespoke governance structures, may (or ought to) become the 'norm' for all other kinds of biobanks. As noted above, many of the biobanks or entities that the respondents in our study worked with had not been established *de novo* with the specific intent of making them available for all researchers, or the world at large. They had been established incrementally and almost as an incidental outcome of the research process, the products of a desire on the part of research teams and PIs to make what they considered to be the best use of samples and data already collected. Consequently, to try to put them together as seamless, integrated resources that can be accessed, for example, under open access policies, requires a substantial change in thinking, collection practices, management and organisation. This will be one of the challenges for the proposed Biobanking and Biomolecular Resources Research Infrastructure (BBMRI) that aims to organise existing clinical and

[9] Ch 4, p 108.

research collections into national 'hubs and spokes' that can be linked into a European and, potentially, global network.[10] It also suggests that it may not be advisable to try to govern this area simply according to the specific characteristics of a biobank—such as its size, the discipline of the collector (whether a clinician, university-based researcher or commercially based researcher), or the locality where the samples and data were collected—for example, within a hospital or university, as part of clinical research practice or during a research project. What is required instead is a focus on the development of protocols, procedures and standards that are universal in nature but which can be interpreted and implemented appropriately within local and specific contexts.[11] Rather than being oriented around a conceptualisation of biobanks as single, isolated entities, then, any new governance mechanisms that are deemed to be required—as well considerations about the appropriate ways in which to apply existing governance arrangements to them—should take into consideration this complexity, and be designed with (multiple) networks in mind.

## DECENTRALISED AND POLYCENTRIC GOVERNANCE

The map of the regulatory space presented in Chapter Three reveals a picture of the governance of medical research as being complex and ever-changing, with a wide range of competing and disparate groups seeking to play a role. In many ways, like the networks of practice we have described, the regulatory space also represents a complex network of associations, relationships and interdependencies, both between 'regulators' or 'governors' and those who are governed. In this regulatory space the state has not provided any tailored legal instrument for medical research, and the oversight processes of formally constituted bodies are far from uniform. There is no one authoritative body that has been given delegated responsibilities by Parliament or the Secretary of State for Health to oversee the medical research context.[12] In this regard the role of the state is minimal and decentralised, rather than being central to the oversight process. Largely, oversight is left to other bodies, such as RECs—which have some delegated authority—and (less directly) funding bodies specific to the context of medical research, whose role in governance is less formalised and less subject to scrutiny.

[10] See BBMRI: <http://www.bbmri.eu/> and ch 2 for more discussion.

[11] This has been the aim of organisations such as P3G—the Public Population Project for biobanks P3G Consortium: http://www.p3gconsortium.org/—that has developed standards and tools for population biobanks and is now engaged in a number of other similar initiatives in genomics.

[12] However the Academy of Medical Sciences Report recommends that this should be established.

This lack of clear authority has opened up a governance 'gap' or lacuna which has been filled by a proliferation of guidance policies, initiatives and materials, issued by an array of bodies, and with overlapping (or even multiplication of) activities and roles. In this polycentric space authority is devolved to many organisations which can exert different kinds of influence. As we observed in Chapter Three, regulatory power is (unevenly) dispersed within the biomedical sphere because the key resources that confer it are fragmented. Which bodies enjoy power, and how much power, are determined by how such resources are distributed. In addition to Scott's classic, four-fold model of the key regulatory space resources that confer power,[13] one further key source of power that enables greater influence in this domain which we have identified is the ability to control professional education, the setting of good practice norms, and the inculcation of professional cultures.[14] Our legal research, our sociological research, and the ethical analysis in Chapter Eleven all suggest that these aspects of practice contribute importantly to shaping individuals' attitudes and day-to-day outlooks and activities, meaning that professional organisations have an important role to play in governance.

The ways in which our respondents engaged with this polycentric governance structure depended upon many things. In no particular order these included: their particular disciplinary perspectives; the professional cultures within which they were embedded; their relationships with their research participants; their areas of responsibility; their levels of knowledge about governance, including their awareness of 'problems' and the need for 'solutions'; their institutional contexts; their personal and professional networks; and the regulatory or advisory bodies (or individual experts) which they happened to encounter along the way, know about or respect, or from whom they deliberately sought information. Therefore, individual respondents' knowledge and experience of governance was incredibly varied. We have documented the 'informality' in some of the processes which determined governance, guidance and oversight structures (this was less evident in the accounts of respondents working in the private sector). In this context we have indicated that emergent practices were shaped by actors such as funding bodies and RECs—the two principal 'gatekeepers' that allow research to proceed; steering, advisory or ethical boards which, in some cases, were attached to particular biobanks or projects; and, members of

[13] C Scott, 'Analysing Regulatory Space: Fragmented Resources and Institutional Design' [2001] *Public Law* 329.

[14] The six key resources that we have identified within the biobanking and biomedical spheres are: (1) formal legal authority; (2) possession and control of information; (3) possession of wealth; (4) organisational capacities; (5) the ability to publish or disseminate one's views and preferences effectively, and to persuade others to accept them as being authoritative or persuasive; and (6) control over professional education, the setting of good practice norms, and the inculcation of professional cultures.

the scientific community themselves within and through these governance bodies, and by virtue of promoting themselves (or being considered by others) as experts. Thus, the importance of professional and social networks cannot be underestimated as a method used by biobanking professionals to share knowledge and to find solutions to governance problems. This leads to the emergence of systems wherein 'trailblazing' projects become the benchmark for standard-setting, and in which individuals become the experts by 'doing'. In this sense, certain key actors appear to become more visible as co-constructors of the emergent governance frameworks.[15] A danger of such ad hoc processes—particularly in terms of offering protection both to practitioners and their patients/research subjects—is that colleagues or those who are seen as 'leaders' or 'experts' in the field may not always in every case be passing on views or practices that are lawful, or which constitute the best ethical practice. Furthermore, the same model does not necessarily suit all types of research project, nor will it adequately take account of the dynamic and interconnected nature of practice. Indeed, some of our respondents were acutely aware of the risks in terms of public trust and damage to professional integrity that they faced in having to determine best practice in the absence of any clear, applicable regulatory framework.

## IS THIS 'SYSTEM' FIT FOR PURPOSE?

Our aim in this book has been not only to understand how biobanks are governed in England and Wales, but also to provide a basis for evaluating this governance structure. The sociological element of the project was carried out with the intention of gathering the views and experiences of members of the biomedical research community with the important caveat that these concerns, which are important and relevant for reflection on future systems of governance, are the opinions of those interviewed for the study and, hence, cannot necessarily be assumed to accord with the opinions of all those working in the biomedical research field. The purpose of this section, then, is to draw some conclusions about the governance system these individuals were working with, based on our analysis of the accounts and opinions provided by our respondents, as viewed through the prism of the 'good governance' or 'better regulation' principles (discussed further below).

The opinions of our respondents varied greatly in respect to the current system of governance, which, as we pointed out in Chapter One, includes but is not restricted to regulation. A number of respondents described having concerns, or reported having had negative experiences of the governance

[15] See M Mayrhofer and B Prainsack, 'Being a Member of the Club: The Transnational (Self-)Governance of Networks of Biobanks' (2009) 12(1) *International Journal of Risk Assessment and Management* 64–81.

system in relation to biobanking. The opinions aired by our respondents are, of course, just that—opinions of the governance landscape as viewed from their own perspectives. Had we conducted interviews with people who are charged with responsibility for legal and ethical oversight we may, of course, have got a very different picture. This said, there was still considerable support among our respondents for having a governance structure—at least, one that would be capable of providing clear and appropriate direction for the emerging field of biobanking. While we recognise our respondents' opinions as partial, then, they are nevertheless a useful and valuable starting point for conducting an analysis of the governance structure based on the good regulation principles.

## The Assessment of Governance by Respondents

If we agree that 'a legitimate authority is one that is recognized as valid or justified by those to who [*sic*] it applies',[16] then understanding the perspectives of those whose activities are subject to and/or constitutive of 'governance' is crucial. As we have demonstrated in this book, our respondents' experiences of governance were extremely varied, depending upon what they were doing, their institutional contexts, their own disciplinary training and those whom they encountered. Some had had negative experiences—notably, where governance measures were seen as unjustifiably impeding or obstructing research, as requiring unnecessary bureaucratic hurdles to be overcome, or as failing to assist with the development of practical solutions to aid the evolution of good practice. Others reported the laborious process involved in finding, interpreting and applying the relevant information or guidance.

In the eyes of some of our respondents the main purpose of the governance system was to confirm that what they were doing—what they considered to be 'good practice'—was 'appropriate' and 'legitimate'. The governance system was seen as a form of validation that would enable them to act confidently; would enable their biobank to be seen as a 'centre of excellence'; would ensure public trust; and would avoid criticism or reputational harm. From this kind of perspective, a governance system that was useful for practitioners would be one that spelt out precisely what they needed to do in order to carry out their work with a minimum of bureaucracy and external interference. Conforming to these requirements could, in turn, be used as a basis to reassure research participants, peers, employers, funders and the general public that biobanking practices were legitimate and of a certain

[16] V Bekker, G Dijkstra, A Edwards and M Fenger, 'Governance and the Democratic Deficit: Introduction' in V Bekker, G Dijkstra, A Edwards and M Fenger (eds), *Governance and the Democratic Deficit: Assessing the Democratic Legitimacy of Governance Practice* (London and New York, Ashgate, 2007) 7.

standard and quality. Specifically, many respondents wanted governance that established clear frameworks—especially frameworks containing the 'nuts and bolts' for how to set up and run databases; manage participant-identifiable data; and design and operate database IT systems in a unified, standardised way—which embodied and clearly identified all of the relevant legal requirements. Thus, we found examples of respondents speaking of the need for robust governance to 'prove' to the world that biobanks and associated practices are legitimate, and for 'path-breaking' initiatives, in particular, to be governed closely from the outset, so that correct benchmarks would be set for the future. However, in the eyes of some respondents, the governance system as it existed at the time failed in these basic requirements.

Some respondents considered that there was a lack of easily accessible information relevant for biobanking; legal requirements (once found) were difficult to understand and apply; licences for holding tissue were expensive; the generic forms used by RECs and NHS clinical governance bodies were inappropriate for new biobanking activities; and the system itself was bureaucratic and time-consuming to navigate, particularly when trying to introduce innovative practices. Others considered there to be considerable duplication in the requirements of different governance bodies, both in the UK and for collaborative projects that involved other national governance systems. Certain respondents reported that some of their key concerns were simply not addressed, or that they were confronted with unfathomable or conflicting interpretations of how laws and guidance should be applied on issues such as ownership of samples and data in multi-institutional collaborations, requirements for IT information handling, cross-jurisdictional issues, and the need for a unified system for dealing with data and samples.

Some of our respondents felt that the underlying presumption or starting point of regulatory and oversight bodies was inherently negative about science—that scientists, doctors and researchers were seen as being potentially harmful to participants and the public, and so potential abuses by them must be prevented. Therefore, regulatory systems and actors were largely perceived as doubting the integrity of researchers, being hostile to research and being resistant to innovation. They were seldom described as being positive, facilitative or enabling. For many respondents, these structures were seen as time-consuming and a burden, taking researchers away from scientific enquiry and wasting precious resources. The requirements of RECs (at the time of the research) were seen by some as being out of kilter with what research participants might expect, and contrary to the aim of facilitating research and 'serendipitous discovery'. It seemed to some of them that there was a mismatch between what the governance framework was trying to protect and what was required to enable research to proceed smoothly. Therefore, as well as being a system that was described by some as being fundamentally hostile to research, some oversight bodies were also perceived to be unhelpful and ineffective because they were unable to assist respondents with their basic requirements. As a result, in a context perceived to be one involving conflicting and ambiguous governance, our analysis revealed instances in which compliance appears 'selective', at least to the extent that some respondents talked positively about applying guidance that they saw as appropriate and were dismissive of that which did not accord with what they believed or were trying to do.

Despite some respondents believing strongly that there were real limitations and problems with the current system, this did not mean that they wanted to do away with governance or regulatory structures per se. In fact, there was a widely shared and positive desire *for* governance—what was wanted was *better* governance rather than no governance at all. However, there were mixed signals as to whether 'better' meant more or less tailored or global; although it almost certainly meant less time-consuming. There were also some respondents who saw governance as being positive, necessary, and, if done in the right way, beneficial. This resonates with the findings of research carried out in the field of stem cell research,[17] which found that clear guidelines and strict rules were perceived by scientists as enabling them to pursue their lab work. This was because the bulk of the ethical responsibility was borne and carried out by the regulatory framework, rather than having to be undertaken by the researchers themselves. In contrast, our analysis suggest that biobanking as an activity did not fit established norms and so individuals were left in many instances having to struggle to develop practices to address the deficiencies in the governance system for themselves. From the perspective of some of our respondents, the regulatory framework for research lacked the capacity to take sufficient account of the many different kinds of activities which were carried out under the banner of 'biobanking'. In this regard the system lacked basic elements needed to govern the field of biobanking at this time, at least to the extent that it was not meeting the needs of some of those operating in this field.

## Good Governance Principles Analysis

The so-called 'good governance' or 'better regulation' principles[18] are a useful tool for structuring our reflections on the regulatory structure for the

[17] SP Wainwright, C Williams, M Michael, B Farsides and A Cribb, 'Ethical Boundary-Work in the Embryonic Stem Cell Laboratory' (2006) 28(6) *Sociology of Health and Illness* 732–48.

[18] As outlined in ch 1 above and developed in ch 3, two key sets of principles are especially pertinent here: those enunciated by the Hampton Review (P Hampton, Final Report, 'Reducing Administrative Burdens: Effective Inspection and Enforcement' (HM Treasury, London, March 2005)); and those identified by the Better Regulation Commission (Better Regulation Task Force, 'Regulation—Less is More: Reducing Burdens, Improving Outcomes: A BRTF Report to the Prime Minister' (London, March 2005)). These principles are now enshrined in the Legislative and Regulatory Reform Act 2006 and accompanying statutory Regulators' Compliance Code (Department for Business, Enterprise and Regulatory Reform, *Regulators' Compliance Code: Statutory Code of Practice for Regulators* (17 December 2007)).

regulation of biomedical research as a whole and the individual regulatory bodies that inhabit the regulatory space. In this section we use the better regulation principles to reflect on both the overall regulatory structure and specific bodies, in order to highlight for discussion some of the deficiencies in the current governance system. As there was no specific governance framework for biobanking in England and Wales at the time of our research, our investigation had to consider the governance of biomedical research as a whole, and how biobanking was being accommodated within that structure. Therefore, we shall argue that the current regulatory system for biomedical research that has implications for biobanking does not comply with all five of the favoured principles of better regulation—namely, consistency, transparency, accountability, targeting and proportionality.[19] In addition, other characteristics of good governance also appear to be lacking—not least, consultation and public/stakeholder participation; predictability; fairness and non-discrimination; systems for overall ongoing review and reform; and a sound legal and empirical basis. These principles and characteristics are important for ensuring that a governance system has legitimacy and credibility. With so many of the better regulation principles being at issue, this means that the legitimacy of the current governance structure for biobanks is brought sharply into question.

### *Consistency*

Our analysis suggests that one of the qualities most lacking in the governance structure that existed in 2005 to 2009 was consistency. This principle requires that:

> [R]egulators should be consistent with each other and work with each other; regulation should be predictable in order to give stability and certainty to those being regulated; enforcement agencies should apply regulations consistently across the country; and new regulations should take account of other existing or proposed regulations (domestic or EU or international).[20]

During the period in which we carried out this study, the system as a whole was experienced by many of the researchers we spoke to as being neither predictable nor stable: there were constant changes to the law, the regulatory bodies, and the ways in which research was regulated (both formally and informally). The polycentric nature of the regulatory space meant that there was no one authoritative body for decision-making, nor any co-ordinated strategy for the regulation of the types of biomedical research that we have labelled as biobanking. This led to a plethora of sometimes conflicting or

[19] Better Regulation Task Force, *Principles of Good Regulation* (revised 2007) available at <http://archive.cabinetoffice.gov.uk/brc/publications/principlesentry.html>.

[20] *Ibid.*

contradictory guidance emerging—none of which was specifically designed for biobanks and the building up of collections of samples and data over successive projects. It could be difficult for researchers to establish what the laws, rules and standards were; and, even where these were identified, there could remain considerable doubts and varying opinions as to how they might apply to biobanking. There was a lack of clear, straightforward, accessible principles that existed nationally that researchers could apply to biobanking activities in order to develop best practice. This has further contributed to variations in how biobanking governance has developed across the country. As the collection and storage of DNA samples and information largely lay outside the legislative frameworks, or sat ambiguously with them, at least some individuals developed their own practices for biobanking or turned to more informal sources of guidance, such as colleagues and advisory bodies. The bodies that regulated research did not tend to work together in a transparent way that was beneficial for researchers. While some regulators did work together, and tried to define the parameters of their regulatory activities, there were a number of bodies in the regulatory space that carried out very similar activities. For instance, well in excess of thirty bodies produced written guidance that could be applied to all aspects of medical research and genetics. Yet very few of these documents had relevance for biobanking. Arguably, based both on the accounts given by our respondents and our own analysis of the regulatory space, regulators failed to communicate to researchers either the law that applied to their work, or what was required for compliance. Both consistency and coherence, then, were lacking.

### *Transparency*

The principle of transparency requires that:

> 'policy objectives should be clearly defined and communicated to interested parties; consultation must take place before proposals are developed to ensure stakeholder views and expertise are taken into account; those being regulated should be made aware of their obligations, with law and best practice distinguished; those being regulated should be given time and support to comply; and the consequences of non-compliance are clear.[21]

Again, with respect to transparency, many of our respondents claimed that it was difficult to find out what applied and when, and who would enforce it. Our legal review indicates that there were no clear, overall policy objectives that were clearly defined and communicated to the research community as to how it should proceed in relation to biobanking. Our analysis of interview data suggests that formally constituted statutory laws, such as statutes, were not universally considered to be a source of guidance. What our research

[21] *Ibid.*

indicates is that many of our respondents adopted a 'need to know' attitude towards knowledge of the law and how it might apply to biobanking activities. The law and regulatory bodies did not provide unambiguous answers in the area of practical application that was needed by practitioners. Additionally, there was a lack of any single, clear legal definitions of much-used (but variously interpreted, construed and applied) concepts and terms.

*Accountability*

For a regulatory system to be regarded as legitimate and to garner respect it must be based on some form of accountability to Parliament. The House of Lords Report on accountability stipulated that the role of Government and Parliament is to be responsible and accountable for designing

> the regulatory framework as a whole, where that framework divides roles and responsibilities in a rational way, focused on the desired outcomes of regulation, and incorporates the necessary checks and balances of accountability within the system.[22]

This should be put into statute. Currently, such a regulatory framework has not been designed, either for medical research in general or for biobanking in particular. While the Report acknowledged that devolved, independent regulators are responsible in many instances for day-to-day regulation, they should still be accountable to Government and to Parliament for their activities: 'The procedures of accountability need to be able to identify this co-responsibility in practice'.[23]

In the case of medical research regulation and governance, the role adopted by Government and Parliament to date has been characterised by a largely 'hands-off' approach. For example, while Parliament does provide some legislative boundaries through the Data Protection Act 1998, the Human Tissue Act 1994, and the GMC requirements, many of the bodies that operate in the regulatory space are not accountable to Parliament as they consist of a combination of professional associations and charitable bodies. Our legal research found that there was no one body that had a democratic mandate to provide a tailored guidance and oversight framework for biobanking activities. The scope of parliamentary delegation in this area is very specific, and is primarily confined to the Human Tissue Authority (for human tissue) and the National Research Ethics Service (NRES) (for research approval).

The sociological and legal research we conducted identified that, in practice, the funders of research and RECs were key de facto oversight bodies

[22] House of Lords Constitution Committee, *Constitution: Sixth Report 2003–4* Ch 8 paras 132–34 (Parliamentary Copyright: London) <http://www.publications.parliament.uk/pa/ld200304/ldselect/ldconst/68/6802.htm>.

[23] *Ibid.*

in this field. While public funding bodies are accountable to the Secretary of State as they are directly accountable to government departments for their funding activities, they are not required to report in their roles as de facto regulators. This is mainly because their regulatory activities come as a secondary outcome of their funding activities, and may be a response to the need for *some* body to regulate to fill the governance 'gap'. Likewise, charitable funding bodies report to the Charities Commission—which is an arm of the state—but this reporting requirement frames their activities as charities, rather than taking account of their role as de facto regulators. RECs are the other key bodies in the regulation of research who do report to the Secretary of State for Health through NRES. Therefore, the scrutiny by the state of the regulation of medical research over recent years has largely been devolved to the funders, RECs and the Human Tissue Authority. Much of the informal governance through project advisory boards—another key category of governance actors identified through our analysis—is carried on without state scrutiny. While the Human Tissue Authority has clear lines of accountability for its regulatory activities, this is not the same for the de facto regulatory activities of the publicly funded Medical Research Council or for the RECs.

NRES delegates its authority to regulate to individual RECs. These committees are independent and consist of volunteer professionals and lay people, who may or may not have direct knowledge of the field. One of the strengths of ethics committees is that they can make tailored decisions on a case-by-case basis. However, this raises questions about the way in which RECs are held accountable for their decision-making. Unlike other bodies in society, ethics committees do not publish the reasons for their decisions, and they are not bound by precedent to follow either their own previous decisions or those of other ethics committees.[24] This is contrary to the better regulation accountability principle. RECs have attracted much criticism for ad hoc decision-making[25] because of this alleged case-by-case decision-making approach, which has also led to a call for the process of assessing research proposals 'to be simplified and made consistent, and the reasons for decisions… [to] be clearly argued and stated'.[26] While there have been considerable changes to address these concerns, at the time of our research decisions and procedures varied considerably, both between and even within regions within the UK. These differences became even more

[24] See RE Ashcroft, AJ Newson and PMW Benn, 'Editorial: Reforming Research Ethics Committees' (2005) 331 *BMJ* 587–88 on whether RECs should meet in public.

[25] J Tully, N Ninis, R Booy, R Viner 'The New System of Review by Multi-Centre Research Ethics Committees: Prospective Study' (2000) 320 *British Medical Journal* 1179–82; P Glasziou and I Chalmers, 'Ethics Review Roulette: What Can We Learn?' (2004) 328 *British Medical Journal* 121–22.

[26] R Souhami, 'Governance of Research that Uses Identifiable Personal Data' (2006) 333 *British Medical Journal* 315–16.

acute when decisions of research ethics committees were compared between countries.

*Targeting*

The 'targeting' principle requires that:

> Regulation should be focused on the problem and minimize side effects. The key considerations in this regard are: regulation should avoid a scattergun approach; regulators should adopt a 'goals-based' approach, with enforcers and those being regulated given flexibility in deciding how to meet targets; enforcers should predominately focus on those whose activities give rise to the most serious risks; and regulations should be reviewed to determine whether they are still necessary and effective.[27]

As the current system stood during the period of our research, it is very difficult to say that it was targeted at all—that is, that the system focused on the problem of biobanking, and sought to minimise side-effects. If anything, our respondents tended to report a lack of attention by regulators to the question of how they should proceed. Consequently, there was little appreciable sense of targets to be met, or goals to be achieved, unless projects had their own advisory boards or a good relationship with their RECs.

*Proportionality*

Proportionality requires that: regulators should only intervene when necessary. Remedies should be appropriate to the risk posed and costs identified and minimised.[28] Our analysis of the interviews conducted in our study revealed three examples whereby regulation was not felt to be proportionate for the goals to be achieved, as it required considerable resource expenditure. The first example was the licensing requirements of the Human Tissue Authority—particularly its requirement to obtain multiple licences for storage facilities held in different buildings within the same institution, which was expensive and did not seem to offer a commonsense approach. The second example was the cost, in terms of both resources and time, involved in checking compliance with the different requirements in each country when working with international collaborators. One respondent confided that the prohibitive time and cost burdens that this would have involved meant that the researchers in that project had made a pragmatic decision not to engage with the laws at all. This was a high-risk strategy that was taken because the costs simply of ascertaining the requirements

[27] Better Regulation Task Force, *Principles of Good Regulation* (revised 2007) available at <http://archive.cabinetoffice.gov.uk/brc/publications/principlesentry.html>.

[28] *Ibid.*

for compliance were too expensive and time-consuming. Thirdly, another respondent highlighted the costs of implementing the EU Clinical Trials Directive 2001/20/EC,[29] which had had a major impact on the way in which they did research due to the costs involved. These three examples of perceived disproportionality show that, in the views of at least some people working in the field, the regulatory requirements for research are at times unnecessarily out of proportion with the aims that they are trying to achieve. From our analysis in general, it is also difficult to say that the system as a whole is proportionate, given that it lacks basic coherence, clarity, transparency and accountability.

*Public Participation*

As we conclude this analysis of our research findings, as viewed through the lens of the better regulation principles, it is worth pausing to draw attention to one of the additional good governance characteristics noted above, as identified by Baldwin and Cave[30]—namely, public/stakeholder participation. In this context, it is illuminating to compare the list of actors presented in the regulatory space analysis in Chapter Three with the list of stakeholders who were seen as being important participants in the biobanking space by our respondents, because they comprise the core target audience for the governance 'message'—namely: participants; potential participants; 'society' or the 'public'; biobank users (such as external researchers); funding bodies; and professional peers. In Chapter Three, aside from funding bodies, none of these stakeholders featured directly in our analysis of the regulatory space as being identifiable 'regulatory actors'. None have any obvious regulatory power or ability to control the six key regulatory space resources that we have identified.[31] Yet, as our sociological findings suggest, such stakeholders have considerable implicit significance to biobanking professionals due to their interrelationships and interdependencies.

It would be going too far to suggest that this renders such stakeholders 'regulatory actors', or gives them an effective 'voice' or power of their own as direct participants in the biobanking governance regulatory space. Instead, as we have shown in previous chapters, the extent of their implicit influence or participation here fundamentally depends upon (and is determined and defined by) what practitioners themselves believe the interests, concerns, perceptions, preferences and so forth of such stakeholders to be, and how they present them. This, in turn, is shaped (and limited) by how far practitioners perceive that their own interests align, elide or coincide

[29] [2001] OJ L/121/34.

[30] R Baldwin and M Cave, *Understanding Regulation: Theory, Strategy, and Practice* (Oxford, Oxford University Press, 1999) ch 6.

[31] See Scott, 'Analysing Regulatory Space', above n 13.

with the (supposed) interests of others. Nevertheless, comparing these two lists of potential 'participants' (our own and that of our respondents) serves to highlight the fact that, for at least some of our respondents—and, perhaps, practitioners more widely—the interests and concerns of key 'silent' stakeholders are important considerations, although such stakeholders are not currently afforded any direct role or status within the regulatory framework.

## LOOKING FORWARD

Our analysis highlights a number of concerns with the governance system for medical research as it existed in 2005 to 2009. It is our view that, when assessed against the good regulation principles, and in the light of the evidence and analysis in the earlier chapters of this book, the current governance framework for biobanks and biobanking research can be said to have important shortcomings. We believe that it is clear that there is room for improvement in the regulation and governance of biobanks and associated research work.

In recent months, there have been a number of other calls for change. An example of this is the recent Academy of Medical Sciences Report of 2011,[32] which makes a number of key recommendations that, if implemented, would have a significant impact on the ways in which medical research would be governed in the future. The experiences of researchers reported by the Academy of Medical Sciences Report accord in many respects with those described in this book.[33] The existing structure for medical research has evolved over time and, as a result, has a number of 'legacy' requirements, structures and processes. Bodies have added their requirements to the list that researchers must fulfil at various stages of the research process. This has resulted in a system which is not necessarily aligned to the needs of researchers or that is appropriate or proportionate, as all types of research (low-risk data analysis or high-risk physical interventions) are subject to the same oversight processes. In this section we extrapolate from our project as a whole to present suggestions for changes that may help to improve and build on the existing strengths of the governance system, and ameliorate some of the deficiencies that have been identified through this research and enable it better to meet the requirements of the principles of good governance.

[32] Academy of Medical Sciences *A New Pathway for the Regulation and Governance of Health Research* (Academy of Medical Sciences London 2011).

[33] One of the authors (MP) was a member of the Academy of Medical Sciences working party which produced this report.

## One Authorative Body

The system of governance that currently exists is polycentric, allowing a number of bodies to exercise their powers to influence the activities in this regulatory space. This has led to duplication of effort, confusion on behalf of some researchers, an inconsistent representation of wider interests in medical research and a lack of clear democratic legitimacy. Some measure of harmonisation, rationalisation and a pruning down (where appropriate) of the number of regulatory actors currently involved might improve the system as a whole. To this end, the Academy of Medical Sciences Report[34] argues that a single authoritative body, which would be able to lay down the requirements for biomedical research, could offer significant benefits. Some elements of our analysis can also be interpreted to suggest that such a body could potentially be beneficial, while others point to the limitations and difficulties that it would face.

Assuming that such a body could be successfully created, constituted, empowered and integrated into the governance framework, it could have the potential to fulfil a number of useful functions that, our analysis of the case of biobanking suggest, were lacking from the system in the recent past. First, there is a need to outline the overarching principles that should be applied to emerging areas of research and new approaches, including biobanking, so as to provide a nationally authoritative, principled governance framework, and policies based on clear normative principles and a comprehensive understanding of medical research. Whether undertaken by a single national body or achieved through other means, having some authoritative articulation of the overall view of the wider issues and benefits of carrying out medical research, coupled with the rationale and justifications for certain courses of action, would help to engender public trust, whilst also helping to formulate a national strategy and co-ordinated vision of where medical research should be focused. Such conceptual frameworks can provide a litmus test by which to evaluate deviations from established norms in the case of new areas, such as biobanking.

Depending on the ambit of its responsibilities and role, having a more centralised, nationally authorised body could help to eliminate (or at least ameliorate) some of the current competition for influence amongst inhabitants of the regulatory space—one notable example being the production of overlapping guidelines by multiple bodies some of which do not have the legal mandate to write or enforce them. Introducing a national body might also enable clearer lines of authority to be established, and a clearer demarcation of areas of activity. This would enable valuable resources that

[34] Academy of Medical Sciences, 'A New Pathway for the Regulation and Governance of Health Research' (London 2011) http://www.acmedsci.ac.uk/p47prid88.html.

currently are expended on producing excessive, overlapping, and poorly targeted guidance to be directed to other areas. Again, whether or not achieved via a national body, such ends would appear to be desirable for the sake of clarity within the system when taken as a whole. Linked in with this, a national body could also perhaps perform a co-ordinating role, liaising with other regulatory actors such as the Information Commissioner's Office, the Human Tissue Authority (or successors to it), and the GMC, to ensure that research issues are appropriately covered. These bodies play an important role in the medical regulatory framework, as they have the direct authority of Parliament to regulate in this area. However, as our findings suggest, they also need to consider in a much more co-ordinated and co-operative manner the relevance of, and interrelationships between, the various guidance materials they produce, as well as the best ways in which to disseminate guidance to the research community. It is these bodies that also have the mandate and the authority to engage with similar authorities in other countries, and which, thus, could address some of the cross-border concerns of carrying out international research that were reported by our respondents.

Our research has highlighted the perceived lack of clear, practical advice about best practice. In the case of biobanking, which spans disciplinary and professional boundaries, the practical advice, which is arguably needed includes clear definitions, template forms, and standards and procedures—perhaps in the form of a 'roadmap' that would show what is expected and what must be done in order to comply, with both ethical and legal requirements. Having such a tool, coupled with authoritative, targeted advice, would make requirements clearer and more certain, as well as encouraging greater uniformity, as one single interpretation could be identified. Such an approach would be able to anticipate new areas of practice wherein guidance was needed, rather than it being left to individual researchers to expend valuable time and resources trying to ascertain what best practice should be. However, the development of best practice (whether by a single oversight body or alternative means) would have to be carried out in a consultative manner, to avoid practitioners experiencing resentment and a reluctance to engage with new policies. Therefore, the development of mechanisms to engage with all stakeholders—including professionals as well as patient groups and participants—in a meaningful and inclusive way about significant changes in policy or practice should be a priority. As our research has shown, it is also crucial that any new governance mechanisms take real account of the variety of research methods and variations in scale in the broad area of 'biobanking' research, as we have described in Chapter Four and Ten.

Were it to be possible to set up a national, authoritative body with appropriate democratic legitimacy—that is, accountable to Parliament through the Secretary of State for the conduct of its activities in the same way as other regulatory bodies—the primary benefit of having such a body to an emergent and inter-disciplinary field like biobanking, then, would be its potential to promote cohesion. While it is an option for the reasons outlined above, it would not necessarily come easily, or without possible drawbacks or difficulties of its own.

A major challenge facing such a body would be how to secure alignment in the face of the diversity and dynamism that characterises a field like biobanking. In addition, as noted above, considerable challenges, difficulties and tensions could also ensue, requiring great care should the option be pursued. Key potential issues could, for example, revolve around the problem of avoiding undue legalism, rigidity and the issuing of 'diktats'; ensuring proper participation and representation of all relevant stakeholders' voices; achieving acceptance from practitioners, the public and other stakeholders; and avoiding co-optation or regulatory 'capture' by the more powerful regulatory space inhabitants and interest groups. It is also questionable whether other bodies that currently enjoy de facto governance powers or engage in regulatory activities would (willingly) withdraw themselves from the field; and whether their withdrawals would be desirable in all cases. Arguably, such withdrawals are both unlikely and potentially damaging to a more flexible, responsive and context-sensitive approach that could be appropriate and useful in innovative or emergent fields of research.

### Acknowledging the Role of Funders as de facto Regulators

Even were there to be a single research regulator, inevitably funding bodies would nonetheless continue to play a significant role in guiding, directing and deciding the research which is funded in the UK. They have such power to influence and control research practice not because they are mandated to do so as regulators, but because they hold the purse strings and can therefore exert direct control over research by laying down the terms under which funding will be given. One leading example of this has been the introduction of open-access data-sharing policies, as discussed in Chapter Nine. This power on the part of funders means that a small number of funding bodies enjoy substantial control over research agendas and the ensuing research practice. While a number of funders are public bodies, there are also a number of charitable bodies that operate independently of state control in the UK. Each of these bodies may have different policies and procedures. It may be beneficial for major UK funding bodies to explore an agreement on a set of common policies, procedures and requirements on key issues—for example open access. However, funders are not currently acknowledged for their important regulatory role. They are an effective regulator as research cannot proceed without funding; they are aware of new techniques and ideas emerging in science and can therefore act as an early warning system for imminent, required changes to governance; and they are in a position to bring about rapid cultural change. However, to acknowledge this role would require that publicly funded research councils be made subject to the same accountability mechanisms as other regulators in terms of how they regulate.[35]

### Research Ethics Committees

RECs, like funding bodies, are one of the key gatekeepers in the research governance system within England and Wales. However, they only have a sound legal footing for the regulation of clinical trials, not for all research activities in this jurisdiction which they oversee. Our interviews were carried out before substantial changes were made to the REC system, notably through the establishment of NRES. Our research findings therefore reflect a specific period in time, and this study did not include more recent interviews with respondents to gauge their opinions of these changes. However, our findings suggest that, at that time—pre-NRES—our interviewees strongly believed that the REC system needed to move to more accountable and transparent decision-making processes, so that researchers were aware of the standards by which they were to be judged and the precedents that would apply. The mandate of RECs is to protect patients. However, according to our respondents, the systems put in place to do this had become very bureaucratic and procedural. Some of our respondents suggested that there may be a 'disconnect' between what RECs regard patients' concerns to be, and what patients themselves might actually want. (Although, for the same reasons as noted in our discussion above of the good governance 'participation' characteristic, and the implicit influence of research participants, how far we accept the presentation of what patients might want from professionals, in the absence of asking patients themselves, is an open question.[36])

## IMPROVING THE ROLE OF LAW

There is need for a foundation document that lays out the principles for research in the jurisdiction of England and Wales. While we have an Act that applies to animal research, there is no equivalent statute for all research

[35] Such as the *Regulators' Compliance Code: Statutory Code of Practice for Regulators* (17 December 2007).

[36] See L Brown, M Parker and M Dixon Woods, 'Whose Interest? British Newspaper Reporting of Use of Medical Records for Research' (2008) 13 *Journal of Health Services Research and Policy* 140–45.

carried out on human beings.[37] To have one piece of legislation may assist researchers to be clear about what is expected of them. However, this would have to be supplemented by ready access to appropriate guidance which is more nuanced for particular kinds of research or biobanking enterprises. There is a need to take all of the references to medical research that are found in a number of different sources of law, to consolidate and streamline them, and to make them easily accessible to researchers. One important shortcoming of the current system is that the lack of clear, consolidated, authoritative legal requirements means that many practitioners do not know what the law is or what their obligations are, and, hence, whether or not they are legally compliant. Any reforms introduced along these lines should include clear guidance on cross-border collaborations and sharing, as currently different legal requirements apply when sharing samples and data with collaborators in different countries. In the case of biobanking, a 'higher-level' framework of requirements should be produced, which could form the basis for more detailed guidance at the 'operational level'. This detailed guidance should explain how to implement the higher-level framework. Such a consolidation and co-ordination of legal requirements and guidance would mean that researchers would know precisely where to go to find the information that they require, and whom to contact to obtain advice. Ideally, the provision of such definitive advice would be carried out by the national body responsible for research (were such a body to be introduced).

## DEVELOPING PROFESSIONAL CULTURE

A clear finding from our research is that informal mechanisms are very important as a way of developing practice in emerging fields such as biobanking. It is clear that no one body or single legal instrument can provide all of the answers to the complex issues or day-to-day questions that surface with new practice developing in different contexts as a result of different relationships,[38] although as noted above, both of these interventions could have benefits over the status quo. Our research has identified various professional influences that shape the decisions made in developing best practice, such as: shared assumptions about what is legally, ethically or scientifically acceptable; and learning from experience, anecdotal evidence from colleagues, and the accumulated shared knowledge within professional or institutional communities (accessed, for example, through training and support). Social networks, as conduits of governance and guidance, can also support and shape evolving practice. Any coherent future governance framework is

[37] Clinical Trial Regulations, as their name suggests cover only specific one part of the biomedical research activity.

[38] This is discussed in some detail in ch 10 above.

going to need to take seriously, encourage, make space for, and account for the use of appropriate, situated judgment in particular cases. This suggests the importance of the role for professional networks in ensuring transparency and accountability.

Professional networks are an important way of disseminating information but also are a form of control which ensures that individuals conform to certain standards of shared professional integrity. Practitioners' sensitivity to perceived 'scrutiny' from participants and peers, and the impact of professional cultural norms, are two obvious illustrations. Professional associations—notably, the GMC, Royal Colleges, Academy of Medical Sciences, and the British Medical Association—can also play an important role in defining professional education, setting good practice norms, and helping to maintain the values and practices that support professional cultures. Currently, many of these bodies also promulgate guidance for their members, which could be seen to take them beyond merely providing support and into the regulatory domain. The merits and disadvantages of this, as well as how this activity inter-relates with other regulatory bodies, all need to be carefully assessed and understood.

## Responsible Researchers

A notable finding was the sense of a 'relationship' of 'responsibility' that some researchers expressed towards patients and funders to make the best use of the resources that each had provided. We have recounted examples of researchers who saw themselves as protectors or champions of participants' interests, and as having developed through their research a sense of custodianship in respect of the data and samples with which they had been entrusted. This sense of responsibility, and the importance placed on moral practice by researchers themselves, is without doubt one of the key factors in the establishment and maintenance of high ethical standards in research. Any effective approach to governance and regulation in biomedical research will need to pay particular attention to safeguarding and fostering conditions conducive to the support and encouragement of these aspects of ethical research practice. Moreover, we found that, through their practices,[39] scientists are important shapers of governance. This resonates with findings from previous studies which show that, through informal networks and through their position as expert users and developers of techniques and technologies, scientists themselves co-produce governance.[40] Chapter Eleven highlights

[39] See ch 9 above.

[40] S Hilgartner and SI Brandt-Rauf, 'Data Access, Ownership, and Control: Toward Empirical Studies of Access Practices (1994) 15 *Science Communication* 355–72: Mayrhofer and Prainsack 'Being a Member of the Club', above n 15.

and explores the complexities of the moral world of the researchers and clinicians involved in the production, use and governance of biobanks.

## Ready Access to Experts

It is important for researchers and those who are responsible for developing and operating biobanks to have easy access to the support, training and advice that is necessary for them to be in a position to identify and analyse the ethical, legal and social issues (ELSI) arising from the new approaches that they are developing, and to have links with experts who can provide tailored advice and support. Such 'expert' sources (mirroring the role of the expert compliance units in pharmaceutical companies) could be along the lines of the Caldicott Guardians, who have been instrumental in providing assistance in cases of confidentiality in clinical care, except that they would be established for new areas of research, such as biobanking. Training and mentoring systems designed to spread knowledge and best practice know-how effectively and quickly would also be valuable for new projects. Currently, while some of the larger biobanking projects and institutions in the public sector, including some of those that we have been calling repositories, have access to colleagues working within the fields of Law and Ethics and the Social Sciences. However, this support is not always formally acknowledged and there are no uniform, formal structures to enable this to happen.

## Training

It is evident from our findings that a number of professionals, such as bioinformaticians, epidemiologists, lab-based scientists, nurses and clinical researchers, all work side-by-side within the biobanking field. However, each of them will have received different disciplinary training, which will not always explicitly have involved an ELSI component on biobanking or the professional practice that should accompany it. Given the importance of this burgeoning emerging field, there ought to be specific courses on biobanking as the practice becomes more established, and recognition of the basic principles that have been developed by organisations such as PHOBEBE, P$^3$G, BBMRI and ISBER. Such specialist training courses—perhaps along the lines of the P$^3$G Summer School and the NOWGEN courses—could bring the various elements for biobanking together. However, there is also a need for an accreditation course. Accreditation systems that specify the technical quality of procedures can be a useful way for biobanks to demonstrate not only excellence of practice but also the fact that their activities are legitimate because they have attained or adhere to certain recognised standards.

### Advisory Boards

Internal advisory bodies currently play a valuable role in advising certain projects on what ultimately will become the benchmarks for future projects. Cutting-edge research projects are often trailblazers in developing innovative approaches and new forms of practices.[41] In emerging areas of practice, advisory boards play a pivotal role that complements and supports the oversight activities of formally constituted regulatory bodies. Many international consortia now have advisory boards or specialist ethics and governance committees that help to guide emerging practice. Principally, this is done through the writing of guidance and project-specific policies. Some projects may have ELSI research embedded in the project as a whole, which can play something akin to this advisory role, allowing for a deeper influence of ELSI considerations over the development of best practice as the research unfolds. In the case of biobanks, advisory bodies may have a particularly significant oversight role—particularly as, over recent years, certain biobanks (those accorded research tissue bank status) have found themselves in a position whereby they are able/required to approve additional research without having to go back to RECs. Internal oversight bodies can be a most effective governance mechanism, as they can target their attention in those areas when and where it is needed most. However, to be most effective, the important role of these informal bodies needs to be acknowledged and there needs to be clarity both about their composition, role, powers and functions, and how they are related to the wider governance context, including any single, overarching regulatory body. Moreover, the individuals involved need to be adequately recognised, supported and, if appropriate, remunerated for their work and expertise.

## THE IMPORTANCE OF INPUT FROM PATIENTS AND RESEARCH PARTICIPANTS

Finally, another important element shaping decision-making that was evident in our research was a perception that decisions about what is to constitute 'good practice' should stand up to scrutiny by patients and research participants, as well as by peers and funders. Patients and research participants may in many cases be willing for some aspects of decision-making to be undertaken by bodies constituted for the purpose of overseeing research, as such bodies may, in many cases, have a better understanding of the issues arising in the field of biomedical research than research participants. However, there is a real need to develop effective and

[41] See Mayrhofer and Prainsack, 'Being a Member of the Club', above n 35.

appropriately inclusive mechanisms for involving patients and participants in research and its governance, and for taking account of their views.

## FINAL THOUGHTS

One motivation for carrying out this project and writing this book was a concern that 'genetic databases' or 'biobanks' may be being developed in an ad hoc manner, without due concern for the law and without proper guidance. Changes in the legal requirements at the time were being driven to some extent by very specific events that had led to public alarm over the inappropriate use of data and biological materials. This significant change in the regulatory framework might not be sufficiently informed by the subtleties and variety of day-to-day practice in genetic research, and had the potential to lead to regulation and governance which would be unfit for purpose and not adequately capture the variety of practice. Another of our concerns was that models of governance appropriate to the development and use of large-scale genetic databases designed to act as community resources might not be appropriate, and might indeed be unethical, in relation to small-scale genetic 'databases' such as those built up by clinician-researchers working in the context of family studies on single gene disorders. In this chapter, we have drawn upon the analysis presented in earlier chapters and have explored some of their implications for the better governance of biobank research.

As we conclude, it is important to highlight again that an important finding of our research is that, just as the regulatory space within which they operate is complex, genetic databases or biobanks and the practices associated with them are themselves far from uniform or static, and the experiences and opinions of those working in this field are variable depending on a host of contextual factors. This complexity and fluidity has implications for what is to count as good practice before one can begin to design mechanisms to ensure that this is encouraged. It requires of those who are involved in the governance of genetic databases or biobanks that they develop not only coherent overarching mechanisms for the governance of research as a whole, but also that they provide the support and resources required to enable researchers themselves to develop processes for the achievement of practical, ethical solutions in particular contexts, which they are able to justify in ways that are reasonable, transparent and accountable. Developing regulatory mechanisms that are responsive and enabling as well as being able to apply in a global context will require a co-ordinated effort that also takes into account the complexity of national biobanking practice and governance frameworks. This study provides some insights into this complexity from the perspective of those undertaking biobanking activities in England and Wales which may inform developments in other countries and at a global level.

# Appendices

# APPENDIX 1

# Selected Formal Governance Sources

*Domestic*

— Access to Health Records Act 1990
— Access to Medical Reports Act 1988
— Copyright, Designs and Patents Act 1988
— Criminal Justice Act 2003
— Criminal Justice and Police Act 2001
— Criminal Justice and Public Order Act 1994
  - Data Protection Act 1998 (as amended)
  - Data Protection (Processing of Sensitive Personal Data) Order 2000 (SI 2000/417)
  - Data Protection (Conditions under Paragraph 3 of Part II of Schedule 1) Order 2000 (SI 2000/185)
  - Data Protection (International Co-operation) Order 2000 (SI 2000/190)
— Freedom of Information Act 2000
— Health and Social Care Act 2008
— Human Fertilisation and Embryology Act 1990 (as amended)
  - Human Fertilisation and Embryology (Research Purposes) Regulations (SI 2001/188)
  - Human Fertilisation and Embryology (Special Exemptions) Regulations 1991 (SI 1991/1588)
  - The Human Fertilisation and Embryology (Quality and Safety) Regulations 2007 (2007/1522)
— Human Fertilisation and Embryology Act 2008
— Human Rights Act 1998
— Human Tissue Act 2004
  - The Human Tissue Act 2004 (Ethical Approval, Exceptions from Licensing and Supply of Information about Transplants) Regulations 2006 (SI 2006/1260)
  - Human Tissue (Quality and Safety for Human Application) Regulations 2007 (SI 2007/1523)
— Legislative and Regulatory Reform Act 2006
  - Legislative and Regulatory Reform (Regulatory Functions) Order 2007 (SI 2007/3544)

- — Medical Act 1983 (as amended)
- — Medicines Act 1968 (as amended)
  - – Medicines for Human Use (Clinical Trials) Regulations 2004 (SI 2004/1031) (as amended)
  - – Medicines for Human Use (Clinical Trials) Amendment Regulations 2006 (SI 2006/1928)
- — Mental Capacity Act 2005
  - – Mental Capacity Act 2005 (Loss of Capacity during Research Project) (England) Regulations 2007 (SI 2007/679)
  - – Mental Capacity Act 2005 (Loss of Capacity during Research Project) (Wales) Regulations 2007 (SI 2007/837)
- — Mental Health Act 1983 (as amended)
- — Mental Health Act 2007
- — National Health Service Act 2006 (especially ss 251–252)
  - – Health Service (Control of Patient Information) Regulations 2002 (SI 2002/1438)
- — National Health Service (Wales) Act 2006
- — Patents Act 1977 (as amended)
  - – Patents Rules 2007 (SI 2007/3291)
  - – Patents Regulations 2000 (SI 2000/2037)
- — Police and Criminal Evidence Act 1984 (PACE) (as amended)
- — Regulatory Enforcement and Sanctions Act 2008
- — Statistics and Registration Service Act 2007

*European Union*

- — Clinical Trials Directive, 2001/20/EC
- — Data Protection Directive 95/46/EC
- — Databases Directive 96/9/EC
- — Directive 98/44/EC of 6 July 1998 on the legal protection of biotechnological inventions
- — Good Clinical Practice Directive, 2005/28/EC
- — Tissues and Cells Directive 2004/23/EC
  - – Commission Directive 2006/17/EC (First technical annexe)
  - – Commission Directive 2006/86/EC (Second technical annexe)

*United Nations and UNESCO*

- — International Covenant on Civil and Political Rights 1966
- — International Covenant on Economic, Social and Cultural Rights 1966
- — Universal Declaration of Human Rights 1948
- — Universal Declaration on Bioethics and Human Rights 2005
- — International Declaration on Human Genetic Data 2003
- — Universal Declaration on the Human Genome and Human Rights 1997

*Pending Bills and Other Significant Areas of Proposed Legislative Reform*

— Coroners and Justice Bill 2008–09
— Data Protection Act 1998—review
— Data Protection Directive 95/46/EC review
— Health Bill 2008–09 (including NHS Constitution and Handbook)
— Human Fertilisation and Embryology Act 2008—draft regulations
— Human Tissue Act 2008—revised Codes of Practice
— National DNA Database
— National Health Service Reform Bill 2008
— NHS electronic Care Record scheme

# APPENDIX 2

# Selected Codes of Practice, Guidelines and Other Guidance Materials

*Statutory Codes of Practice*

- Human Tissue Act 1998—notably:
  - Code 1—Consent
  - Code 5—Disposal of human tissue
  - Code 8—Import and export of human bodies, body parts and tissue
  - Code 9—Research
- Human Fertilisation and Embryology Act 1998—HFEA Code of Practice (as updated)
- Legislative and Regulatory Reform Act 2006—Regulators' Compliance Code: Statutory Code of Practice for Regulators
- Mental Capacity Act 2005—Code of Practice

*Other Domestic Codes of Practice and Guidance Materials*

- British Medical Association (BMA)—wealth of guidance documents, webpages, working papers, toolkits, joint guidance documents and so forth, covering a range of topics including confidentiality, consent, capacity, genetics, health records, human rights, mental health and new technologies. Examples include:
  - Access to health records by patients (2007)
  - Access to medical reports (2007)
  - Advance decisions and proxy decision-making in medical treatment and research
  - *Assessment of Mental Capacity* (2nd edn, 2004)
  - Clinical genetics factsheet
  - Confidentiality and disclosure of information tool kit (2008)
  - Confidentiality and people under 16 (1994)
  - Confidentiality as part of a bigger picture—a discussion paper (2005)
  - Consent tool kit—4th edn (2008)
  - Human Tissue Legislation: Guidance from the BMA (2006)

- Impact of the Human Rights Act (2007)
- Information technology glossary
- Medical information and insurance—BMA/ABI (2008)
- Mental Capacity Act 2005—Guidance for health professionals
- Mental Capacity Act 2005 tool kit (2008)
- Remotely held records and centralised servers (2002)
- Secondary uses of patient information (2007)

— Department of Health
  - Directions from the Secretary of State (under various statutes)
  - Host of letters, circulars, policy documents, guidance and good practice documents, health service guidelines, regular bulletins, guides, toolkits and other resources

— General Medical Council (GMC)—detailed guidance on good practice and ethics, covering a wide range of topics including:
  - 0–18 years: guidance for all doctors
  - Confidentiality: Protecting and Providing Information (2004)
  - Confidentiality FAQs (2004)
  - Conflicts of interest (2008)
  - Consent: patients and doctors making decisions together
  - *Good Medical Practice* (2006)
  - Research: the role and responsibilities of doctors (2002)

— Human Fertilisation and Embryology Authority (HFEA)
  - Directions
  - Chair's Letters
  - Chief Executive's Letters

— Human Genetics Commission—numerous consultations, reports, information sheets and policy documents on a range of topics

— Human Tissue Authority (HTA)—various non-statutory codes of practice, Guidance Notes, position statements, and other publications and webpages addressing a range of topics, including:
  - Definition of relevant material under the Human Tissue Act
  - Expected Standards (Directions)
  - Guidance Notes
  - Licensing guidance
  - Model consent forms
  - Non-consensual DNA analysis
  - Regulation of acellular material
  - Regulatory enforcement policy

— Information Commissioner and Information Commissioner's Office (ICO)—a wealth of publications and online tools, forms, leaflets, checklists, toolkits and other resources, aimed at both the public and organisations. Relevant materials include:
  - A Strategy for Data Protection Regulatory Action
  - Advice Notices
  - Awareness Guidance

- Data Protection—Protecting People: A Data Protection Strategy for the Information Commissioner's Office
- Data Protection Act 1998 Legal Guidance
- Data Protection Act 1998: The Eighth Data Protection Principle and International Data Transfers
- Dealing with subject access requests involving other people's information (Technical Guidance Note)
- Determining what is personal data (Technical Guidance)
- Framework Code of Practice for Sharing Personal Information
- Freedom of information Act Practical Guidance: Information about the deceased (v 2, 31 July 2008)
- Glossary of terms
- Guidance on data security breach management (Guidance Note)
- International transfers of personal information—General advice on how to comply with the 8th data protection principle (Data Protection Guidelines)
- It's Your Information guides (various)
- NHS Electronic Care Records—The Information Commissioner's View
- Personal Information Toolkit
- Providing personal information to a third party (Good Practice Note)
- Releasing information to prevent or detect crime (Good Practice Note)
- Sharing personal information: our approach
- Use and Disclosure of Health Data: Guidance on the Application of the Data Protection Act 1998 (May 2002)
- What is data for the purposes of the DPA (Technical Guidance Note)
- What is personal data?—A quick reference guide

— Medical Research Council (MRC)—numerous ethics, research governance, and best practice publications, statements, reports and guidelines, including:
- Clinical Trials Toolkit
- Tissues and Data Toolkit
- Personal information in medical research (as updated)
- Human tissue and biological samples for use in research—operational and ethical guidelines (plus Addendum)
- Research involving human participants in developing societies
- Medical research involving adults who cannot consent
- Medical research involving children
- *Good Research Practice* (2000, as updated)
- Research regulation and ethics—MRC Position

— Medical Research Council (MRC) Regulatory Support Centre—wealth of guidance (particularly online, including a dedicated Advice

Service) provided to help scientists implement good practice and meet legal and ethical requirements. Topics covered in detail include:
- Clinical research governance
- Consent to take part in research
- Data access
- Data sharing initiative
- Ethics regulation and public involvement committee
- Global bioethics
- Good research practice
- Open access to published research
- Regulatory guidance (including human tissue; Research Governance Framework)
- Research and the Human Tissue Act 2004—Consent (March 2007)
- Research and the Human Tissue Act 2004—Licensing (March 2007)
- Training
- UK Biobank
- UK Stem Cell Bank
- Use of human tissue
- Global Bioethics Regulatory Support Centre

— National Research Ethics Service (NRES) (formerly COREC)
- Bulletins (information and guidance for research ethics committees)
- Defining research (Leaflet)
- Explaining research (Leaflet)
- *Governance Arrangements for NHS Research Ethics Committees* (GAfREC)
- Guidance on information sheets and consent forms
- Guidance on research involving adults unable to consent for themselves
- *Standard Operating Procedures for Research Ethics Committees in the United Kingdom* (SOPs)
- Various information sheets and guidance sheets for research ethics committees

— Nuffield Council on Bioethics—numerous reports and recommendations, including:
- Human Tissue: Legal and Ethical Issues (April 1995)
- Public Health: Ethical Issues (November 2007)
- The Ethics of Patenting DNA (July 2002)

— Research Councils UK (RCUK)
- Expectations for Societal and Economic Impact
- Mission for Economic and Societal Impact
- Numerous guidance documents and websites

— Royal Colleges—including the Royal College of Physicians, Royal College of Pathologists, Royal College of General Practitioners, and Joint Committee on Medical Genetics

  - Numerous reports
  - Various guidance materials
- UK Biobank
  - UK Biobank Ethics and Governance Framework (version 3, October 2007)
- UK Research Integrity Office (UKRIO)
  - Code of Practice for Research
- Wellcome Trust—various guidelines, policy and position statements, reports and other publications. Also, detailed web-based professional resources (biomedical, professional, education, and researcher support). Matters covered include:
  - Access to collections of data and materials for health research
  - Grant Conditions: UK and overseas (as amended, August 2007)
  - *Guidelines on Good Research Practice* (as revised, November 2005)
  - Harmful misuse of research
  - Managing risks of misuse associated with grant funding activities: A joint Biotechnology and Biological Sciences Research Council (BBSRC), Medical Research Council (MRC) and Wellcome Trust policy statement
  - Policy on access to bioinformatics resources by Trust-funded researchers (2001)
  - Policy on data management and sharing (2007)
  - Policy on intellectual property and patenting
  - Policy on relationships between Trust-funded researchers and commercial organisations
  - Policy on stem cell research
  - Position statement in support of open and unrestricted access to published research (as updated, February 2008)
  - Publisher's guide and FAQ: Open and unrestricted access to published research (as updated, 24 May 2007)
  - Statement on genome data release
  - Statement on human participants in research
  - Statement on the Handling of Allegations of Research Misconduct (as revised, November 2005)

*European Union Guidance Materials*

- Article 29 Data Protection Working Party (Art 29 DPWP)—various opinions and working documents laying down interpretative views of the EU Data Protection Authority on genetic issues.
  - Art 29 DPWP Opinion 6/2000 on the Human Genome and Privacy (WP 34)
  - Art 29 DPWP Working Document on Genetic Data (WP 91)

  - Art 29 DPWP Working Document on the processing of personal data relating to health in electronic health records (EHR) (WP 131)
  - Art 29 DPWP, Opinion 4/2007 on the concept of personal data (01248/07/EN—WP136)
  - Art 29 DPWP, Opinion on the review of the Directive 2002/58/EC on privacy and electronic communications (ePrivacy Directive) (00989/08/EN—WP150)
- European Research Council (ERC)
  - ERC Scientific Council Guidelines for Open Access
  - ERC Scientific Council Statement on Open Access
  - European Standards on Confidentiality and Privacy in Healthcare

*Council of Europe Recommendations*

- Recommendation 1160 (1991) on the preparation for a Convention on Bioethics
- Recommendation 1240 (1994) on the protection and patentability of materials of human origin
- Recommendation 1468 (2000) on biotechnologies
- Recommendation 1512 (2001) on the protection of the human genome
- Recommendation R(81)1 the Committee of Ministers to member states on Regulations for automated medical data banks
- Recommendation R(83)10 on the protection of personal data used for scientific research and statistics
- Recommendation R(89)4 the Committee of Ministers to member states on the collection of epidemiological data on primary health care
- Recommendation Rec(90)3 of the Committee of Ministers to member states concerning medical research on human beings
- Recommendation R(94)1 of the Committee of Ministers to member states on Human Tissue Banks
- Recommendation R(97)5 of the Committee of Ministers to member states on the protection of medical data
- Recommendation R(97)18 of the Committee of Ministers to member states on the protection of personal data collected and processed for statistical purposes
- Recommendation Rec(2006)4 of the Committee of Ministers to member states on research on biological materials of human origin

*Other Leading International Organisations and Internet Resources*

- Council for International Organizations of Medical Sciences (CIOMS)
  - International ethical guidelines for biomedical research involving human subjects (Geneva, 2002)

  - International guidelines for ethical review of epidemiological studies (1991)
- European Science and Technology Observatory (ESTO)
- Human Genome Organization (HUGO) and HUGO Ethics Committee
  - Statement on benefit sharing (2000)
  - Statement on DNA sampling: control and access (1998)
  - Statement on human genomic databases (2002)
  - Statement on the patenting of DNA sequences (2005)
  - Statement on the patenting of DNA sequences, in particular response to the European Biotechnology Directive (2000)
  - Statement on the principled conduct of genetics research (1996)
- HumGen International—extensive web-based resource, providing links to national and international laws, guidelines, policies and literature on the ethical, legal and social issues in human genetics. Includes:
  - Databases of laws and policies (national and worldwide)
  - Databases of selected literature
  - PopGen (database on the legal and socio-ethical aspects of population genetics)
  - StemGen (database on the ethical, legal and social issues around stem cell research)
  - IPGen (database concerning intellectual property law and policy)
- OECD: various publications, declarations, guidelines, working papers, working party and committee reports, best practice guidelines and reports, including on biotechnology, health, science and innovation
  - *Best Practice Guidelines for Biological Resource Centres* (Paris, OECD, 2007)
  - *Creation and Governance of Human Genetic Research Databases* (Paris, OECD, 2006) (OECD Biotechnology Division)
  - Declaration on Access to Research Data from Public Funding (OECD Committee for Scientific and Technological Policy) (adopted 30 January 2004)
  - Draft Guidelines for Human Biobanks and Genetic Research Databases (Paris, 1 April 2008)
  - Guidelines on the Protection of Privacy and Transborder Flows of Personal Data (1980)
  - Principles and Guidelines for Access to Research Data from Public Funding (2007)
  - *The Management and Governance of Human Genetic Research Databases (HGRDs): Options and Way Forward* (OECD Directorate for Science, Technology and Industry Committee for Scientific and Technological Policy Working Party on Biotechnology)
- $P^3G$ Observatory (Public Population Project in Genomics ($P^3G$) Consortium)
- UNESCO Global Ethics Observatory (GEObs)

— UNESCO International Bioethics Committee (IBC)
  - Report of the International Bioethics Committee of UNESCO (IBC) on Consent (SHS/EST/CIB08-09/2008/1) (5 December 2008)
  - Bioethics and human population genetics research (1995)
  - Report on confidentiality and genetic data (2000)
  - Human genetic data: preliminary study by the IBC on its collection, processing, storage and use' (revision 2, 2002)
  - Guide No 1, Establishing Bioethics Committees (2005)
  - Guide No 2, Bioethics Committees at Work: Procedures and Policies (2006)
  - Guide No 3, 'Educating Bioethics Committees (2007)
  - Bioethics Committee Guidebooks (February 2006)

— World Health Organization (WHO)
  - A Declaration on the Promotion of Patients' Rights in Europe (ICP/HLE 121) (28 June 1994)
  - Genetic Databases: Assessing the Benefits and the Impact on Human and Patient Rights (Geneva, 2001)
  - Genomic Resource Centre (GRC)
  - Guideline for Obtaining Informed Consent for the Procurement and Use of Human Tissues, Cells, and Fluids in Research (2003)
  - Handbook for Good Clinical Research Practice (GCP): Guidance for Implementation (2002)
  - International Digest Of Health Legislation
  - Operational Guidelines for Ethics Committees That Review Biomedical Research (TDR/PRD/ETHICS/2000.1) (2000)
  - Proposed international guidelines on ethical issues in medical genetics and the provision of genetic services (1997)
  - The Tallinn Charter: Health Systems for Health and Wealth (aka European Health Charter) (Tallinn, 27 June 2008)

— World Medical Association
  - Declaration of Helsinki: Ethical Principles for Medical Research Involving Human Subjects (1964, 2000, as amended, Seoul, October 2008)
  - Declaration on Ethical Considerations regarding Health Databases (2002)
  - Declaration on the Human Genome Project (1992)
  - Statement on Genetics and Medicine 2005

*Other Significant Recent Reports, Papers and Publications (in chronological order)*

— House of Lords Select Committee on Science and Technology, 'Report on Human Genetic Databases: Challenges and Opportunities' (HL 57) Fourth Report of Session 2000–01 (March 2001)

— Department of Health, 'Human Bodies, Human Choices: The Law on Human Organs and Tissue in England and Wales: A Consultation Report' (July 2002)
— Human Genetics Commission, 'Inside Information: Balancing Interests in the Use of Personal Genetic Data' (2002)
— WW Lowrance, 'Learning from Experience: Privacy and the Secondary Use of Data in Health Research'—Report to the Nuffield Trust (2002)
— Better Regulation Task Force, 'Scientific Research: Innovation with Controls' (2003)
— Department of Health Genetics White Paper, 'Our Inheritance, Our Future: Realising the Potential of Genetics in the NHS' (Cm 5791-II, June 2003) (and subsequent progress review reports)
— House of Commons Library Research Paper 04/04, 'The Human Tissue Bill, Bill 9 of 2003–04' (January 2004)
— Healthcare Industries Task Force, 'Final Report: Better Healthcare Through Partnership: A Programme for Action' (November 2004)
— Better Regulation Task Force, 'Regulation—Less is More: Reducing Burdens, Improving Outcomes: A BRTF Report to the Prime Minister' (March 2005)
— Academy of Medical Sciences, 'Personal Data for Public Good: Using Health Information in Medical Research: A Report of the Academy of Medical Sciences' (January 2006)
— House of Commons Library Research Paper 06/06, The Legislative and Regulatory Reform Bill: Bill 111 of 2005–06 (February 2006)
— WW Lowrance, 'Access to Collections of Data and Materials for Health Research: A Report to the Medical Research Council and the Wellcome Trust' (March 2006)
— Joint Committee on Medical Genetics, 'Consent and Confidentiality in Genetic Practice—Guidance on Genetic Testing and Sharing Genetic Information: A Report of the Joint Committee on Medical Genetics' (April 2006)
— National Patient Safety Agency/COREC, 'Building on Improvement: Implementing the Recommendations of the Report of the Ad Hoc Advisory Group on the Operation of NHS Research Ethics Committees' (August 2006)
— Care Records Development Board, 'Information Governance in the Department of Health and the NHS' (rev September 2006)
— Department of Health White Paper, 'Trust, Assurance and Safety—The Regulation of Health Professionals in the 21st Century' (Cm 7013, February 2007)
— Healthcare Industries Task Force, 'Innovation for Health: Making a Difference. Report of the Strategic Implementation Group (SIG)' (March 2007)

— Medical Research Council/Ipsos MORI, 'The Use of Personal Health Information in Medical Research General Public Consultation: Final Report' (June 2007)
— Ipsos MORI, 'Human Tissue Authority Stakeholder Evaluation: General Public Qualitative and Quantitative Research' (June 2007)
— House of Commons Justice Committee, 'Protection of Private Data' (HC 154) First Report of Session 2007–08 (January 2008)
— R Thomas and M Walport, 'Data Sharing Review Report' (July 2008)

# APPENDIX 3

# Selected Common Law Doctrines and Areas Governed by Judicial Rulings

*Civil Law*

— Statutory interpretation and application (especially under the Data Protection Act 1998 and Human Rights Act 2004)
— Tort of battery (relevant to consent to the taking of biosamples from living persons, although courts now prefer to use negligence law)
— Defence of necessity
— Tort of negligence (most relevant to informed consent to the taking of biosamples and participation in research, but possibly also to returning research results and incidental findings to participants)
— Equitable doctrine of confidence (as expanded under the Human Rights Act 1998, art 8, in respect of privacy protection)
— Capacity and competence
— Consent
— Intellectual property rights (including patents and database rights)
— Property law (eg, regarding the human body and body parts: the no-property principle; the work and skill exception to the no-property principle)

*Criminal Law*

— Offences against the person (assault; causing actual bodily harm)
— Offences and sentences under relevant legislation—including the Human Tissue Act 2004 and Data Protection Act 1998 (as amended)

# APPENDIX 4

# Selected Regulatory Space Governance Actors

*Domestic and UK-based*

— Academy of Medical Royal Colleges
— Academy of Medical Sciences
— Association of Chief Police Officers (ACPO)
— Association of British Healthcare Industries (ABHI)
— Association of British Insurers (ABI)
— Association of the British Pharmaceutical Industry (ABPI)
— Association of Charitable Foundations
— Association of Medical Research Charities (AMRC)—and its members (>100)
— Association of Research Ethics Committees (AREC)
— Better Regulation Commission (BRC)
— Better Regulation Executive (BRE)
— Better Regulation Task Force (BRTF)
— BioIndustry Association
— Biosciences Federation (BSF)
— Biotechnology and Biological Sciences Research Council (BBSRC)
— British Association for Tissue Banking
— British Heart Foundation
— British Medical Association (BMA)
— British Society for Human Genetics (BSHG) (incorporating the Clinical Genetics Society and Clinical Molecular Genetics Society)
— British Standards Institute (BSI)
— Caldicott Guardians
— Cancer Research UK
— Care Quality Commission
— Charities (various, including over 100 medical charities)
— Chief Medical Officer
— Chief Scientific Adviser
— Commission for Equality and Human Rights
— Confederation of Cancer Biobanks (CCB)
— Council for Healthcare Regulatory Excellence (CHRE)

— Council for Science and Technology (CST)
— Courts and Tribunals
— Department for Business, Enterprise and Regulatory Reform (BERR)
— Department for Innovation, Universities and Skills (DIUS)
— Department of Health, Social Services and Public Safety (aka Department of Health)
— Economic and Social Research Council (ESRC)
— Engineering and Physical Sciences Research Council (EPSRC)
— Gene Therapy Advisory Group (GTAC)
— Genetics Commissioning Advisory Group (GenCAG)
— Genetics and Insurance Committee (GAIC)
— Genetics Interest Group
— GeneWatch
— General Medical Council (GMC)
— Global Forum on Bioethics in Research
— Government Office for Science (within DIUS)
— Health and Safety Executive
— Health Protection Agency
— Healthcare Industries Task Force (HITF)
— Health Innovation Council
— Home Office
— House of Commons Health Committee
— House of Commons Innovation, Universities, Science and Skills Committee
— House of Commons Justice Committee
— House of Commons Library
— House of Commons Public Accounts Committee
— House of Commons Public Administration Committee
— House of Commons Regulatory Reform Committee
— House of Commons/House of Lords Joint Committee on Human Rights
— House of Lords Science and Technology Committee
— Human Fertilisation and Embryology Authority (HFEA)
— Human Genetics Commission (HGC)
— Human Tissue Authority
— Independent Complaints Advisory Service (ICAS)
— Information Centre for Health and Social Care
— Information Commissioner
— Information Commissioner's Office
— Information Tribunal
— Innovation Research Centre
— Institute for Public Policy Research
— Institute of Biomedical Science (IBMS)
— Joint Committee on Medical Genetics (of The Royal College of Physicians, The Royal College of Pathologists and the British Society for Human Genetics) (JCMG)

— Joint Information Systems Committee (JISC)
— Liberty
— Lord Chancellor
— Medical Devices Agency (MDA)
— Medical Education England (MEE—aka NHS:MEE)
— Medical Research Council (MRC)
— Medical Research Council (MRC) Regulatory Support Centre
— Medicines Healthcare Products Regulatory Agency (MHRA)
— Medicines Control Agency (MCA)
— Ministry of Justice (MoJ)
— National Audit Office (NAO)
— National Cancer Research Institute (NCRI)
— National Clinical Assessment Service (formerly the National Clinical Assessment Authority)
— National DNA Database Ethics Group
— National DNA Database Strategic Board and Custodian
— National Endowment for Science, Technology and the Arts (NESTA)
— National Information Governance Board for Health and Social Care (NIGB)
— National Institute for Health Research (NIHR)
— National Institute for Mental Health in England
— National Institute of Health and Clinical Excellence (NICE)
— National Institute for Biological Standards and Control (NIBSC) (aka National Biological Standards Board (NBSB))
— National Patient Safety Agency
— National Research Ethics Service (NRES) (formerly the Central Office for Research Ethics Committees (COREC))
— NHS Care Record Development Board
— NHS Care Records Service
— NHS Connecting for Health
— NHS Connecting for Health—the Secondary Uses Service
— NHS Foundation Trusts
— Monitor
— NHS Litigation Authority
— NHS National Genetics Education and Development Centre
— NHS National Knowledge Service
— NHS Research and Development (R&D) Forum
— NHS Research and Development (R&D) Offices
— NHS Primary Care Trust Data Protection Officers
— Nuffield Council on Bioethics
— Nuffield Foundation
— Nuffield Trust
— Office of the Health Professions Adjudicator (OHPA)
— Office for National Statistics (ONS)
— Office for Strategic Coordination of Health Research (OSCHR)

— Office of Science and Innovation (OSI)
— onCore UK
— PACE Review Board
— Parliamentary and Health Service Ombudsman
— Parliamentary Office of Science and Technology (POST)
— Patient Information Advisory Group (PIAG)
— Prescription Medicines Code of Practice Authority (PMCPA)
— Primary Care Research Network (PCRN)
— Progress Trust
— Public Health Genetics Unit (PHGU)
— Research Base Funders' Forum (BERR)
— Research Councils UK (RCUK)
— Research Ethics Committees (RECs) (various)—including within the NHS, universities, and private sector
— Research Governance teams/bodies within institutions (eg, NHS and other hospitals, universities)
— Risk and Regulation Advisory Council
— Royal Colleges (various)—including the Royal College of Pathologists, Royal College of Obstetricians and Gynaecologists, Royal College of General Practitioners, Royal College of Physicians, Royal College of Nursing, Royal College of Surgeons of England, Royal Society of Medicine
— Science and Technology Facilities Council
— Secretary of State for Health
— Social Research Association (SRA)
— Statistics Board
— Technology Strategy Board (TSB)
— The National Archive (TNA)
— UK Accreditation Service (UKAS)
— UK Biobank
— UK Biobank Ethics and Governance Council
— UK Clinical Research Collaboration (UKCRC)
— UK Clinical Research Collaboration Regulatory and Governance Advice Service
— UK Clinical Research Network (UKCRN)
— UK Council of Caldicott Guardians (UKCGC)
— UK Data Archive
— UK Ethics Committees Authority (UKECA)
— UK Genetic Testing Network
— UK Good Laboratory Practice Monitoring Authority (GLPMA)
— UK Intellectual Property Office (formerly the UK Patent Office)
— UK Parliament
— UK Research Integrity Office (UKRIO)—including the UK Panel for Research Integrity in Health and Biomedical Sciences

— UK Statistics Authority
— Universities UK
— Wellcome Trust
— Welsh Assembly
— Welsh Assembly Department for Public Health and Health Professions

*European*

— Council of Europe
— Committee of Ministers of the Council of Europe
— Council of Europe Steering Committee on Bioethics (CDBI)
— Council of the European Union
— EuroGenGuide project
— EuroGenTest project
— European Association of Health Law (EAHL)
— European Association of Tissue Banks
— European Commission (EU)
— European Commission Health and Consumer Protection Directorate-General (DG SANCO)
— European Committee for Standardization (CEN)
— European Court of Human Rights
— European Court of Justice (ECJ) (formally, the Court of Justice of the European Communities)
— European Data Protection Supervisor (EDPS)
— European Federation of Pharmaceutical Industries and Associations (EFPIA)
— European Group on Ethics in Science and New Technologies to the European Commission (EGE)
— European Heads of Research Councils (EuroHorcs)
— European Institute of Technology
— European Medical Research Councils (EMRC)
— European Medicines Agency (EMEA)
— European Network of Research Tissue Banks
— European Parliament (EU)
— European Patent Office (EPO)
— European Research Council (ERC)
— European Science Foundation (ESF)
— European Science Foundation Standing Committee for the European Medical Research Councils
— European Society of Human Genetics (ESHG)
— European Society of Human Genetics Professional and Public Policy Committee (PPPC)
— European Union (EU)

— EU Working Parties—including the Article 29 Working Party on Data Protection, Working Party on Research, Working Party on Public Health, and Working Party on Pharmaceuticals and Medical Devices
— Parliamentary Assembly of the Council of Europe
— PHOEBE (Promoting Harmonisation Of Epidemiological Biobanks in Europe)
— Public Health Genomics European Network (PHGEN)

*Foreign, International, Supranational and Multinational*

— American Society of Human Genetics (ASHG) (USA)
— American Society of Law, Medicine and Ethics (USA)
— Council for International Organisations of Medical Sciences (CIOMS)
— EQUATOR Network (Enhancing the QUAlity and Transparency Of health Research)
— Food and Drug Administration (FDA) (USA)
— Genomic Resource Centre (GRC)
— GRaPH-*Int* (Genome-based Research and Population Health International Network)
— HuGENet™ (Human Genome Epidemiology Network)
— Human Genome Organisation (HUGO)
— Human Genome Organisation (HUGO) Ethics Committee
— Human Genome Variation Society (HGVS)
— International Association of Bioethics
— International Cancer Genome Consortium
— International Conference on Harmonisation of Technical Requirements for Registration of Pharmaceuticals for Human Use (ICH)
— International Federation of Human Genetics Societies (IFHGS)
— International Society for Biological and Environmental Repositories (ISBER)
— International Organisation for Standardisation (ISO)
— International Working Group on Data Protection in Telecommunications (IWGDPT)
— National Bioethics Advisory Committee (USA)
— National Human Genome Research Institute (NHGRI) (USA)
— National Institutes of Health (NIH) (USA)
— OECD Directorate for Science
— OECD Internal Coordination Group for Biotechnology
— OECD Technology and Industry Committee for Scientific and Technological Policy Working Party on Biotechnology
— Organisation for Economic Co-operation and Development (OECD)
— Privacy International
— Public Population Project in Genomics ($P^3G$) Consortium

- UNESCO Division of Ethics of Science and Technology Sector for Social and Human Sciences
- United Nations (UN)
- United Nations Economic and Social Council (ECOSOC)
- United Nations Educational, Scientific and Cultural Organisation (UNESCO)
- UNESCO Intergovernmental Bioethics Committee (IGBC)
- UNESCO International Bioethics Committee (IBC)
- US President's Council on Bioethics (previously the National Bioethics Advisory Commission (NBAC))
- World Health Organisation (WHO)
- World Medical Association (WMA)

# APPENDIX 5

## Interview Subjects

| *Identifier* | *Professional background* | *Organisational affiliation* |
|---|---|---|
| R1 | Bioinformatician | University |
| R2 | Clinical Scientist | University linked to NHS Hospital |
| R3 | Research Scientist | University |
| R4 | Clinical Scientist | University |
| R5 | Research Scientist | University |
| R6 | Epidemiologist | University |
| R7 | Research Scientist | University |
| R8 | Research Scientist | Repository |
| R9 | Clinical Scientist | University linked to NHS Hospital |
| R10 | Bioinformatician | Repository |
| R11 | Clinical Scientist | University linked to NHS Hospital |
| R12 | Clinical Scientist | University linked to NHS Hospital |
| R13 | Health Professional (Clinician) | NHS Trust or Hospital |
| R14 | Clinical Scientist | University linked to NHS Hospital |
| R15 | Clinical Scientist | University linked to NHS Hospital |
| R16 | Clinical Scientist | University |
| R17 | Clinical Scientist | University linked to NHS Hospital |
| R18 | Epidemiologist | University |

(*continued*)

| | | |
|---|---|---|
| R19 | Health Professional (Public Health) | NHS Trust or Hospital |
| R20 | Bioinformatician | Repository |
| R21 | Clinical Scientist | Pharmaceutical Company |
| R22 | Research Scientist | Pharmaceutical Company |
| R23 | Research Scientist | Pharmaceutical Company |
| R24 | Epidemiologist | University |
| R25 | Research Scientist | Repository |
| R26 | Clinical Scientist | University linked to NHS Hospital |
| R27 | Research Scientist | University linked to NHS Hospital |
| R28 | Clinical Scientist | University linked to NHS Hospital |
| R29 | Bioinformatician | Repository |
| R30 | Bioinformatician | University |
| R31 | Clinical Scientist | University linked to NHS Hospital |
| R32 | Health Professional (Public Health) | NHS Trust or Hospital |
| R33 | Health Professional (Genetic counsellor) | NHS Trust or Hospital |
| R34 | Research Scientist | Pharmaceutical Company |
| R35 | IT Specialist | Repository |
| R36 | Bioinformatician | Repository |
| R37 | Research Scientist | Repository |
| R38 | Bioinformatician | Repository |
| R39 | Bioinformatician | Repository |
| R40 | IT Specialist | Repository |
| R41 | Research Scientist | University linked to NHS Hospital |
| R42 | Clinical Scientist | University linked to NHS Hospital |
| R43 | Research Scientist | Repository |

(*continued*)

| | | |
|---|---|---|
| R44 | Research Scientist | Repository |
| R45 | Clinical Scientist | University linked to NHS Hospital |
| R46 | Epidemiologist | University linked to NHS Hospital |
| R47 | Clinical Scientist | University linked to NHS Hospital |
| R48 | Clinical Scientist | University linked to NHS Hospital |
| R49 | Clinical Scientist | University |

| *Professional background* | *Number in our sample* |
|---|---|
| Bioinformatician | 8 |
| Clinical Scientist | 18 |
| Research Scientist | 13 |
| Epidemiologist | 4 |
| IT Specialist | 2 |
| Other Health Professional | 4 |

| *Organisational affiliation* | *Number in our sample* |
|---|---|
| University | 11 |
| University linked to NHS Hospital | 17 |
| Repository | 13 |
| NHS Trust or Hospital | 4 |
| Pharmaceutical Company | 4 |

# Index


**Academy of Medical Sciences Report (2011)**, 317, 318
**Access to data**
applications for access
criteria, 236
good scientific proposals, 236, 237
lack of standardised procedures, 237
negotiation, 238
persuasion, 238
professional status, 237, 238
scientific advice, 237
scientific quality, 237
biological material/data distinction, 193
*see also* **Biological material/data distinction**
changing cultures
dominant cultures, 257
obligatory passage points, 251, 257, 258
redefinition of existing objects, 258
changing models
common resources, 242
Fort Lauderdale Agreement, 242
Human Genome Project, 242, 243
open-access model, 241–4
sources of tension, 242, 243
Wellcome Trust, 242, 243
collaboration, 240, 241
contingency of enacted access
conflicting influences, 253
consent issues, 253
contextual factors, 253
desire to share materials, 253
ethical concerns, 253
formalised governance, 254
multiple influences, 253
need to share materials, 253
notion of data-stream, 253
privacy issues, 253
public good, 251, 253
controversy
connections between controversies, 254
distribution of benefits, 255–7
enacting access, 233
ethical issues, 255, 257
exploitation, 257
inequity, 257
self-protection, 257
data-access agreements, 291
data-sharing, 232–6
*see also* **Data-sharing**
decisions on access
decision-making process, 235, 236, 241
designated committees, 235
documentation, 236
informal networks, 236
informed consent agreements, 236
responsibility, 235
sources of authority, 236
degrees of access, 233, 240
ethical factors, 233, 253
exchange regimes
co-operative, 240
gift, 240
sub-contracting, 240
exchange relationships, 251, 253, 254, 257
governance of access
contextual motivations, 253, 254
enacting governance, 233
formalised governance, 253, 254
multiple interactions, 253
social practices, 253
governing mechanisms, 241, 242
importance, 232
improvements to governance, 220
*see also* **Improvements to governance**
limitations on access, 233
motivations, 254
*see also* **Motivations for data-sharing**
open-access policies, 232
*see also* **Open-access policies**
plurality of access practices, 233, 240
preferential access, 240
regulatory factors, 233
restriction of access
access procedures, 238
constraints on usage, 239
contractual restrictions, 239, 241
decisions, 239
managing access, 239
protecting research/expertise, 248
security practices, 239, 241
technical constraints, 240, 241
unauthorised access, 239
social aspects, 240, 241
technical/social factors, 233
**Actor Network Theory (ANT)**
deployment of controversies, 232

**Advisory Boards**
ethical, legal and social issues, 325
function, 325
internal advisory boards, 325
**Attitudes to governance**
*see also* **Positive desire for governance**
access issues, 164
*see also* **Access to data**
analysis, 163, 164
balancing of interests
expectations/techno-scientific capabilities, 195, 196, 199, 200
over-protection of data, 195, 197, 200
researchers' needs/participants' wishes, 200
behaviour and practice, 140
beliefs and assumptions
*see* **Beliefs and assumptions**
biological material/data distinction
*see* **Biological material/data distinction**
common themes
lack of consistency, 184
lack of organisation, 184
lack of understanding, 184, 185
obstacle to research, 171–3, 184, 198, 275
stakeholder's real interests, 185
wastefulness, 184, 200
concerns
competency issues, 226
consequences for governance, 225
genetic exceptionalism, 185, 188, 189, 197
guidance, 225
importance, 201
inappropriate requirements, 200
inconsistent regulation, 90, 201
jurisdictional differences, 226
lack of legal clarity, 185–7, 197, 200, 204
legal compliance, 198, 226
legal treatment of biological material, 200
legal treatment of data, 200
multiple governance sources, 200
multiple governance systems, 185, 188, 197, 198, 200
perceptions of hostility, 201, 202, 309
practical problems, 226
purpose of governance, 221
regulatory actors, 225
research implications, 200
resource implications, 185, 189, 190, 197, 198
specific laws, 225, 226
stifling of research, 185, 186, 197, 200, 224, 230
wastefulness, 184, 200, 226
data protection 165, 172, 180
*see also* **Data protection**
disregard for governance, 224, 225
governance sources, 141, 226
*see also* **Governance sources**
governance systems, 164, 225, 226
Human Tissue Act 2004
*see* **Human Tissue Act 2004**
implicit guidance/influence
dynamics of governance, 148
educational bodies, 149, 150
individual contacts, 150
influence, 223, 225
interaction with other stakeholders, 142, 146–8, 150, 161, 222
participants' interests, 150, 161
peer pressure, 150, 161
practical experience, 142, 144–6, 148, 222, 272
professional background, 141–4, 148, 222
professional bodies, 149
professional cultures, 149, 150, 222, 225
professional training, 272
public trust, 150, 273
social relationships, 150
implicit influences, 140, 141
intellectual property rights, 46, 119, 127, 164, 171–3, 226
interaction with other stakeholders
attitudes/behaviour, 146
funding bodies, 148
future participants, 147
mandatory data-sharing, 148
meetings with colleagues, 148
patients, 147
peer review, 147
professional community, 147
public scrutiny, 147
Research Ethics Committees, 148
research participants, 147
sense of duty, 147
universities, 148
interviews, 140
jurisdictional differences
concerns, 226
conflicting legislation, 191, 192, 198
effect, 278, 279
exclusionary techniques, 192, 198
harmonisation, 191, 228
knowledge requirements, 191, 198
legal awareness, 191
legal variations, 192
national laws, 191
source of difficulty, 198
strategic response, 192, 198
messaging function, 156

mismatches
competing concepts, 200, 228, 229
conflicting legislation, 224
data regulation, 228, 229
governance framework, 200
parallel legal regimes, 229
stakeholder needs, 200
stifling of research, 200, 224, 230
negative attitudes, 176, 179, 181, 183, 223, 225
obstacles to research, 171–3, 181, 198, 275
practical experience
common practice, 145, 146
embedded practice, 145, 146
experience through learning, 144, 146
exposure to liability, 224
implicit guidance, 142, 144
individual abilities, 145, 146
specific people, 145
working practice, 144, 145
practitioner requirements, 141
practitioner understanding, 141
professional background
disciplinary issues, 142
ethical matters, 142, 143
etiquette, 144, 222
influencing norms, 142, 222
interdisciplinary approaches, 142, 143
professional boundaries, 142
professional values, 143, 144
regulation of data
anonymity, 195–7
confidentiality, 195, 197
conflicting interests, 195, 198, 199
expectations/techno-scientific capabilities, 195, 196, 199, 200
over-protection, 195, 197, 200
patient protection, 197
regulatory actors, 163, 164, 200, 211, 225
regulatory bodies
common attitudes/themes, 171–3, 184, 185, 198
funding bodies, 164, 181–3
National Health Service, 164, 179–81, 183, 184, 223
Research Ethics Committees, 164, 175–9, 183, 211
selective compliance, 224, 310
specific laws, 163, 164, 172, 174, 200

**Beliefs and assumptions**
core attitudes, 221, 222
decision-making, 222
disregard for regulation, 156, 159
doing good, 153, 154, 158–60, 210, 223
governance perspective, 160
importance, 160
interpretative judgments, 154, 155, 158, 159, 223
knowing what to do, 151, 152, 158, 160, 209, 210, 223, 230
mental benchmarks, 151
positive self-image, 159
reassuring others, 156, 203, 206, 207, 209, 210
regulatory compliance, 156
selective compliance, 157–9, 310
self-confidence, 159, 223
stretching the rules, 155, 159
trust in practitioners, 158, 159
violating ethical requirements, 155, 160
*see also* **Ethics**
**Bermuda Agreement**
data-sharing, 297
open-access policies, 33
**Better Regulation Commission**
recommendations, 26, 27, 58
**Better Regulation Principles**
*see* **Good governance principles**
**Better Regulation Taskforce**
guidelines, 23, 26
**Biobanking and Biomolecular Resources Research Infrastructure (BBMRI)**
infrastructure development, 43, 44, 304, 305
**Biobanks**
boundary objects, 22, 56, 87, 88, 112, 113, 303
complexity, 289, 299, 300, 305
corporeal biomaterials, 63
data-sharing, 10
*see also* **Data-sharing**
definitional problems, 53–5
differing forms of expertise, 300
differing obligations, 300
differing standards, 5
disease-specific, 3, 7
diversity, 28, 288, 289
emergence
funding strategies, 282
genome sequencing, 283
incremental development, 303, 304
policy initiatives, 282
research developments, 282
establishment, 29, 30
ethical principles, 8, 289, 300, 301
*see also* **Ethics**
governance, 8, 9, 28, 29
*see also* **Governance framework**
heterogeneity, 288, 289, 299
informational material, 63
large population biobanks, 3, 6, 7
legal framework, 4, 5
legal principles, 5–8

linking research collections, 30
networks, 43
origin of samples, 299
privacy, 5, 6
proliferation, 28
public support, 91
regional/national differences, 4
regulation, 7, 8, 10, 11, 20
research networks, 302
rhizomatic character, 300, 303
security, 4
terminology, 4, 5
**Biological material/data distinction**
accessibility, 193
bifurcated statutory system, 192, 193, 198, 199
clear differentiation, 193
custodial arrangements, 193
emotional factors, 194, 199, 228
equivalent material, 193
harmonisation, 195, 199, 228
no qualitative difference, 193
practical issues, 194
practitioner views, 192–4, 199
professional expectations, 228, 229
public sensitivity, 193, 198
**Biomedical research framework**
*see also* **Medical research**
accountability, 91
coherence, 91
consistency, 91
proportionality, 92
transparency, 91
uncertainty, 91
**Boundary objects**
biobanks, 22, 56, 87, 88, 112, 113, 303

**Caldicott Guardians**, 122, 179, 180, 270, 324
**Codes of practice/guidance materials**
*see* **Guidance**
**Common diseases**
aetiology, 37
environmental contribution, 37
familial clustering, 37
genetic basis, 37
polygenic diseases, 37
research efforts, 37
sequencing projects, 37
**Common law doctrines**
civil law, 342
criminal law, 342
governance framework, 51, 62, 66, 68
*see also* **Governance framework**
legal framework, 263
**Computer technology**
advances, 3
**Consent**
access to data, 253
adequate information, 290, 291
capacity, 262
change to governing body, 291
data-access agreements, 291
data-sharing, 290
DNA analysis, 71
future research, 291
good practice, 293
governance in practice, 119
historic data collections, 292
human tissue (research), 166, 167, 170
individual decision-making, 274
informed consent agreements, 236, 252
legal changes, 291
medical research, 63
open-access policies, 290
*see also* **Open-access policies**
process, 291
purpose, 291
re-consent, 292
Research Ethics Committees, 71
re-use of samples, 292
valid consent
competence, 290
informed consent, 45, 290, 291
voluntary consent, 290

**Data protection**
guidance materials, 74, 75
legal regime, 74
living individuals, 74
medical research, 55
personal data, 74
practitioner attitudes, 165, 172, 180
sanctions, 72
**Data-release**
Bermuda Agreement, 297
condition of funding, 297
Fort Lauderdale Agreement, 297, 299
Human Genome Project, 297
implications for populations, 298
managed approaches, 299
open-access policies, 33, 297–9
*see also* **Open-access policies**
scientific collaboration, 299
sustainability of research, 299
**Data-sharing**
access to data, 232, 290
*see also* **Access to data**
Bermuda Agreement, 297
condition of funding, 297
consent issues, 290
data-sharing agreements, 46
Fort Lauderdale Agreement, 297, 299

Human Genome Project, 297
managed approaches, 299
mandatory data-sharing, 148
methods
collaboration, 235
IT systems, 235
technical arrangements, 235
motivation
*see* **Motivations for data-sharing**
open-access policies, 297–9
*see also* **Open-access policies**
scientific practice, 44, 45, 298
what is shared with whom
actors, 234
aggregated data, 234
anonymised data, 234
artefacts, 234
content/quality of materials, 234
exchange relationships, 234, 240
forms of data, 234
overlapping projects, 233
**Decentralised/polycentric governance**
areas of responsibility, 306
awareness of problems, 306
delegated authority, 305
disciplinary perspectives, 306
governance gap, 306
levels of knowledge, 306
networking, 306, 307
professional cultures, 306
professional integrity, 307
public trust, 307
relationships with participants, 306
role of the State, 305
**Decision-making**
*see* **Individual decision-making**
**Dissemination of information**
media coverage, 66
professional bodies, 66
regulatory space, 66, 89, 90
*see also* **Regulatory space**
**DNA**
analysis
living donors, 71
qualifying consent, 71
crime prevention, 47, 55
data, 14
National Database, 47, 55
samples
collections, 3
legislative provisions, 166–70, 174, 278
regulation, 265
sequencing
high-throughput advances, 3
sequencing technology, 34–6
structure, 36, 37
theft, 71
**Documentary governance sources**
best practice norms, 66
changing rules, 88
common law doctrines, 62
electronic sources, 60
formal sources
codes of practice, 61, 64
EU Directives, 61, 62
legislation, 61, 62
governance in practice, 117
*see also* **Governance in practice**
guidelines, 61
inconsistent documents, 88
informal sources, 64, 65
informal stakeholders, 64, 65
international sources, 61
internet-based resources, 61
judicial rulings, 62, 66, 68
legal/ethical guidance, 60
multiple sources, 64, 65, 88
overlapping documents, 88
publication/dissemination, 66, 89
regulatory authorities, 64, 65
**Duties of care**
conflicting obligations, 295
differing obligations, 295
feedback findings, 295–7
future use of samples, 295
nature of obligations, 295
origin of samples, 295
personal risk, 296
**Dynamic networks of practice**
boundary objects, 303
development of protocols/ procedures, 305
differing disciplines, 303
incremental development, 303, 304
multiple collaborations, 303
multiplicity of sources, 303
reconfiguration, 303
rhizomatic character, 303
social relationships, 303

**Education**
*see also* **Training**
professional education, 83, 89, 212, 223
**ELSAGEN Project**, 4, 8
**Ethical analysis**
bioethics, 21
Governing Genetic Databases (GGD) Project, 21, 22
*see also* **Governing Genetic Databases (GGD) Project**
practical experience, 21, 22
practice issues, 21, 22
research method, 11, 21, 22
social science methods, 21

**Ethics**
*see also* **National Research Ethics Service**
access to data, 233, 246, 253, 255, 257, 294
accountability, 301
anonymity, 298
autonomy, 298
biomedical ethical framework, 213
consent issues, 290, 291, 298
*see also* **Consent**
duties of care, 295–7
*see also* **Duties of care**
ethical governance, 282
future research use, 298, 299
genomic research, 282
*see also* **Genomic research**
genuine ethical concerns, 246
good practice, 300, 301
governance
governance mechanisms, 298
purpose-built governance, 301
HUGO Ethics Committee, 81
importance, 289
potential harms
commercialisation, 298
confidentiality, 293, 294
cultural beliefs, 294
discrimination, 294
employment issues, 294
individuals, 293, 294
insurance issues, 294
loss of trust, 293, 294
marriage prospects, 294
personal information, 293
physical harm, 293
populations, 293, 294
privacy, 293, 298
removal of identifiers, 294
security concerns, 293, 294, 298
public health ethical framework, 213
research/protection balance, 298
support/training, 301
sustainability of databases, 298
third-party use, 298, 299
transparency, 301
trust, 298
violations, 78, 155, 160
**EUROGENEBANK Project**, 8
**Experts**
access to experts
Caldicott Guarians, 122, 179, 180, 270, 324
ethical, legal and social issues, 324
training and mentoring systems, 324
advisory panels, 22, 248
expert advice, 279
expert knowledge, 262
in-house experts, 126, 271
protecting expertise, 248
**External oversight bodies**
*see also* **Governance actors; Internal oversight bodies**
forms of oversight, 125
funding bodies, 125, 136
licensing functions, 125
open-access policies, 125
private sector respondents, 125
professional accreditation, 125, 136
public sector respondents, 125
relevant bodies, 133
Research Ethics Committees, 125, 136
voluntary/non-voluntary activity, 125

**Fort Lauderdale Agreement**
access to data, 242
data-sharing, 297, 299
open-access policies, 33
**Funding bodies**
accountability, 85, 267, 321
charitable funding, 86
common policies, 320
common procedures, 320
consultation, 279
entry controls, 138, 161, 222
ethical guidelines, 86
funding applications, 84
gatekeeping function, 84
government departments
Department of Health, 85
Department of Innovation, Universities and Skills, 85
importance, 85
influence, 85, 127, 128, 266, 267, 270, 305, 320
interaction, 148
open-access policies, 85, 181–3, 270, 290, 320
*see also* **Open-access policies**
oversight
external oversight, 125, 136
mechanisms, 85, 86
peer review, 267
policy preferences, 138, 161
practitioner attitudes
control issues, 182
insufficient guidance, 181, 183
multi-funder problem, 182, 213
negative comments, 181
open-access requirements, 181–3
positive comments, 181
prescribed norms/conditions, 84, 85
priorities/policies, 84
publicly funded research councils, 85
regulatory role, 320
research governance, 267
self-review, 279
sources of advice, 270

transparency, 85
UK Biobank, 86, 87

**General Medical Council**
influence/role, 69, 70, 89
registered practitioners, 265
**Genetic database(s)**
characteristics, 93, 94, 99
classification
age of study, 99–101
materials held, 104–5
observed characteristics, 99
research method/objective, 93–6
size categories, 99, 101–4
collection of samples, 14
commonly used term, 14
concept, 94
constitution, 113, 114
definition, 93, 94, 97, 98, 113
differing features, 95, 96
diversity, 15
DNA data, 14
dynamism, 94, 111, 114
genomic sequence data, 14
*see also* **Genomics**
genotype data, 14
governance, 15, 95
*see also* **Governance framework**
heterogeneity of practices, 95
ideal database, 95, 96
interaction, 94
IT system, 14, 95, 96
multiplicity, 111
organisational setting, 97, 108, 110
*see also* **Organisational setting**
phenotype data, 14
professional/scientific interests, 97
rhizomatic system, 96, 112, 113, 114
structural dimension, 94
terminology, 97
typology, 93–5, 113, 114
**Genetic information**
*see also* **Genetic databases**
access to data, 7
*see also* **Access to data**
benefit-sharing, 6
consent, 6
*see also* **Consent**
decision-making, 7
differing standards, 5
familial nature, 5, 6
feedback, 6
intellectual property, 7
*see also* **Intellectual property rights**
legal dimension, 5, 6
ownership, 7
privacy concerns, 5–7
public interest, 5, 6
**Genetic research**
increase, 3
resource collections, 3
**Genetically modified organisms**
medical research, 55
**Genome-wide Association Studies (GWAS)**
1000 Genomes Project, 39, 42, 43
clinical applications, 41
co-ordinated resources, 38, 39
development of treatments, 38
disease aetiology, 40
genetic risk variants, 38
genomic research, 283, 284
*see also* **Genomic research**
genotyping chips, 38
heritability, 40
individual's genomic data, 40
methodological problems, 40
new methodologies, 39, 45
risk assessment, 41, 42
sampling, 38–40
single nucleotide polymorphisms, 38, 39, 283
validity, 41
**Genomic research**
array-based genotyping, 283
best practice, 300, 301
development, 283
diagnostic/predictive tests, 283, 284
different disciplines, 285
differing research/scientific principles, 285, 288
epidemiological studies, 285
Genome-wide Association Studies (GWAS), 283, 284
*see also* **Genome-wide Association Studies (GWAS)**
genomic diversity study, 284
genotyping facilities, 284
insertions/deletions relationship, 283, 284
institutional setting, 288
knock out mice, 285
research networks, 285
samples/data
collection/storage, 284
contents of biobank, 287
differing research/scientific principles, 285, 288
history, 285, 286
multiplicity, 285, 286, 288, 289
recruitment methods, 287
size of biobank, 285, 286
types of participant, 287
single gene disorders, 284
small-scale studies, 285
whole-exome sequencing, 283

**Genomics**
*see also* **Genomic research**
development of disease, 36
functional genomics, 36
global dimension, 48
governance, 9
*see also* **Governance framework**
high throughput data, 36
Human Hap Map, 37, 39, 42
linkage disequilibrium, 38
open-access policies, 33, 34
*see also* **Open-access policies**
sequencing, 283
single nucleotide polymorphisms, 37–9, 283
supportive infrastructure, 30
technology intensive approaches, 36, 37
**Genotyping**
array-based genotyping, 283
genotyping facilities, 284
individuals, 36
**Good governance principles**
*see also* Improvements to governance
accountability, 27, 60, 91, 138, 311, 313, 323
application, 59, 87
authoritative advice, 64
benefits, 59
Better Regulation Commission, 26, 27
coherence, 59
consistency, 27, 28, 91, 311, 312
consultation, 59, 311
co-ordination, 59
differing views, 59
efficiency, 60
European Union, 59
fairness, 59, 311
governance system, 307, 308
Hampton Review, 26, 58, 60
implementation, 26
importance, 310, 311
legal/empirical basis, 59
legal integration, 59
legislation, 27, 58, 60
legitimacy, 60, 138
non-discrimination, 59, 311
omissions, 59
on-going review, 59
Organisation for Economic Co-operation and Development, 59
predictability, 59, 311
proportionality, 27, 64, 91, 92, 311, 315, 316
public authorities, 26
public/stakeholder participation, 59, 311, 316
quality control, 138
reforms, 139
Regulators' Compliance Code, 27, 59, 73, 91
targeting, 27, 91, 311, 315
transparency, 27, 28, 64, 76, 91, 138, 311–4, 323
**Governance actors**
domestic (UK based), 343–7
formal actors
Department of Health, 69, 89
European institutions, 68, 347, 348
General Medical Council, 69, 70, 89
government, 68, 69, 89
Human Fertilisation and Embryology Authority, 69, 73
Human Tissue Authority, 70–3, 89
Information Commissioner's Office, 69–76, 89
judiciary, 68, 69, 89
legislators, 68, 69, 89
Office of the Health Professions Adjudicator, 69, 78
recognised actors, 67, 68
Research Ethics Committees (RECs), 76–9
HUGO Ethics Committee, 81
informal actors
discernible governance function, 67
funding bodies, 80, 82, 84–7
guidelines, 90
hidden governance actors, 68, 75, 76, 79
implicit symbiosis, 90
industry groups, 80
international bodies, 80, 81
interrelationships, 67
intrinsic interdependency, 90
multiplicity, 79, 80
networks, 80
non-governmental organisations, 80, 81
power to influence, 67, 80
professional associations, 80, 82–4
prominent actors, 80
regional bodies, 80, 81
status/authority, 67, 79, 80
sub-categories, 80
international/multinational, 80, 81, 348, 349
National Health Service, 81, 82
oversight/governance bodies
*see* **External oversight bodies; Internal oversight bodies**
UNESCO, 81, 330
United Nations, 81, 330
World Health Organisation, 81
World Medical Association, 81

**Governance framework**
*see also* **Good governance principles; Governance sources; Governance system**
analysis, 51–3, 55
beliefs and assumptions
*see* **Beliefs and assumptions**
benefit-sharing, 63
biomedical research, 88
biomedical scandals, 55
bio-samples, 63
boundary objects, 22, 56, 87, 88, 112, 113, 303
clinical trials, 88
codes of practice, 51
common law authorities, 51, 66, 68
complexity, 52, 53, 262, 263, 267–9, 279, 280, 301
corporeal materials, 88
credibility, 92
decentralised polycentric governance
*see* **Decentralised polycentric governance**
decision-making
*see* **Individual decision-making**
deficiencies, 91, 92, 317
definition, 114
delegated responsibilities, 279
designated supervisory authority, 78
draft laws, 63
duplication, 318
dynamic networks
*see* **Dynamic networks of practice**
electronic resources, 51
ethical issues, 78, 326
*see also* **Ethics**
governance in practice
*see* **Governance in practice**
governance structure(s)
assessment, 23
education/training, 223
entry controls, 138, 161, 222
gatekeeping role, 222
guidelines, 23
legitimate authority, 23
new methodologies, 29
new technologies, 29
oversight systems, 222
policy preferences, 222
professional networks, 222
purpose-built, 301
role of individuals, 222
steering committees, 222, 306
trailblazers, 222, 271, 306
government databases, 55
guidance materials, 51
harmonisation, 54, 89
improvements to governance, 212, 214, 215, 226, 302
*see also* **Improvements to governance**
inadequacy, 62, 63, 67
incomplete nature, 90
inconsistency, 90, 201
independence, 91
interdependency of organisations
*see* **Interdependency of organisations**
lack of coherence, 261, 262, 267
legacy requirements, 317
legal framework
*see* **Legal framework**
legal research, 60
*see also* **Legal research**
legitimacy, 23–5, 92, 266, 318
*see also* **Legitimacy**
medical research, 67
*see also* **Medical research**
multiple sources, 52, 53
negative perceptions, 221
negotiated space, 268, 279, 280
new forms of research, 55
non-corporeal materials, 88
operation, 51
oversight
formal oversight, 270
informal oversight, 279, 306, 307
internal oversight bodies, 270
State intervention, 268, 279
ownership/property rights, 63
polycentric nature, 318
practice
best practice, 262, 326
current practice, 261
interactions, 302
interdependencies, 302
practical problems, 53
practitioners' experience, 261, 302
voluntary compliance, 262
prevention of harm, 224
public authorities, 51, 52
public trust, 55
reforms, 317, 326, 331
regulation
legislation, 51, 63, 329, 330
regulatory bodies, 51–3, 264–6
regulatory gaps, 90, 117, 120, 135
regulatory requirements, 261
self-regulation, 262, 279
scientific collaboration, 53
social networks, 262, 272, 273, 276
*see also* **Social relationships/networks**
social sanctions, 262
sources of advice, 270
sources of authority, 266
standard operating procedures, 265, 266
statutory bodies, 264, 265
strengths/weaknesses, 302
systematic symbiosis, 78, 90
terminology, 54, 55, 89

therapeutic applications, 88
transparency, 91
**Governance in practice**
*see also* **Governance actors; Governance system**
actors
domestic (UK based), 343–7
European, 347, 348
formal actors, 122
informal actors, 122
international, 348, 349
official bodies, 121, 122, 128, 133
quasi-formal actors, 122
relevant actors, 119, 120
best practice principles, 118, 119, 134
confidentiality, 119
consent issues, 119
criminal/civil liability, 136
cross-border dimension, 121, 135
data analysis, 116, 117
documentary governance sources, 117
*see also* **Documentary governance sources**
empirical research, 116
entry controls, 138, 161, 222
formal laws
broad spectrum, 117, 134
clarity of expression, 120, 134
EU Directives, 117, 118, 120, 134
foreign national laws, 121
freedom of information, 119
guidelines, 117
intellectual property laws, 119
legal forms/processes, 117
legal reference points, 117, 119
legislation, 117, 119–21, 134, 329, 330
non-UK law, 118
practitioner awareness, 133, 134
regulatory gaps, 117, 120
responses, 120
funding bodies, 127, 128
governance reforms, 116, 139
improvements to governance
*see* **Improvements to governance**
informal oversight methods, 127, 138
legal analysis, 116
material transfer agreements, 119
oversight/governance bodies
*see* **External oversight bodies; Internal oversight bodies**
policy preferences, 138, 222
practitioner uncertainty, 137
provision of information, 137
public/private sector differences, 125–8
seeking guidance
*see* **Seeking guidance**
sources of governance
*see* **Governance sources**
support networks, 137
**Governance sources**
*see also* **Documentary governance sources**
accreditation bodies, 122, 123
advisory bodies, 122
Caldicott Guardians, 122, 179, 180, 270, 324
domestic sources, 329, 330
European Union, 330
formal regulators, 122–4
funding bodies, 122, 134, 135
government departments, 122
improvements to governance, 212, 218–21, 227
*see also* **Improvements to governance**
industry associations, 122
lack of engagement, 133, 136, 138
pending legislation, 331
penetration/use, 123, 135
practitioner attitudes, 115, 116, 124, 135
practitioner awareness, 115–7, 120, 121, 123, 133, 134
professional backgrounds, 124, 135
professional bodies, 122
National Health Service, 122
open-access policies, 115
*see also* **Open-access policies**
practitioner engagement, 115, 116, 133
Research Ethics Committees, 122, 134
restricted engagement, 124
sociological research, 11
sources cited, 124
voluntary engagement, 124, 138
**Governance system**
*see also* **Governance framework**
ambiguity, 277, 278, 310
assessment of governance
bureaucracy, 308
clear frameworks, 309
conflicting/ambiguous governance, 310
duplication of requirements, 309
good practice, 308
hostility to research, 309
interpretation of information, 308
lack of accessible information, 309
negative views, 308–10
practitioners' experience, 308
protection/research balance, 309
public trust, 308
selective compliance, 310
avoidance, 154
best practice, 275
credibility, 311
detrimental consequences
administrative wastage, 226
avoidance strategies, 226
duplication, 226
exclusionary techniques, 226
undermining practitioner trust, 226

failure to meet practitioner needs, 276, 277
generic approval, 280
good governance principles, 307, 308
governance in practice
*see* **Governance in practice**
lack of relevance, 276
legal impact
conflicting laws, 279
EU Directives, 278
Human Tissue Act 2004, 278
intellectual property rights, 278
jurisdictional differences, 278, 279
legitimacy, 311
multiplicity, 185, 188, 197, 198
new initiatives, 280, 281
obstacle to research, 275
practitioner attitudes, 164, 275, 308
practitioner needs, 280
public trust, 281
single authoritative body
authority, 318
beneficial effect, 318, 320
best practice, 319
democratic legitimacy, 319
liaison role, 319
limitations, 318, 320
national strategy, 318
public trust, 318
research principles, 318
responsibilities, 318
sources of guidance, 276
*see also* **Guidance**
**Governing Genetic Databases (GGD) Project**
age of study
distinction between stages, 100, 101
incremental collections, 101
legacy samples, 99, 100, 107
range of ages, 99, 100
basic premise, 93
discussion
boundary objects, 112, 113
common structural dimensions, 111
continuing changes/development, 110–2
determining boundaries, 111
nature of genetic databases, 111
new technology, 110–2
repositories, 111
rhizomaztic approach, 112
ethical analysis, 21, 22
genetic databases, 93–105
*see also* **Genetic databases**
governance practices, 232, 253
individual participants, 105
legal research, 52
materials held
available biological material, 104, 105
biological samples, 104
data, 104
differing types of data, 104
digitised data, 104
methodology
methodology employed, 110
recruitment, 98
respondents, 98, 99, 113
sampling, 98
organisational setting, 108, 109
*see also* **Organisational setting**
recruitment
assessment centres, 106
clinical interaction, 105–7
collaborations, 107, 108
criteria, 105
direct recruitment, 106
healthcare records, 108
hospital databases, 106
opportunistic approach, 105, 106
partial representation, 107
recruitment practices, 107
research clinics, 106
regulatory space analysis, 58
*see also* **Regulatory space**
size of categories
epidemiological studies, 103
evolution, 103
individual research participants, 102
larger collections, 102, 103
methodological approach, 102
projections, 103
units of measurement, 101, 102
variability, 103
**Guidance**
codes of practice/guidance materials
domestic codes, 332–6
EU materials, 336, 337
guidance materials, 332–7
international organisations, 337–9
internet resources, 337, 338
recent reports, 339, 340, 341
statutory codes, 332
conflicting guidance, 311, 312
implementation guidance, 227
implicit guidance/influence
dynamics of governance, 148
educational bodies, 149, 150
individual contacts, 150
influence, 223, 225
interaction with other stakeholders, 142, 146–8, 150, 161, 222
participants' interests, 150, 161
peer pressure, 150, 161
practical experience, 142, 144–6, 148, 222, 272
professional background, 141–4, 148, 222
professional bodies, 149
professional cultures, 149, 150, 222

professional training, 272
public trust, 150, 273
social relationships, 150
seeking guidance
*see* **Seeking guidance**

**Hampton Review**, 26, 58, 60
**Hap Map Project**, 37, 39, 42, 283
**Human Fertilisation and Embryology Authority**
influence/role, 69, 73
**Human Genome Project (HGP)**
bioinformatics, 30
collaborative approach, 32, 33
data-sharing, 297
ethos of sharing, 33
influence, 48
international dimension, 33
international interactions, 32
key principles
achievement, 31, 32, 36
data release, 31
importance, 31
inclusion, 31
new scientific information, 36
open-access policies, 30, 33
*see also* **Open-access policies**
origins, 31
purpose, 31
resources, 32
results, 3
self-interest, 32
sequencing
effort, 31–3
technologies, 30, 34–6
**Human rights**
EU law, 45, 46
European Convention on Human Rights, 45
medical records, 47
right to life, 45
**Human Tissue Act 2004**
access issues, 168, 170, 173
cell cultures, 169
consent issues, 166, 167, 170
criticism, 166, 167
culture change, 166, 226, 278
different kinds of tissue
failure to differentiate, 174, 226
ontological status, 169, 174
quantity of tissue, 169, 170, 174
sample type, 169, 174
DNA samples, 166–70, 174, 278
excluded materials, 167, 168, 173, 174, 278
expert advice, 279
implications for practice, 166
importance, 165, 278
lack of clarity, 166
licensing requirements, 166, 167, 168, 174, 278
post-mortem tissue
access for research, 170, 173
consent, 170
practitioner attitudes, 164–7
relevant materials, 166–9, 173, 226
storage licence exception, 168, 169, 173
transplantation material, 168, 169
**Human Tissue Authority**
authority, 7, 20, 70–2, 89
civil interventions, 73
codes of practice, 7, 70, 71
collaboration, 72, 73, 89, 268, 269
consent procedures, 71
DNA samples, 265
enforcement, 72
funding, 72
licences, 7, 70, 71, 265
powers, 265
role, 70, 71
sanctions, 70–3
stakeholder influence, 72
statutory powers, 71

**Improvements to governance**
*see also* **Good governance principles; Positive desire for governance**
access to data, 220
*see also* **Access to data**
alternatives to governance, 212, 219, 227
common regulatory framework, 214, 215
consultation, 213, 214
data security standards, 214
dynamic networks
*see* **Dynamic networks of practice**
framework plus guidance, 215
generally, 202, 203, 221
generic matters, 227
genetic research, 217
governance framework, 212, 214, 215, 226
*see also* **Governance framework**
governance sources/dissemination, 212, 218–21, 227
*see also* **Governance sources**
implementation guidance, 227
international collaboration, 218
legal instruments, 212
multi-jurisdictional issues, 227
oversight/governance bodies, 212, 217–8, 220, 227
*see also* **External governance bodies; Internal governance bodies**
political/policy steps, 212–4, 226
*see also* **Political/policy steps**
practitioner experience, 219, 220, 302

practitioner views, 212
regulatory gaps, 219
specific legal changes
definition of terms, 216
generally, 212, 227
genetic discrimination, 217
material transfer agreements, 217
ownership, 216
property rights, 216
standardised guidelines, 216
spectrum of suggestions, 220
steering committees, 218, 222
systemic matters, 227
volume of complaints, 220
working relationships, 227
**Individual decision-making**
anonymisation issues, 274, 275
appropriate course of action, 273
collaborations, 274
consent issues, 274
custodian responsibilities, 273–5
material transfer agreements, 275
memoranda of understanding, 275
negotiated process, 273
open-access policies, 273
*see also* **Open-access policies**
patent licences, 275
professional networks, 274
protecting participants, 274
source of guidance, 273
standard operating procedures, 275
third-party access agreements, 275
**Information Commissioner's Office**
authority, 269
bio-medical sphere, 75
collaboration, 89
complaint procedure, 265
enforcement, 74–6
expert advice, 279
formal undertakings, 76
functions, 73, 76
funding, 75
guidance standards, 75
influence, 75, 89
medical research, 76
negotiation, 76, 89
personal data, 73
reforms, 75
registered data controllers, 265
regulatory action, 76
resources, 267, 268
sanctions, 73–5
status, 73
**Infrastructure development**
Biobanking and Biomolecular Resources Research Infrastructure (BBMRI), 43, 44, 304, 305
collaborative funding, 42–4
collaborative projects, 44
governance structures, 43
information repositories, 43
networked biobanks, 43
P3G, 44
**Intellectual property rights**
access to materials, 278
copyright, 7
impact, 278
lack of clarity, 171, 173
material transfer agreements, 46, 119, 127, 171, 172
negotiation, 171, 173
obstacle to research, 171–3
patents, 7, 171, 173, 226, 262, 275, 278
practitioner attitudes, 164
protection, 7
sharing materials, 172
transaction costs, 171, 173
**Interdependency of organisations**
collaboration, 268, 269
complexity, 267, 268
distribution of power, 268
enforcement powers, 268
negotiated space, 268, 269
overlapping guidelines, 267
research approvals, 269
self-regulation, 268
**Internal oversight bodies**
*see also* **External oversight bodies; Governance actors**
company management, 126
company meetings, 271
control systems, 270
in-house experts, 126, 271
institutional governance, 126, 136, 138, 161
internal audit, 270
material transfer agreements, 127
monitoring, 270
private sector respondents, 126
relevant bodies, 133
sources of advice, 270
standard form template documents, 126
standard operation procedures, 127, 136
steering committees, 126, 127, 136, 138, 161, 218, 222, 306
supervisory actors, 126, 136
team meetings, 271
trailblazers, 271
written agreements, 127
**Interview subjects**, 350–2

**Legal framework**
application, 264
best practice principles, 263
capacity to consent, 262

clinical trials, 262
codes of practice, 263
common law doctrines, 263
data protection, 262
*see also* **Data protection**
expert knowledge, 262
guidelines, 263
human rights, 262
human tissue, 262
interpretation, 264
lack of clarity, 263
lack of clear definitions, 264
medical research, 262, 264
patents, 262
regulations, 263
sources of guidance, 263
statutes, 263
**Legal regime**
authoritative legal requirements, 322
consolidation, 322
coordination, 322
reforms, 321, 322
single overriding statute, 321, 322
**Legal research**
documentary evidence, 60, 61
*see also* **Documentary governance sources**
existing legal framework, 20, 21
governance framework, 60
*see also* **Governance framework**
governance theory, 20
purposes, 20
regulatory bodies, 20, 21
regulatory space, 20, 29
*see also* **Regulatory space**
**Legitimacy**
accountability, 23, 24
definition, 23
due process, 23, 24
efficiency, 23, 25
expertise, 23, 25
governance system, 311
*see also* **Governance system**
legislative mandate, 23, 24
legitimate authority, 23
legitimation, 23
regulatory bodies, 266

**Material transfer agreements**, 46, 119, 127, 171, 172, 217, 275
**Medical research**
*see also* **Biomedical research framework**
changing rules, 63
clinical trials, 47
consent, 63
data protection, 55
draft laws, 63
genetically modified organisms, 55
governance, 46–8, 67
*see also* **Governance framework**
legislation, 46, 47
legislative reforms, 63
medical records, 47
misconduct, 55
organ retention, 47
privacy, 55
public trust, 47, 55
regulatory framework, 29, 58
stem cells, 55
transgenic embryos, 55
**Motivations for data-sharing**
*see also* **Access to data; Data-sharing**
consent agreements, 252
distribution of resources/power/influence, 251
emerging culture of practice, 251
exchange relationships, 251
lack of status, 251
normative disincentives
confidentiality, 249, 250
credit for input, 248, 249
free riders, 248
inappropriate disclosure, 249
original context of research, 249, 250, 252
private sector involvement, 250
relationships of responsibility, 249
normative motivations
consent agreements, 252
expectations of patients/participants, 245, 246, 252
genuine ethical concern, 246
good intentions of others, 245
moral arguments, 246
social good, 246, 247, 252
patterns of exchange, 252, 257
practical disincentives
access costs, 247
access request systems, 247
confidentiality, 248
depletion of finite resources, 247
protecting expertise, 248
protecting financial benefits, 248
protecting research niches, 248
practical motivations
access to clinical trials, 244
company involvement, 244
depth of resources, 245
necessary exchange, 244
organisational stability, 244
particular knowledge/technology, 244
sharing as core business, 244, 245
status, 244, 245
professional norms, 251
public good, 251, 253
scientific advances, 251

self-interest, 251
territorialism, 251

**National Health Service (NHS)**
biobank research, 81, 82
governance sources, 122
organisational setting, 108, 109
practitioner attitudes
Caldicott Guardians, 179, 180
data protection, 180
incompetence, 184
lack of centralisation, 184
negative attitudes, 179, 183, 223
research and development approvals, 180, 181, 183
research and development offices, 180, 181, 184, 270
research governance requirements, 180
**National Research Ethics Service**
delegation of authority, 314
establishment, 321
governance sources, 65, 78, 355
parliamentary delegation, 313
research regulation, 314
standard operating procedure, 266

**Office of Health Professions Adjudicator**
influence/role, 69, 78
**Open-access policies**
access to data, 232
*see also* **Access to data**
Bermuda Agreement, 33
clinical data, 34
community resources, 33, 34
consent issues, 290
data-release, 33, 297–9
*see also* **Data release**
Fort Lauderdale Agreement, 33, 297, 299
funding bodies, 85, 181–3, 270, 290, 320
*see also* **Funding bodies**
genome sequence data, 33, 34
Human Genome Project, 30, 33
*see also* **Human Genome Project**
individual decision-making, 273
*see also* **Individual decision-making**
scientific practice, 44
sources of governance, 115
Toronto Agreement, 34
tripartite responsibility, 33
**Organisational setting**
charity-funded repositories, 108, 109
collaboration, 109
cross-sectoral connections, 108, 109
multiple settings/different research projects, 108
pharmaceutical companies, 108
public health system, 108
university settings, 108, 109

**Oversight/governance bodies**
*see* **External oversight bodies; Internal oversight bodies**

**Patient/research participant input**
good practice, 325
inclusive mechanisms, 326
**Political/policy steps**
biomedical ethical framework, 213
common policies, 213
consultation, 213, 214
data coding, 214
improvements to governance, 226
*see also* **Improvements to governance**
multi-funder problem, 213
open-access policies, 213
*see also* **Open-access policies**
public health ethical framework, 213
security standards, 213
**Positive desire for governance**
*see also* **Attitudes to governance; Improvements to governance**
analysis, 202
assuring participants, 204
benefits of governance, 203, 208, 210, 211, 310
better governance, 310
demand for clarity, 209, 211
demonstrating legitimacy, 203, 206, 207, 209, 210
ensuring appropriateness, 203, 205, 206, 209–11, 230
fostering participation, 203, 207, 208
generally, 202, 203, 209, 221
increased level of governance, 203
management of identifiable data, 204
personal risk, 204, 205
professional support, 203–5, 209, 210, 230
projects subject to governance, 203, 204, 209
public concerns, 211
public trust, 210
reassuring others, 203, 206, 207, 209, 211, 212, 230
self-created governance arrangements, 205
sensitivity concerns, 211, 230
source of protection, 204
**Preferences for governance**
*see also* **Attitudes to governance; Positive desire for governance**
governance needs
appropriate governance, 229, 230
desire for governance, 229, 230
implementation, 231
personal governance, 229
purpose of governance, 229

professional expectations
data regulation, 228, 229
emotional approach, 228
ethical reasoning, 228
genetic exceptionalism, 228
harmonisation, 228
key expectations, 231
ontological approach, 228

**Privacy**
access to data, 253
protection of privacy, 5–7, 45, 55

**Professional associations**
advocacy work, 83
authority, 83
guardians of professional culture, 83, 149
*see also* **Professional cultures**
guidance materials, 83
influence, 82, 90
lobbying, 83, 84
medical research, 82
regulatory space resources, 83
standard-setting, 83
status, 82, 83

**Professional cultures**
accountability, 322
anecdotal evidence, 322
best practice standards, 149, 322
common experience, 222, 322
educational bodies, 149, 150
etiquette, 144, 222
guidance materials, 149
inculcation, 83, 89, 90, 149, 212, 222
influence, 149, 151, 159, 161, 225, 322
professional bodies, 83, 149
professional integrity, 323
professional networks, 323
professional training, 149, 323
protection, 83
setting good practice norms, 323
shared practice, 222, 322
social networks, 322
transparency, 322

**Public interest**
protection, 5, 6, 91
public interest theory, 56

**Public trust**
desire for governance, 210
*see also* **Positive desire for governance**
governance framework, 55, 281, 308, 318
*see also* **Governance framework**
implicit influence, 150
medical research, 47, 55
social relationships/networks, 273

**Regulators' Compliance Code**, 27, 59, 73, 91

**Regulatory space**
actors, 58, 60, 61
ambiguity, 300, 310
complexity, 300, 326
conflicting guidance, 311, 312
control over professional education, 83, 89, 149, 212, 306
distribution of power, 57
elements
assertion of power/influence, 57, 67
key resources, 57
occupant of regulatory domain, 57
resource distribution, 57
governance actors
*see* **Governance actors**
governance sources
*see* **Documentary governance sources**
holistic approach, 57
inculcation of professional cultures, 83, 89, 90, 149, 212
informal authority, 57
interdependencies, 305
key resources
dissemination of information, 66, 89, 90, 212
formal legal authority, 57, 65, 89, 90, 212
organisational capacities, 57, 65, 89, 90, 212
possession/control of information, 57, 65, 66, 89, 90, 212
possession of wealth, 57, 65, 89, 90, 212
regulatory gaps, 90, 117, 120, 135
uneven distribution, 90
lack of coherence, 305
mapping, 56, 139
metaphor, 56, 57, 87
networks of associations, 305
occupants, 57, 58
polycentric nature, 311, 318
professional cultures, 306
reforms, 139
regulatory theory
*see* **Regulatory theory**
setting good practice norms, 89, 149, 212, 306, 323
uncertainty, 300

**Regulatory structures**
challenges, 45
data-sharing agreements, 46
*see also* **Data-sharing**
EU law, 45, 46
global activity, 45
human rights, 45, 46
material transfer agreements, 46
national orientation, 45, 46
research governance, 46, 47
*see also* **Governance framework**

**Regulatory system**
explicit avoidance, 154

**Regulatory theory**
institutionalist theory, 56
network theories, 56
private interest theory, 56
public interest theory, 56
regulatory space analysis, 56, 57
*see also* **Regulatory space**
systems theories, 56
**Research Ethics Committees (RECs)**
approvals, 76–9, 138, 265, 266, 269, 270, 278
authority, 77, 265, 266, 269
clinical trials, 77
consent procedures, 71
criticisms, 227, 321
decision-making, 265
dependence, 227
enforcement, 71
explicit avoidance, 154
gatekeeping role, 127, 128, 138, 161, 222, 270, 306, 321
genetic research, 78
geographical distribution, 76
governance sources, 122, 134, 305
guidelines, 79
influence, 77, 79, 89, 127, 270
interaction, 148
lack of uniformity, 78
mandate, 321
monitoring, 78
powers, 77, 127, 266, 269
practitioner attitudes
application forms, 177
appropriate governance, 211
bureaucratic procedures, 176, 183
duplicated functions, 177
inconsistent decision-making, 179
lack of knowledge, 177, 178, 183
negative attitudes, 176, 179, 183, 223
positive comments, 175, 176, 183
understanding participant's needs, 178, 179, 184
wasteful procedures, 176, 177, 183
practitioner interaction, 183
quality assurance, 78
quasi-formal nature, 77, 89, 127
research progress reports, 78
resources, 267, 268
role, 77
sources of advice, 270
staffing, 76
status, 77
voluntary system, 77
**Research funding**
*see* **Funding bodies**
**Research method**
definitional issues
biobank, 15
genetic database, 14
governance, 11–4
regulation, 11–3
ethical analysis, 11, 21, 22
*see also* **Ethical analysis**
interdisciplinary approach
boundary objects, 22
credibility, 22
expert advisory panel, 22
legal distinctions, 22
jurisdictional focus, 11
legal research, 11, 14, 20, 21, 29
*see also* **Legal research**
sociological research, 11, 14–20
*see also* **Sociological research**
**Research projects**
ELSAGEN Project, 4, 8
EUROGENEBANK Project, 8
GeneBanC, 8
PRIVILEGED, 9
Tiss.EU, 9
UK Biobank, 8, 86, 87
**Research tissue banks (RTBs)**
autonomy, 78
powers, 78
research projects, 78
**Researchers**
ethical research practice, 323
moral practice, 323, 324
responsibility, 323

**Sanctions**
compliance requirements, 73
data protection, 72
designated regulators, 73
disciplinary action, 77
flexibility, 73
Human Tissue Authority, 70–3
Information Commissioner's Office, 73–5
legislation, 72–4
monetary penalties, 73
restoration requirements, 73
stop notices, 73
**Scientific practice**
data generators, 45
data-sharing, 44, 45
*see also* **Data-sharing**
data users, 45
global collaborations, 44
informed consent, 45
open-access policies, 44
*see also* **Open-access policies**
privacy, 45
scientific endeavour
acknowledgment, 45
attribution, 45
reward, 45
technological developments, 45

**Seeking guidance**
biobanking conferences, 130, 138
breadth of information, 131, 132
diverse forms, 129, 133
formal approaches, 129
individual contacts, 129, 130, 132, 133, 136–8, 161, 222
informal approaches, 129
in-house specialists, 129
lack of engagement, 133, 136, 138
lack of motivation, 133
meetings/briefings, 130
proactive approaches, 129, 133, 136, 138
reading materials, 130, 131, 136
trailblazing projects, 130, 136, 161, 222, 271, 306
training courses, 130, 136, 223
websites, 131, 136
**Sequencing technology**
accuracy of genome assembly, 36
bioinformaticians, 36
costs of sequencing, 35
data analysis, 35, 36
data storage, 36
Human Genome Project (HGP), 34–6
lack of software, 35
quality of sequencing data, 36
sequencing facilities, 34, 35
specialisation, 34
**Single nucleotide polymorphisms**, 37–9, 283
**Social relationships/networks**
implicit guidance, 272
influence, 276
opinions of other stakeholders, 272, 273
practical experience, 272
professional background, 272, 276
public trust, 273
**Sociological research**
coding scheme, 18
consent form, 20
ethical approval, 19
exploratory qualitative methodology, 15, 18
genetic databases, 15, 16
interview schedule, 16, 17
organisational categories, 17
professional categories, 17
professional practice, 15
purposive sampling strategy, 16
recruitment strategy, 16, 18
respondents, 17, 19, 29
samples, 19
scoping interviews, 16
semi-structured interviews, 16
**Standard operating procedures**
decision-making, 275
governance framework, 265, 266
internal oversight, 127, 136
**Stem cells**
medical research, 55

**Toronto Agreement**
open-access policies, 34
**Training**
accreditation courses, 324
different disciplinary training, 324
ethical, legal and social issues, 324
professional training, 130, 136, 149, 223, 272
specialist courses, 324

**UK Biobank**
biosamples, 86
Ethics and Governance Framework, 86, 87
influence, 86, 87
procedures, 86, 87
research project, 8
standards, 86, 87
**UNESCO**
governance role, 81, 330
**United Nations**
governance role, 81, 330

**World Health Organisation**
governance role, 81
**World Medical Association**
governance role, 81

www.ingramcontent.com/pod-product-compliance
Ingram Content Group UK Ltd.
Pitfield, Milton Keynes, MK11 3LW, UK
UKHW020132290926
15471UKWH00003B/443

9 781841 139050